Transforming Agriculture in Southern Africa

This book provides a synthesis of the key issues and challenges facing agriculture and food production in Southern Africa.

Southern Africa is facing numerous challenges from diverse issues such as agricultural transformations, growing populations, urbanization and climate change. These challenges place great pressure on food security, agriculture, water availability and other natural resources, as well as impacting biodiversity. Drawing on case studies from Lesotho, Malawi, Mozambique, Namibia, South Africa, Swaziland, Tanzania, Zambia and Zimbabwe, the chapters in this book consider these challenges from an interdisciplinary perspective, covering key areas in constraints to production, the most important building blocks of good farming practices, and established and emerging technologies. This book will be a valuable support for informing new policies and processes aimed at improving food production and security and developing sustainable agriculture in Southern Africa.

This informative volume will be key reading for those interested in agricultural science, African studies, rural studies, development studies and sustainability. It will also be a valuable resource for policymakers, governmental and nongovernmental organizations, and agricultural practitioners.

Richard A. Sikora is Emeritus Professor of the University of Bonn, Germany; former chairman of the Department of Plant Pathology and the Section, Soil Ecosystem Phytopathology and Nematology. He was Chair of the CGIAR System-wide programme on integrated pest management and is presently a Fellow of Stellenbosch Institute of Advanced Study (STIAS) and Convener of the Forum on Sustainable intensification. Richard has received numerous awards including the University of Illinois Alumni Association Award of Merit, the International Association for the Plant Protection Sciences Award of Distinction, the American Phytopathological Society International Service Award and the 2017 German Phytopathological Society Anton de Bari medal for excellence in phytopathology.

Eugene R. Terry was Director General of WARDA, now AfricaRice, and the Founding Director of the African Agricultural Technology Foundation (AATF). He currently serves as the Chair of the Advisory Panel for Biosciences for East and Central Africa (BecA) – ILRI Hub, and the Chair of the Advisory Board of the West Africa Centre for Crop Improvement (WACCI) hosted by the University of Ghana in Accra. Dr Terry was awarded the AfricaRice Distinguished Service Award in 2010, for exemplary leadership in rice research and development in Africa, and The Macdonald College, McGill University Distinguished Alumni Award in 2012, for outstanding contributions to society and humanity.

Paul L.G. Vlek is Emeritus Professor of the University of Bonn, Germany, and former Director of the Centre for Development Research ZEF as well as the West African Science Service Center on Climate Change and Adapted Land Use WASCAL. He devoted many years of service to a number of journal editorial boards and national and international scientific committees. In 2014, he was awarded the World Agriculture Prize for his lifetime contribution to agricultural development.

Joyce Chitja is Senior Lecturer in Food Security at the University of KwaZulu-Natal, South Africa. Her expertise and present research programme include food security and water use security, smallholder farmer development and market access. She has received a Women in Science (rural development of women) award and a Wonder Women in Science (WWS) award in 2018. She is a Cornell University and STIAS Visiting Scholar.

Earthscan Food and Agriculture Series

Other books in the series include:

Farming Systems and Food Security in Africa
Priorities for Science and Policy Under Global Change
Edited by John Dixon, Dennis P. Garrity, Jean-Marc Boffa, Timothy Olalekan Williams, Tilahun Amede with Christopher Auricht, Rosemary Lott and George Mburathi

Consumers, Meat and Animal Products
Policies, Regulations and Marketing
Terence J. Centner

Gender, Agriculture and Agrarian Transformations
Changing Relations in Africa, Latin America and Asia
Edited by Carolyn E. Sachs

A Global Corporate Trust for Agroecological Integrity
New Agriculture in a World of Legitimate Eco-states
John W. Head

Geographical Indication and Global Agri-Food
Development and Democratization
Edited by Alessandro Bonanno, Kae Sekine and Hart N. Feuer

Multifunctional Land Uses in Africa
Sustainable Food Security Solutions
Elisabeth Simelton and Madelene Ostwald

Food Security Policy, Evaluation and Impact Assessment
Edited by Sheryl L. Hendriks

Transforming Agriculture in Southern Africa
Constraints, Technologies, Policies and Processes
Edited by Richard A. Sikora, Eugene R. Terry, Paul L.G. Vlek and Joyce Chitja

For further details please visit the series page on the Routledge website: www.routledge.com/books/series/ECEFA/

Transforming Agriculture in Southern Africa

Constraints, Technologies, Policies and Processes

Edited by Richard A. Sikora, Eugene R. Terry, Paul L.G. Vlek and Joyce Chitja

LONDON AND NEW YORK

First published 2020
by Routledge
2 Park Square, Milton Park, Abingdon, Oxon OX14 4RN

and by Routledge
52 Vanderbilt Avenue, New York, NY 10017

Routledge is an imprint of the Taylor & Francis Group, an informa business

British Library Cataloguing-in-Publication Data
A catalogue record for this book is available from the British Library

Library of Congress Cataloging-in-Publication Data
Names: Sikora, Richard A., editor. | Terry, E. R., editor. | Vlek, Paul L. G., editor. | Chitja, Joyce, editor.
Title: Transforming agriculture in southern Africa : constraints, technologies, policies and processes / edited by Richard A Sikora, Eugene R. Terry, Paul L Vlek, Joyce Chitja.
Description: Abingdon, Oxon ; New York, NY : Routledge, 2020. | Includes bibliographical references and index.
Identifiers: LCCN 2019039355 (print) | LCCN 2019039356 (ebook) | ISBN 9781138393530 (hardback) | ISBN 9780429401701 (ebook)
Subjects: LCSH: Agricultural innovations—Africa, Southern. | Sustainable agriculture—Africa, Southern. | Food security—Africa, Southern.
Classification: LCC S494.5.I5 .T743 2020 (print) | LCC S494.5.I5 (ebook) | DDC 630.968—dc23
LC record available at https://lccn.loc.gov/2019039355
LC ebook record available at https://lccn.loc.gov/2019039356

ISBN: 978-1-138-39353-0 (hbk)
ISBN: 978-0-429-40170-1 (ebk)

Typeset in Bembo
by Apex Covantage, LLC

MIX
Paper from
responsible sources
FSC
www.fsc.org   FSC® C013056

Printed and bound in Great Britain by
TJ International Ltd, Padstow, Cornwall

Contents

Contributors

Cornel Adler, Senior Researcher, Julius Kühn-Institut (mJKI) Federal Research Centre for Cultivated Plants, Germany.

Juliet Akello, International Institute of Tropical Agriculture, Zambia.

Ranajit Bandyopadhyay, International Institute of Tropical Agriculture, Nigeria.

Frédéric Baudron, CIMMYT (International Maize and Wheat Improvement Centre), Zimbabwe.

Heike Baumüller, Department of Economic and Technological Change, Centre for Development Research (ZEF), Germany.

Janos Bogardi, Emeritus Professor and Senior Fellow, Centre for Development Research (ZEF), University of Bonn, Germany.

Michael Brüntrup, Senior Researcher, Deutsches Institut für Entwicklungspolitik (DIE), Germany.

Katati Bwalya, National Institute for Scientific and Industrial Research, Zambia.

Mariama Camara, START International, Burkina Faso.

David Chikoye D., International Institute of Tropical Agriculture, Zambia.

Joyce Chitja, Senior Lecturer, University of KwaZulu-Natal, South Africa.

Pauline Chivenge, International Rice Research Institute (IRRI), Philippines.

Ralph D. Christy, Professor of Emerging Markets, Dyson School of Applied Economics and Management, Cornell University, USA.

Jennifer Cockerill, Master of International Affairs Graduate Student, Pennsylvania State University, USA.

Tracy Davids, Bureau for Food and Agricultural Policy and Department of Agricultural, University of Pretoria, South Africa.

Joseph D. DeVries, President, Seed Systems Group, Kenya.

Thomas Dubois, International Centre of Insect Physiology and Ecology (*icipe*), Kenya.

Kennedy Dzama, Deputy Dean and Professor, Faculty of AgriSciences, University of Stellenbosch, South Africa.

Barbara Gemmill-Herren, Senior Associate, World Agroforestry Centre, Kenya.

Mario Giampietro, ICREA Research Professor, Institute of Environmental Science and Technology, Universitat Autònoma de Barcelona & Catalan Institution for Research and Advanced Studies (ICREA), Spain.

Evan Girvetz, International Center for Tropical Agriculture (CIAT), Kenya.

Delia Grace, Co-Leader, Animal and Human Health, International Livestock Research Institute (ILRI), Kenya.

E.T. Gwata, Department of Plant Production, School of Agriculture, University of Venda, South Africa.

Tinyiko Edward Halimani, Department of Animal Science, University of Zimbabwe, Zimbabwe.

Rhett D. Harrison, World Agroforestry (ICRAF), Zambia.

Gisèle L. Herren, Research Associate in Nematology, University of Ghent, Belgium.

Jeremy K. Herren, International Centre for Insect Physiology and Ecology, Kenya.

Thomas S. Jayne, Professor, Department of Agriculture, Food and Resource Economics, Michigan State University, USA.

Muhammadou M.O. Kah, Professor of IT & Computing, Vice President for Academic Affairs & Provost, American University of Nigeria, Nigeria.

Hannah N. Kamau, World Agroforestry Center (ICRAF), Kenya.

Mohammad Karaan, Professor, Department of Agricultural Economics, University of Stellenbosch, South Africa.

Reiner Klingholz, Director, Berlin Institute for Population and Development, Germany.

Mark D. Laing, Professor, Department of Plant Pathology, University of KwaZulu-Natal, South Africa.

Christine Lamanna, World Agroforestry Center (ICRAF), Kenya.

Moses Herbert Lubinga, Senior Economist: Markets and Economic Research Centre (MERC), National Agricultural Marketing Council (NAMC), South Africa.

Giuseppe Maggio, Econometric Specialist, FAO Food and Agriculture Organization of the United Nations, Italy.

George Mahuku, Senior Plant Pathologist for Eastern, Southern and Central Africa International Institute of Tropical Agriculture (IITA), Tanzania.

Paul Mapfumo, Professor, University of Zimbabwe, Zimbabwe.

Tlou S. Masehela, South African National Biodiversity Institute, South Africa.

Dorcas Matangi, CIMMYT (International Maize & Wheat Improvement Center), Zimbabwe.

Ferdi Meyer, Associate Professor, Bureau for Food and Agricultural Policy and Department of Agricultural Economics, University of Pretoria, South Africa.

Thulasizwe Mkhabela, Agricultural Research Council (ARC), Agricultural Economics, Impact & Partnerships, South Africa.

Gabisile Mkhize, Lecturer and Coordinator Gender Studies, The School of Social Sciences, University of KwaZulu-Natal, South Africa.

Annelin Molotsi, Department of Animal Sciences, University of Stellenbosch, South Africa.

Phetogo I. Monau, Department of Animal Science & Production, Botswana University of Agriculture and Natural Resources, Botswana.

Florence Mtambanengwe, Professor, Department of Soil and Agricultural Engineering, University of Zimbabwe, Zimbabwe.

Milu Muyanga, Assistant Professor, Department of Agriculture, Food and Resource Economics, Michigan State University, USA.

Caroline Mwongera, International Center for Tropical Agriculture (CIAT), Kenya.

Raymond Nazare, University of Zimbabwe, Zimbabwe.

Edson Ncube, Agricultural Research Council, Grain Crops, South Africa.

Simphiwe Ngqangweni, Senior Manager: Markets and Economic Research Centre (MERC), National Agricultural Marketing Council (NAMC), South Africa.

Thibault Nordey, CIRAD, France and The World Vegetable Center, Eastern and Southern Africa, Tanzania.

Pheneas Ntawuruhunga, International Institute of Tropical Agriculture (IITA), Zambia.

Ingrid Öborn, University Professor, Department of Crop Production Ecology, Swedish University of Agricultural Sciences, Sweden, and World Agroforestry (ICRAF), Kenya.

Erich-Christian Oerke, University Professor, Institute of Crop Science and Resource Conservation (INRES), Department of Plant Pathology and Protection, University of Bonn, Germany.

Sylvester O. Oikeh, African Agricultural Technology Foundation (AATF), Kenya.

Umezuruike Linus Opara, Distinguished Professor and SARChI Chair in Postharvest Technology, Postharvest Technology Laboratory, Faculty of AgriSciences, Stellenbosch University, South Africa.

Jon Padgham, Director of Partnerships, START International, USA.

Anusuya Rangarajan, Professor and Director, Cornell Small Farm Program, Cornell University, USA.

Elizabeth Ransom, Associate Professor, School of International Affairs and Senior Research Associate, Rock Ethics Institute, Pennsylvania State University, USA.

Hussein Shimelis, University Professor, African Centre for Crop Improvement, University of KwaZulu-Natal, South Africa.

Richard A. Sikora, Emeritus Professor, Phytopathology, Institute for Crop Science and Resource Conservation (INRES), University of Bonn, Germany.

Idah Sithole-Niang, Department of Biochemistry, University of Zimbabwe, Zimbabwe.

Nicholas J. Sitko, Food and Agriculture Organization (FAO) of the United Nations, Italy.

Philip C. Stevenson, Professor and Head of Chemical Ecology, Natural Resources Institute (NRI), University of Greenwich, UK.

Lulseged Tamene, International Centre for Tropical Agriculture (CIAT), Ethiopia.

Ndiadivha Tempia, Lecturer in Economics, Tshwane University of Technology, South Africa.

Eugene R. Terry, Implementing Director, The African Agricultural Technology Foundation, USA.

Jennifer A. Thomson, Emeritus Professor, Department of Molecular and Cell Biology, South Africa.

Leena Tripathi, Plant Biotechnologist, International Institute of Tropical Agriculture (IITA), Kenya.

Johnnie Van den Berg, Programme Manager: IPM, North West University South Africa (NWU), North-West University, South Africa.

Bernard Vanlauwe, Central Africa and Natural Resource Management, International Institute of Tropical Agriculture (IITA), Kenya.

Altus Viljoen, University Professor and Chair, Department of Plant Pathology, Stellenbosch University, South Africa.

Nick Vink, University Professor and Chair, Department of Agricultural Economics, University of Stellenbosch.

Paul L.G. Vlek, Emeritus Professor and Senior Fellow, Centre for Development Research (ZEF), University of Bonn, Germany.

Elizabeth R. Weatherly, Master of International Affairs Graduate Student, School of International Affairs, Pennsylvania State University, USA.

Sileshi G. Weldesemayat, Senior Scientist and Consultant, Ibex Meanwood, Zambia and University of KwaZulu-Natal, South Africa.

Shamie Zingore, International Plant Nutrition Institute (IPNI), Kenya.

About the editors

Richard A. Sikora is Emeritus Professor of the University of Bonn, Germany; former chairman of the Department of Plant Pathology and the Section, Soil Ecosystem Phytopathology and Nematology. He was Chair of the CGIAR System-wide programme on integrated pest management and is presently a Fellow and Convener of the STIAS Forum on Sustainable Intensification. Richard has received numerous awards including the: University of Illinois Alumni Association Award of Merit; International Association for the *Plant Protection* Sciences Award of Distinction; American Phytopathological Society International Service Award and the 2017 German Phytopathological Society *Anton de Bari* medal for excellence in phytopathology.

Eugene R. Terry was the Director General of WARDA, now AfricaRice, and the Founding Director of the African Agricultural Technology Foundation (AATF). Currently serves as the Chair of the Advisory Panel for Biosciences for East and Central Africa (BecA) – ILRI Hub, and the Chair of the Advisory Board of the West Africa Centre for Crop Improvement (WACCI) hosted by the University of Ghana in Accra. Dr. Terry was awarded the AfricaRice Distinguished Service Award in 2010, for exemplary leadership in rice research and development in Africa; and also, The Macdonald College, McGill University Distinguished Alumni Award in 2012, for outstanding contributions to society and humanity.

Paul L.G. Vlek is Emeritus Professor of the University of Bonn and former Director of the Centre for Development Research ZEF as well as the West African Science Service Center on Climate Change and Adapted Land Use WASCAL. He devoted many years of service to a number of journal editorial boards and national and international scientific committees. In 2014, he was awarded the World Agriculture Prize for his lifetime contribution to agricultural development.

Joyce Chitja is a Senior Lecturer in Food Security at the University of Kwa-Zulu-Natal. Her expertise and present research programme include food security and water use security, smallholder farmer development and market access. She has received a Women in Science (rural development of women) award and a Wonder Women in Science (WWS) award in 2018. She is a Cornell University and STIAS Visiting Scholar.

Foreword

Much has been written and said about African food security and agricultural development. Results from economic and policy studies have provided input into a multitude of conferences and workshops. Problems have been diagnosed, challenges have been identified and recommendations for action have been made. The evidence created by these efforts is essential to guide action by policymakers.

So why is it that African policymakers are still confronted with huge challenges, and why are so many African children still malnourished and natural resources being degraded? Maybe lack of evidence is not the binding constraint to achieve food security and sustainability goals, or maybe the evidence is either not reaching the decision makers or is irrelevant to the situation within which decisions are made. Policymakers are busy people. They are under much pressure from various interest groups, trying to achieve a variety of goals, of which agricultural development and food security may or may not take priority. Policy recommendations based on an understanding of the policy process within which they will be received are more likely to translate into action than those that are not. First-best solutions from the sole perspective of food security and agricultural development may not be feasible with the policy space of the decision maker. Pragmatism, which may include second- or third-best solutions, is called for when trying to influence policy decisions.

Is this book going to make a difference? I believe so for at least three reasons. First, the editors and chapter authors are among the most knowledgeable experts on African agricultural development and food security. Equally important, they provide recommendations for action that take into account the pressures on policymakers from other interest groups. Second, each chapter is short enough to entice the policymaker or policy advisor to read it and sharply focused on how to turn the most relevant existing evidence into policy action, and third, the book provides an integrated, wholistic set of policy recommendations focused on some of the most critical challenges facing Southern Africa, including rapidly increasing population and urbanization, continued malnutrition and household food insecurity as well as land pressures and climate change.

Large productivity gaps in Southern African agriculture provide opportunities for expanded food production and improved productivity. As stressed by the editors and several of the chapter authors, these opportunities are best exploited by accelerated use of existing technology and agricultural research to fill the gaps in the existing knowledge and technology. In particular, there is, in my opinion, an urgent need for more research to help African agriculture adjust to drought, floods, strong winds, new plant and animal diseases and other biotic and abiotic factors resulting from climate change and land pressures. Increasing food production and reduced risks do not, by themselves, solve the food security and nutrition problems in the region, but they are important components of a wholistic solution, particularly if they are oriented to a diversified portfolio of foods to meet both nutrient and calorie needs. As urbanization proceeds in Southern Africa and urban food demands increase, a closer look at land ownership and the future of smallholdings seems appropriate, along with an increasing emphasis on urban food security, nutrition and related health problems.

I congratulate the editors and chapter authors for an excellent book, which, I believe, will make a difference in both human and environmental health and well-being.

Per Pinstrup-Andersen
Professor Emeritus, Cornell University, USA, and
Adjunct Professor, Copenhagen University, Denmark

Preface

The Stellenbosch Institute of Advanced Study (STIAS) initiated a select number of long-term programmes in 2013, with the broad objective of making an impact on African development. One of the projects selected for long term support under the broader research theme of Sustainable Agro-Ecosystems was the "The impact of sustainable intensification of agriculture on food security, the environment and human well-being in the rural urban continuum of Southern Africa".

The STIAS Forum on Sustainable Intensification that materialized devoted particular attention to those challenges relevant to "Strategic directions for Agricultural Transformation in Southern Africa" towards the year 2050, which was the topic of a roundtable held at STIAS with regional representatives in 2015.

This book attempts to capture all the ideas, visions, strategies and lessons generated through the various discussion fora, seminars, individual research and analysis, as well as the visions of other stakeholders, in a consolidated publication on agricultural transformation.

The editors of the book are STIAS Fellows with a great deal of practical experience across Africa both in the field of education but also in applied research. They were the core fellows of the Forum and were responsible for developing the research programme and organizing the discussion fora.

Richard A. Sikora, is emeritus Professor of the University of Bonn, Germany; former chairman of the Department of Plant Pathology and Head of the Research Programme – Soil Ecosystem Phytopathology and Nematology. He is Convener of the STIAS Forum on Sustainable Intensification.

Eugene R. Terry was the Director General of WARDA, now AfricaRice, and the Founding Director of the African Agricultural Technology Foundation (AATF).

Paul L.G. Vlek is Professor emeritus University of Bonn and former Director of the Centre for Development Research ZEF as well as the West African Science Service Center on Climate Change and Adapted Land Use WASCAL.

Joyce Chitja is a Senior Lecturer at the University of KwaZulu-Natal in Food Security. Her expertise and present research programme include food security and water use security, smallholder farmer development and market access.

•

The authors of the chapters on technology and policy/processes were either from Southern Africa or had experience in other regions of Africa. Many of the authors were invited to the STIAS Forum on Sustainable Intensification to discuss their expertise as it relates to improving sustainable agricultural production.

The chapters are divided into distinct parts that include: an introduction to Southern African agriculture; discussion of the drivers and constraints influencing change; description of both current and proven as well as emerging technologies that can improve sustainable intensification; and finally a section on policies and processes that are needed to implement transformation.

The book directly targets decision makers, or those who have the greatest influence on agricultural transformation and make decisions that directly impact food production and food security. The authors of the chapters, therefore, were asked to include a section on policy recommendations for this target group.

The short and concise format of the chapters basically represents expanded science- and/or policy-briefs, which are used to ensure that the expert analyses and critical reviews of the factors important in driving future transformation are readable and useful for decision makers. The goal is to stimulate the development of government programmes that will lead to meaningful and substantive improvement of agriculture at the small- to medium-size family farm level. These farmers are underproducing at the present time, and they are an integral component of the responses to food security challenges and a key to solving future food security issues in the region. Many of the technologies and policy recommendations will also have importance for larger family and commercial farms.

We believe the findings presented in the chapters in this book are relevant to other agricultural regions of the world where transforming agricultural systems is needed and important for future food security.

Acknowledgements

The editors want to acknowledge STIAS for the excellent support given throughout the Forum's five year research program. We also want to thank them for their generous financial support for the open-access form of publication. The facilities they offered the Fellows and the financial support for workshops and expert visitation were outstanding.

We want to thank the Wallenberg Foundation for their support of the important Roundtable meeting held in 2015 on "Agricultural transformation: Sustainable intensification of food production in the Southern African region".

The STIAS Fellows want to express our special thanks to the following:

Professor Heinrich Geyer, then-STIAS Director for his scientific and strategic and financial support.

Dr Christoff Pauw, Programme Manager for his advice and support in secure funding for final publication.

Professor Eugene Cloete, Vice Rector, Research, Innovation and Postgraduate Studies, Stellenbosch University for offering financial support for publication

Professor Peter Stehle, Dean, Faculty of Agriculture, University of Bonn for financial support for the book's publication.

Ms Nel-Mari Loock, Programme and IT administrator for her energy giving technical help when needed.

Ms Michelle Galloway, Media officer for her effort to open the forum up to the press at large.

Professor Johann Groenewald, Coordinator: Strategic initiatives, who helped in the organization of the Roundtable Meeting.

Gratitude is also given to the STIAS Fellows Janos Bogardi, former acting Vice Rector of the United Nations University Campus, Bonn, Germany, and Lucas Gakale, Former Permanent Secretary, Ministry of Agriculture, Botswana, who supported the Forum in the first year and who were influential in making the roundtable meeting a great success.

We also greatly appreciate all those who actively participated in the Round-table on Strategic Directions for Agricultural Transformation in Southern Africa, and who expressed their vision for the future.

The support of the editors at Taylor & Francis were essential in ensuring the high quality of the book, and we thank them all.

Part I

Overview of Southern African agriculture

1 Introduction

Richard A. Sikora, Eugene R. Terry,
Paul L.G. Vlek and Joyce Chitja

Global population is expected to reach an estimated nine billion by the year 2050. Addressing the global food demand that this represents will require significant investments and policy reforms to transform important key agricultural production systems. Specifically, these investments should lead to higher returns from sustained productivity growth, from infrastructure development, from institutional reforms and from the environmental services generated by sustainable resource management (FAO, 2008, 2009; IFPRI, 2013).

Nearly all of this future increase in population will take place in those parts of the world comprising today's developing countries. This trend in growth is very evident in Southern Africa where population growth rates are among the highest in the world. In 1960, there were only 43 million people living in the region. It is projected, however, that by the year 2050, the population will increase to about 350 million people (Chapter 3 on population growth rates in this volume) with most of the growth in rural areas.

Rural growth will outpace opportunities for employment in primary agriculture. Therefore, in Southern Africa, there is an urgent need for the creation of a transition to non-agriculture employment (see Chapter 31 on small-scale enterprise in this volume).

More than 70% of the world's population is expected to be urban by 2050 due to rural-urban migration and this will significantly influence lifestyles, income levels and it will change food consumption patterns (FAO, 2009). Africans living in urban areas increased from 28% in 1980 to 40% today and are projected to grow to 50% by 2030.

Africa is the world's most food insecure continent, with relatively low levels of agricultural productivity, low rural incomes, high rates of malnutrition and a significantly worsening food trade balance (see Chapter 2 of this volume). However, Africa possesses 60% of the world's arable land and 70% of its water as well as human capital. These natural resources are important components needed for improved production. Conversely, African small-scale farmers, who are in the majority, lack access to modern technology to improve food production. These farmers are an important element in the transformation of food system in Southern Africa as outlined in chapters in this volume. Southern Africa could contribute significantly to the growing global demand for food, as well as, to energy markets through a sustainable transformation of agriculture.

Agriculture accounts for about 40% of GDP, 15% of exports and 60–80% of Africa's employment. Therefore, the transformation to a more efficient and environmentally sound agriculture system will be fundamental in achieving agriculture economies that: 1) create growth, 2) deliver opportunities for a growing youthful population, 3) tackle malnutrition and food insecurity and 4) simultaneously protect and sustain the natural resource base.

These objectives will not be realized without significant investments and radical improvements in access to modern agricultural technology and a simultaneous improvement in agricultural policy that favours the small-scale landowners. This shift will require substantial private and public investment as well as more efficient public investment (NEPAD, 2003).

The countries in Southern Africa mainly targeted in this book includes Lesotho, Malawi, Mozambique, Namibia, South Africa, Swaziland, Tanzania, Zambia and Zimbabwe, but countries bordering this area are also considered in the book chapters. Agriculture in Southern Africa is at present labour intensive and inefficient, especially with regard to the small size landholders. An important element of the agricultural transformation is therefore improvement of agricultural productivity by improving land tenure issues, access to knowledge, improve credit, access to mechanization and better market access, to name a few of the modern technologies of production used in other areas of the world as outlined in the chapters in this volume. Initiatives in this direction would provide multiple benefits, including increased food availability and improve food security, while freeing up labour to participate in future non-agriculture enterprises (SADC, 2018).

Evidence suggests that the future of Southern Africa is an urban one, and that urban food insecurity is therefore a large and growing challenge to the agricultural community of the population. The causes, determinants and solutions for food insecurity vary between rural areas where food is produced and the urban settings that are basically domestic food importers. It is important therefore that urban food insecurity be addressed as an integral component of the food security agenda of Southern Africa (Crush and Frayne, 2010).

The editors provided the authors writing the chapters with a set of hypotheses as they relate to agricultural transformation in the region as follows:

1   The effect of continuous population pressure and adverse effects of climate change resulting in agriculture being practiced today on land that is often unsuitable for sustainable food production and where the cost in ecosystem services exceeds the meagre returns in any form of agricultural production. Therefore, land-use policies, reforms and appropriate community interventions will be needed to address this issue.
2   The land that can be managed sustainably for agricultural production in Southern Africa is extremely diverse as is the cultural and institutional

environment in which farmers operate. Community-based land-use planning is needed to ensure the optimal use of land, be it for plantations, pastures, agricultural or horticultural crops.

3   The myriad of management and technological interventions to augment agricultural production systems in a profitable fashion need to be tailored to these conditions and adjusted as these conditions change.

4   In contrast with most other parts of the world, the productivity gap in most African agricultural regions is well in excess of 50% even in many areas that can be considered favourably endowed biophysically as well as institutionally.

5   In the quest to produce the food needed for the Southern Africa population of the future, efforts to intensify agriculture sustainably should target those areas where the biophysical as well as institutional conditions are conducive to socioeconomic success and advancement.

6   The interventions and technologies that are proposed for sustainable intensification should aim at reaching the economically feasible production potentials of the targeted region by eliminating genetic and other biophysical constraints, avoid major losses due to pest or diseases either pre- or postharvest and avoid any cost due to loss of ecosystem services.

7   Pre and postharvest losses in Southern Africa amount to 30% or more. Areas where food production systems are intensified and serve commercial markets such losses need to be eliminated through proper pest management, extension and infrastructural investments.

8   Enabling conditions for sustainable intensification will require secure land rights, public investments in infrastructure, market, storage and value chain development, credit and other services such as research and extension, education and safety nets to cope with crop failures as well as an active engagement with the private sector.

9   Given the rapidly ageing farmer and rural population of Southern Africa, there is an urgent need to provide strong motivation and incentives to retain more tech-savvy and entrepreneurial youths to both farming and business in the food and value chain economy in the rural communities who can be the agents of sustainable intensification.

The book provides policy recommendations derived from analysis of the most relevant elements of these hypothetical considerations and predicated on the following assumptions:

1   That most of the technologies required to transform agriculture and significantly increase crop production are already available and effective

2   That farmer access to these technologies is often impeded by a lack of appropriate agricultural policies and/or processes and market access

3   Bringing technology and policy together is required to successfully transform agricultural systems in the region.

Thus, the book is divided into five sections that present relevant content on:

1   The drivers of change towards agricultural transformation, such as population growth, climate change, globalization and urbanization; constraints to improving and increasing agricultural productivity, such as land and water resources, as well as biological constraints and economic barriers
2   A section on current technologies and a second on emerging technologies available to address biophysical and other production constraints, such as germplasm improvement, pest and disease management and improved crop and animal production tools
3   An additional section is devoted to policy issues, capacity building, land tenure, infrastructure and markets that will be needed to facilitate the agricultural transformation towards sustainable intensification
4   The last section defines the outlook and where we hope the book is helpful in the future for transforming agriculture in Southern Africa.

The book should be used as a handbook, or solutions cookbook, whereby decision makers attempting to improve agricultural production, select those chapters (technologies, policies, processes) that fit the problems facing the farmers and agricultural production systems in their own country. This is basically a book of high-quality ingredients and not of recipes – this aspect is being left to the decision makers.

We hope the book will help decision makers develop the policies that will improve agricultural production on the farms of Southern Africa and the adoption of processes that will improve food production and access to food and thus enhance human nutrition and well-being in the region.

References

Crush, J. and Frayne, B. (eds.) (2010). *Pathways to Insecurity: Food Supply and Access in Southern African Cities. African Food Security Urban Network (AFSUN) Series 3.* Cape Town: Unity Press: 48.

FAO (2008). *Food Security in the World.* Rome, Italy: FAO: 56.

FAO (2009). How to Feed the World in 2050. *Proceedings of the Expert Meeting, FAO Headquarters.* Rome, Italy: FAO: 514.

IFPRI (2013). Global Food Policy Report. *International Food Policy Research Institute.* Washington, DC: USDA: 142.

NEPAD (2003). *New Partnership for Africa's Development. Comprehensive Africa Agriculture Development Programme (CAADP) - New Partnership for Africa's Development.* Midrand, South Africa: NEPAD: 102.

SADC (2018). *Promoting Infrastructure Development and Youth Empowerment for Sustainable Development. 38th SADC Summit.* Namibia, Gaborone, Botswana: SADC Secretariat: 108.

2 The strategic role of agriculture in the economic space of the Southern Africa region

Elizabeth Ransom, Jennifer Cockerill and Elizabeth R. Weatherly

The 15 countries under discussion in this volume are culturally, geographically and socioeconomically diverse, with the one commonality that the majority of people in these countries rely on agriculture for their livelihoods. While the contribution of agriculture to GDP is relatively small, approximately 70% of the population in this region relies on agriculture, most as smallholders. Climate change when coupled with the prevalence of smallholders' dependence upon rain-fed agriculture in this region creates a daunting challenge for policymakers, leaders and citizens moving forward.

Overview

Apart from Mauritius, Namibia, Seychelles and South Africa, all other countries under discussion in this book have a majority of their population reliant upon agriculture (World Bank, 2019a).[1] Even among the four countries just listed, food and agriculture remain critically important. For example, food and nutritional security is a growing concern for the majority of Seychelles inhabitants, as they are highly dependent upon food imports, agriculture in the country has received limited investment, and there is limited arable land available for production (AfDB, 2016). Moreover, while all the countries under discussion have seen a decline in the percentage of GDP earned from agriculture (average 17% of GDP contribution), job creation in other sectors, particularly the service sector, has been slow and poverty remains high. For example, the percentage of the population living on less than US$1.90 per day in Democratic Republic of Congo (DRC), Lesotho, Madagascar, Malawi, Mozambique and Zambia ranges between 58 and 78% (World Bank, 2019b). This dependence on agriculture among most of the population is ever more problematic due to changing weather patterns related to climate change, as approximately 94% of all agriculture in this region is rain-fed, with South Africa having the most irrigated agriculture (SADC, 2018).

Southern Africa has been described as a climate hotspot, as the region is expected to experience increasing aridity with low adaptive capacity (Nhamo et al., 2019). Of the countries under discussion, South Africa is by far the largest, both economically and in terms of agricultural production and trade in the region. However, the drought that hit the Southern Africa region in 2016–17, and the prevalence of animal diseases, particularly highly pathogenic avian influenza (H5N8) that decimated egg production in South Africa, contributed to higher than normal imports and revealed the vulnerability of not only South Africa's agricultural sector but also much of Southern Africa to extreme weather events and animal diseases (Friedenberg, 2018).

Trade in agriculture and food: economic partnerships agreement negotiations

Following the global trend, the number of regional trade agreements (RTAs) that the aforementioned countries are a part of has grown exponentially over the past few decades and for very good reasons. One outgrowth of the creation of the World Trade Organization in 1995 has been an explosion in RTAs, whereby countries and regions negotiate terms of trade with other countries and regions. The existence of RTAs has been found to significantly impact agricultural trade flows, which means not having a free trade agreement with a specific market is considered a disadvantage (OECD, 2015). Today, WTO member countries have an average of 13 RTAs, with some WTO members having as many as 20 RTAs (OECD, 2015).

Within our region of focus the number of individual country RTAs is much lower and individual countries differ in their RTA memberships, but all are members of SADC (WTO, 2019). SADC created a Protocol on Trade (signed in 1996; entered into force 2001) and SADC is currently negotiating with the Common Market for Southern and Eastern Africa (COMESA) and East African Community (EAC) to finalize a Tripartite FTA, which is intended to lay the groundwork for the Africa Continental Free Trade Agreement (AfCFTA). Both the Tripartite FTA and AfCFTA are designed to increase trade internal to the continent of Africa. AfCFTA was signed by 55 African countries July 7, 2019, and once implemented, is considered a possible "game changer" due to the sheer size of coverage, both in terms of population (1.3 billion people) and economies (US$3.4 trillion) (Balima, 2019).

These new RTAs are important for expanding the import/export opportunities in agriculture for the 15 countries under discussion, as many of these countries have had an overreliance on the EU and individual European countries, which has become an increasingly difficult market to access for agricultural commodities (Ransom, 2015). However, the degree to which either trade agreement will benefit smallholders is questionable, as most trade policies tend to favour systems of production that are well capitalized, thereby ensuring farm upgrades to meet export specifications (i.e., disease or pest control requirements).

Key issues impacting the transformation of Southern African agriculture

In what follows we discuss three interlocking issues that, when combined, create an uncertain future for Southern African agriculture moving forward.

Specifically, smallholder production and food insecurity, land tenure and gender inequality are three important and complex topics. There is not the space for a comprehensive discussion of each, but we will identify the ways in which these three topics have bearing on whether agriculture can successfully be transformed in Southern Africa.

Smallholders and food insecurity

The countries under discussion with a few exceptions, have relatively high levels of poverty and food insecurity and a large percentage of smallholders. Among rural populations, approximately 16% of people have been consistently designated as food insecure in the past five years (SADC, 2018). The DRC, Malawi, Madagascar, South Africa and Zimbabwe were identified by SADC's RVAA report (2018) as having the most food insecure people in the previous year. For all five of these countries, agricultural production was impacted by weather events (droughts, cyclones) or human and animal disease outbreaks and most of these countries' populations consist of smallholders.

With the preponderance of smallholders, there are several trends to note. First, smallholders are an increasingly diverse group, with evidence that there may be growing inequality in landholdings, access to water resources and livestock (Jayne et al., 2010; Moyo, 2014; Swatuk, 2008). Second, the size of most land holdings among smallholders is shrinking, and third, as mentioned previously, economic trade agreements tend not to benefit smallholders. This contributes to a situation where most small farms in Africa are "becoming increasingly unviable as sustainable economic and social units", which leads Jayne et al. (2010, p. 1394) to conclude that "unless government policy is changed radically, the world may see increasingly frequent and severe economic and social crises in Sub-Saharan Africa".

Land tenure and the financialization of agriculture

Land tenure systems in Southern Africa continue to reflect the complicated histories of most countries in the region. There is a combination of private (freehold and leasehold), communal/customary and state-owned lands, with private ownership the least likely form of ownership in most countries (see Table 2.1). In cases where private landownership is higher – Zimbabwe, Namibia, South Africa – the governments continue to grapple with extremely inequitable land holdings.

For the most part, smallholder farms in Southern Africa are getting smaller and less sustainable for smallholder livelihoods (Lowder et al., 2016). Even in areas where land has been (i.e., Namibia, Zimbabwe) or could be redistributed (i.e., Angola), simply giving land to rural people, does not ensure agricultural productivity, as rural people may use lands in a variety of ways (Ferguson, 2013). Moreover, while increasing smallholders' landholdings is needed to improve livelihoods, more land without increases in investment, such as animal traction, irrigation and fertilizers will limit productivity (Foley, 2007).

Table 2.1 Types of land tenure in Southern Africa

Country	Area (ha)	Population[1]	Ownership (% of land)		
			Public	Private	Communal/ customary
Angola[2]	124,670,00	30,809,760			Approx. 85–90%
Botswana	58,200,000	2,254,130	25%	4%	71%
DRC[2]	226,700,000	84,068,090			Unknown
Eswatini (formerly Swaziland)	1,736,400	1,136,190	19%	25%	56%
Lesotho	3,035,500	2,108,130	5%	5%	90%
Madagascar[3]	58,704,100	26,262,370	45%	15%	40%
Malawi	9,400,000	18,143,310	22%	12%	66%
Mauritius	204,000	1,265,300		80%	
Mozambique	80,159,000	29,495,96	7%	0%	93%
Namibia	82,400,000	2,448,260	20%	44%	36%
Seychelles[4]	45,000	96,760	Over 60%		0%
South Africa	122,103,700	57,779,620	14%	72%	14%
Tanzania	94,508,700	56,318,350	15%	1.50%	84%
Zambia	75,261,400	17,351,820	30%	6% (Leasehold)	64%
Zimbabwe	39,100,000	14,439,020	16%	41%	42%

1 2018 Population estimates (World Bank)
2 Land tenure in Angola and DRC are not known, as reliable data has not been collected and land reform to date has largely failed. For both countries, the majority of citizens are said to exist on land that has been passed down through families or falls under customary, local rules. For example, an estimated 85–90% of Angolans hold their land without any recognized rights under formal law. In both countries, technically, the government controls the majority of land. In the case of Angola, Cain (2013) notes that legislators have historically and in the present demonstrated a tendency to contain or circumscribe the land rights of the country's rural and poor peri-urban populations.
3 Madagascar went through land reform in 2005, with the goal of converting most land to titled private property, including communally managed lands. However, only 10–15% is estimated to have successfully been titled in subsequent years.
4 Actual percentage not available, but at least 60% of land is protected for environmental reasons according to FAO.

Source: Amended from SADC (2010)

Another dimension to land tenure in Southern Africa is the increase in large-scale land acquisitions by financial actors. Land grabs, as they have been dubbed by critics, are part of a broader shift to the financialization of agri-food systems, whereby financial actors and their financial logics are transforming agri-food systems (Bjørkhaug et al., 2018). According to The Land Matrix (2019), the only country in the region of our focus that does not have a reported land deal is Seychelles. With the vast majority of smallholders situated on communal lands that have weak legal status in these countries, the opportunity arises for unscrupulous governing leaders to take undue liberties with their citizen's lands (Wily, 2011). Thus, there are calls for a formalization of land ownership in Southern Africa to protect citizens from land dispossession. However, formalization of land tenure is not so straightforward within the diverse populations of Southern Africa – close to half the countries in our focus have pastoralist or agro-pastoralist populations. Formalization can often serve as a trojan horse for sedentarization of pastoralists who have historically not fit formal legal systems (Basupi et al., 2017).

Gender inequality

Land tenure and gender inequality are intimately linked in agriculture in Southern Africa. For the most part, even when legal frameworks have been amended to recognize women's right to land, as in the case of Madagascar, Mozambique and Zimbabwe, customary laws and traditions continue to block women from ownership (Kimani, 2008; USAID, 2019). Most women can only access land through male relatives (i.e., husband), which makes their land tenure insecure and vulnerable to losses of livelihoods for themselves and their children should they lose their husbands (e.g., HIV/AIDS related deaths). In addition, due to their limited access to not only land but also inputs and extension services, production levels are lower (FAO, 2011; Galiè et al., 2018). Moreover, women in agriculture often spend many more hours working, both in their own fields to grow food for household consumption and in men's fields to assist with cash crop production. A more general measure of gender inequality, not limited to women in agriculture, is the Gender Inequality Index, which reveals that much of the Southern Africa region scores poorly for gender equality along the three domains measured – reproductive health, empowerment and economic status (UNDP, 2019).

Gender inequality has important implications for the transformative potential of agriculture (Perez et al., 2015). Because men and women fulfill different roles and expectations and have access to different resources within agricultural systems, understanding gender dynamics are important for thinking about adaptation strategies in the region. Too often, proposed adaptions do not consider the gendered dimensions to agricultural systems, which inevitably contributes to the failure of these interventions (Farnworth and Colverson, 2015).

Additional factors

There are two additional issues that will shape agricultural policies, practices and institutional resources moving forward. First, there is the prevalence of HIV/AIDS in the region. Eastern and Southern Africa have the highest rates of HIV/AIDS infected populations in the world, with eSwatini having the highest percentage of the population infected (27.4%) and South Africa containing the most individuals infected in the world (7.7 million; 20% of the population) (UNAIDS, 2018). While the impact on the available labour force for agriculture has not been as significant as predicted (Jayne et al., 2010), the budgetary costs for providing much needed treatment will inevitably put a financial strain on national governments' budgets. Moreover, as mentioned earlier, there is a gender dimension to the consequences of HIV/AIDS deaths in rural areas in terms of women and girls' access to land.

A second factor that impacts the transformative potential of agriculture are the extreme inequalities in income and wealth in this region of the world. Approximately half of the countries under analysis in this book ranked among the most inequitable in the world – South Africa, Namibia, Botswana and Zambia rank first through fourth in 2018 (Beaubien, 2018). The populations most in need of agricultural transformations in Southern Africa, are also the populations that are most vulnerable to the effects of climate change, including water scarcity and increased pests and diseases in agriculture. Extreme inequality will exacerbate the ability of vulnerable communities to participate in the transformation of agriculture.

Note

1 Botswana and eSwatini while officially having lower numbers employed in the agriculture sector, unofficially the FAO estimates 50 and 70%of the population respectively rely on agriculture for food and income.

References

AfDB. (2016). *African Development Bank Country Strategy Paper 2016–2020: Republic of Sey-chelles*. Nairobi, Kenya. Retrieved from www.afdb.org/fileadmin/uploads/afdb/Docu ments/Project-and-Operations/SEYCHELLES_-_CSP_2016-2020_-_FINAL.pdf.

Balima, B. (2019). Economic 'Game Changer'? African Leaders Launch Free-Trade Zone. *Reuters*. Retrieved from www.reuters.com/article/us-africa-trade/economic-game-changer-african-leaders-launch-free-trade-zone-idUSKCN1U20BX.

Basupi, L. V., Quinn, C. H., & Dougill, A. J. (2017). Pastoralism and Land Tenure Trans-formation in Sub-Saharan Africa: Conflicting Policies and Priorities in Ngamiland, Bot-swana. *Land, 6*(4), 89.

Beaubien, J. (2018, April 2). The Country with the World's Worst Inequality Is: By World Bank Estimate, South Africa Is the World's Most Unequal Country. *Goats and Soda: Stories of Life in a Changing World*. Retrieved from www.npr.org/sections/goatsandsoda/2018/04/02/598864666/the-country-with-the-worlds-worst-inequality-is.

Bjørkhaug, H., Magnan, A., & Lawrence, G. (Eds.). (2018). *The Finacialization of Agri-Food Systems: Contested Transformations*. New York: Routledge.

Cain, A. (2013). Angola: Land Resources and Conflict. In J. Unruh & R. C. Williams (Eds.). *Land and Post-Conflict Peacebuilding*. New York, NY, Earthscan: 173–200.

FAO. (2011). *The State of Food and Agriculture, Women in Agriculture: Closing the Gender Gap for Development*. Rome: Food and Agriculture Organization of the United Nations (FAO).

Farnworth, C. R., & Colverson, K. E. (2015). Building a Gender-Transformative Extension and Advisory Facilitation System in Sub-Saharan Africa. *1*, 20–39. Retrieved from http://ageconsearch.umn.edu/record/246040/files/MAR%202015%20Vol%201%20issue%201%20Paper%202.pdf.

Ferguson, J. (2013). How to Do Things with Land: A Distributive Perspective on Rural Livelihoods in Southern Africa. *Journal of Agrarian Change, 13*(1), 166–174. doi:10.1111/j.1471-0366.2012.00363.x.

Foley, C. (2007). *Land Rights in Angola: Poverty and Plenty*. Humanitarian Policy Group Working Paper. London: Overseas Development Institute.

Friedenberg, J. (2018). *Southern Africa: A Promising Region for U.S. Agricultural Exports*. Washington, DC: USDA.

Galiè, A., Teufel, N., Korir, L., Baltenweck, I., Webb Girard, A., Dominguez-Salas, P., & Yount, K. M. (2018). The Women's Empowerment in Livestock Index. *Social Indicators Research*. doi:10.1007/s11205-018-1934-z.

Jayne, T. S., Mather, D., & Mghenyi, E. (2010). Principal Challenges Confronting Smallholder Agriculture in Sub-Saharan Africa. *World Development*, *38*(10), 1384–1398. https://doi.org/10.1016/j.worlddev.2010.06.002.

Kimani, M. (2008). Women Struggle to Secure Land Rights. *African Renewal*. Retrieved from www.un.org/africarenewal/magazine/april-2008/women-struggle-secure-land-rights.

The Land Matrix. (2019). *Africa Map*. Retrieved from https://landmatrix.org/.

Lowder, S. K., Skoet, J., & Raney, T. (2016). The Number, Size, and Distribution of Farms, Smallholder Farms, and Family Farms Worldwide. *World Development*, *87*, 16–29. https://doi.org/10.1016/j.worlddev.2015.10.041.

Moyo, S. (2014). *Land Ownership Patterns and Income Inequality in Southern Africa*. Background paper prepared for World Economic and Social Survey.

Nhamo, L., Mathcaya, G., Mabhaudhi, T., Nhlengethwa, S., Nhemachena, C., & Mpandeli, S. (2019). Cereal Production Trends Under Climate Change: Impacts and Adaptation Strategies in Southern Africa. *Agriculture*, *9*(2), 30.

OECD. (2015). *Regional Trade Agreements and Agriculture*. Paris: OECD.

Perez, C., Jones, E. M., Kristjanson, P., Cramer, L., Thornton, P. K., Förch, W., & Barahona, C. (2015). How Resilient Are Farming Households and Communities to a Changing Climate in Africa? A Gender-Based Perspective. *Global Environmental Change*, *34*, 95–107. https://doi.org/10.1016/j.gloenvcha.2015.06.003.

Ransom, E. (2015). The Political Economy of Agriculture in Southern Africa. In A. Bonanno & L. Busch (Eds.), *Handbook of the International Political Economy of Agriculture and Food*. Northampton, MA: Edward Elgar Publishing.

SADC. (2010). *Land Policy in Africa: Southern Africa Regional Assessment*. Addis Ababa, Ethiopia. Retrieved from www.uneca.org/sites/default/files/PublicationFiles/regionalassessment_southernafrica.pdf.

SADC. (2018). *SADC Regional Vulnerability Assessment & Analysis (RVAA) Synthesis Report on the State of Food and Nutrition Security and Vulnerability in Southern Africa*. Gaborone, Botswana: SADC.

Swatuk, L. A. (2008). A Political Economy of Water in Southern Africa. *Water Alternatives*, *1*(1), 24.

UNAIDS. (2018). *AIDSinfo*. Retrieved from http://aidsinfo.unaids.org/#.

UNDP. (2019). *Gender Inequality Index (GII), 2017*. Retrieved from http://hdr.undp.org/en/composite/GII.

USAID. (2019). *Madagascar-Lan Tenure and Property Rights Profile*. Washington, DC. Retrieved from www.land-links.org/wp-content/uploads/2010/12/USAID_Land_Tenure_Madagascar_Profile-2019.pdf.

Wily, L. A. (2011). 'The Law is to Blame': The Vulnerable Status of Common Property Rights in Sub-Saharan Africa. *Development and Change*, *42*(3), 733–757.

World Bank. (2019a). *Employment in Agriculture*. Retrieved from https://data.worldbank.org/indicator/SL.AGR.EMPL.ZS.

World Bank. (2019b). *Poverty Headcount Ratio at $1.90 a Day (% of Population)*. Retrieved from https://data.worldbank.org/indicator/SI.POV.DDAY?view=chart.

WTO. (2019, July 23). *Regional Trade Agreements Database*. Retrieved from http://rtais.wto.org/UI/PublicMaintainRTAHome.aspx.

Part II

Major drivers and constraints impacting agricultural transformation

Here the major drivers and constraints that are limiting effective crop production in Southern Africa are discussed. The technologies, both current and emerging, that could be used to reduce or offset the impact of these drivers and constraints are then outlined in Parts III and IV.

3 Twice as many people in 2050

The need for agricultural transformation in Southern Africa

Reiner Klingholz

One of the biggest challenges for Africa is its ongoing population growth. According to the medium variant of the United Nations projections the continent as a whole will double its population until 2050 to an estimated 2.6 billion people. Worldwide, the population is expected to grow only by 30%. Africa will account for half of the global population growth in that period (UN DESA, 2017). The continent has the youngest and least urbanized population of the world, the lowest education and health as well as the highest poverty levels. All these factors favour further population growth.

The southern part of the continent[1] is not an exception to this trend, with a population expected to grow from 176 million in 2017 to nearly 350 million by mid-century. In 1960 the population was 43 million (UN DESA, 2017). Nevertheless, in terms of demographic development Southern Africa is very heterogeneous. Countries that score higher in the Human Development Index (HDI) like South Africa and Botswana already have lower fertility rates and less population growth, whereas women in countries in the lower category of the HDI like Angola, Mozambique and Zambia still have many children. These countries face the challenge to provide anything from schools to housing and meet the increasing demands for ecosystem products like fresh water, food, timber, fibre and fuel for a fast growing population (UN DESA, 2017).

The projected growth might worsen the already existing food insecurity. It will speed up the urbanization trend and will make the countries' various problems more difficult to manage, from extreme poverty to the lack of basic services. Only with investments into the crucial sectors for development and the political will for family planning programmes the countries of Southern Africa will be able to manage the transition from high to low birth rates and to slower population growth.

This socioeconomic process that all countries pass through during their development is called the demographic transition. It starts in every country in the preindustrial age, when women give birth to a large number of children. Fertility rates are high. Mortality rates are high as well, based on limited environmental and economic potentials. Population growth is only minor or absent in this stage. In stage 2, death rates decline as living standards improve through better food supply, improved hygiene and health services, safe drinking water, etc. Since fertility levels remain high in this stage this results in rapid population growth. With a certain time lag, in stage 3 fertility rates decline, catalyzed by social change like industrialization and urbanization, better (women's) education, women's rights and economic progress. Finally, in stage 4, both mortality and fertility rates reach a low level, population growth fades out and population might eventually decline in numbers when fertility falls below mortality in stage 5.

Mortality, especially of children and mothers, has already declined considerably in Southern Africa. Today in Angola 90 out of 1000 children die before the age of five, compared to 167 in 2005, in South Africa the respective numbers are 38 and 74. But whereas fertility in South Africa, the only industrialized country on the continent, has already declined to 2.4 children per woman, fertility decline in the less developed countries is delayed, resulting in a persisting high population growth. Women in Angola, Zambia, Malawi and Mozambique still give birth to an average of 4.2 (Malawi) to 6.2 (Angola) children (Population Reference Bureau, 2018). Replacement fertility that stands for a stable population in the absence of inward migration is 2.1 children per woman. But even if this level is reached, it takes decades to stop population growth because due to previous high fertility rates there is a large number of girls that have not yet entered their reproductive age (Weeks, 2018). This is why South Africa, which is already close to replacement level, is projected to grow from today's 58 million to 82 million in 2050 (UN DESA, 2017). The country will then be equalled by Angola that will add 52 million to today's population of 30 million in that period, giving Angola one of the fastest growing populations of the world. The pace of fertility decline in Southern Africa is much slower

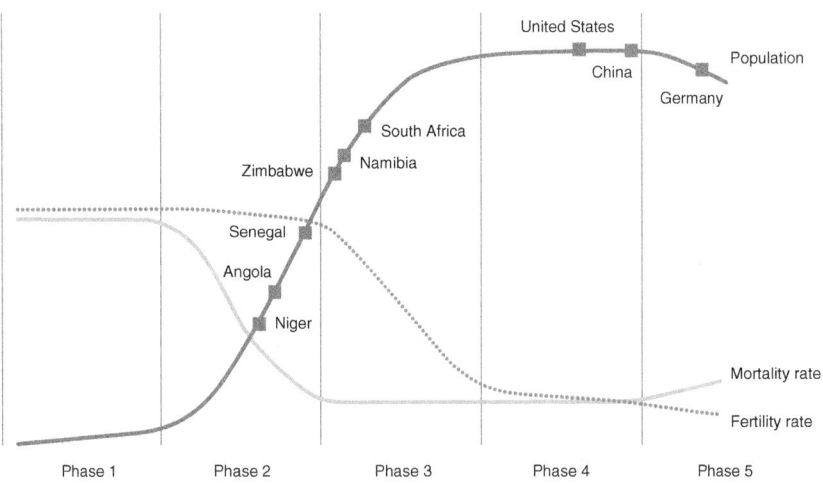

Figure 3.1 Demographic transition

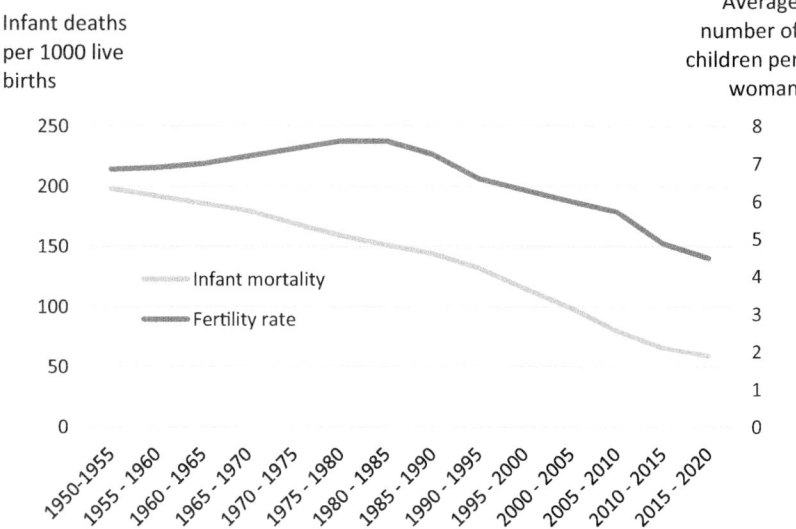

Figure 3.2a Infant mortality and total fertility rate in Malawi

Source: UN DESA, 2017

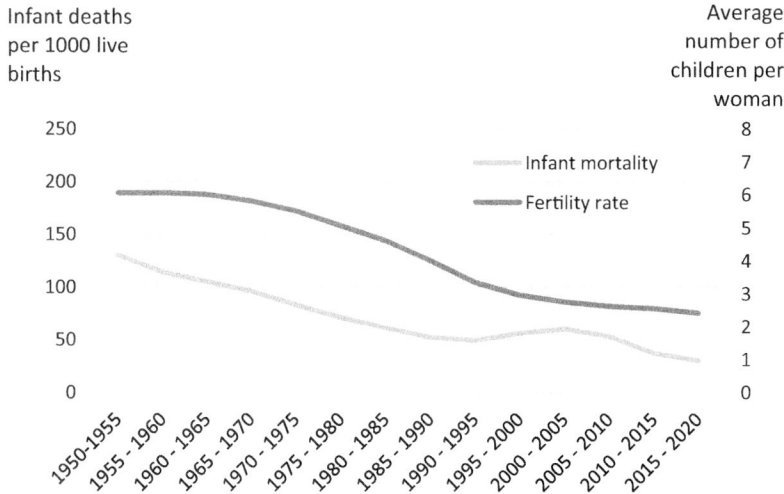

Figure 3.2b Infant mortality and total fertility rate in South Africa

Source: UN DESA, 2017

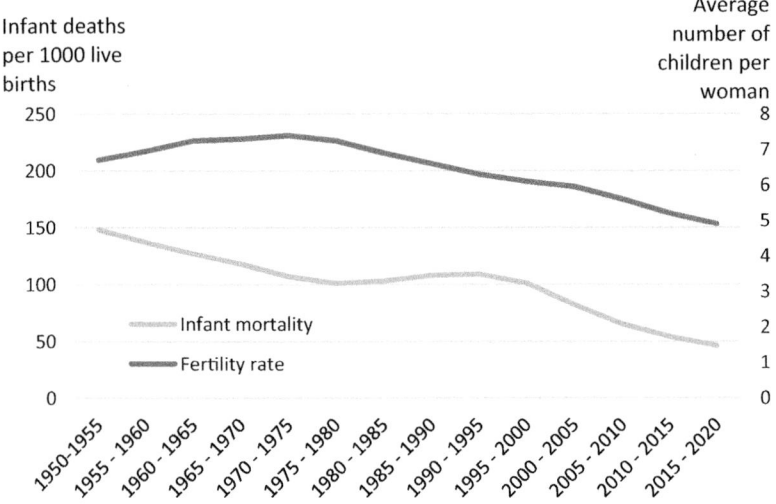

Figure 3.2c Infant mortality and total fertility rate in Zambia

Source: UN DESA, 2017

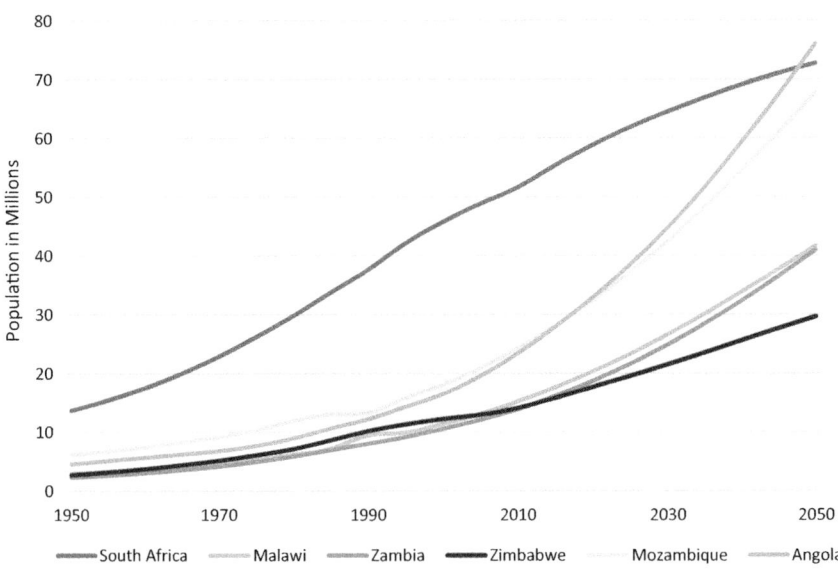

Figure 3.3a Population development in Southern Africa

Source: UN DESA, 2017

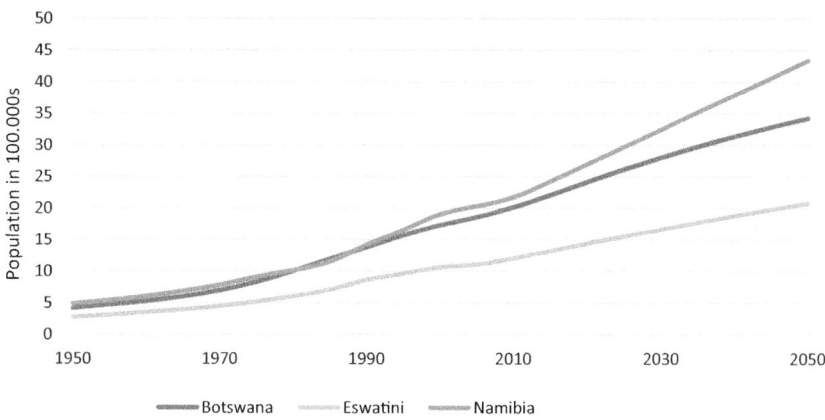

Figure 3.3b Population development in Southern Africa (small states)
Source: UN DESA, 2017

than it was in Asia or Latin America during their demographic transition in the 1970s, forcing the United Nations Population Division to repeatedly correct their population estimates for African countries upward in the past (UN DESA, 2015).

Dividend or disaster?

Whether a growing population is a benefit or a burden to society depends largely on its age structure. As high fertility rates result in large cohorts of children and young adults, this structure causes high costs for families as well as for the public health and education sectors. More problems arise when young adults in great numbers, a so-called youth bulge, enter the workforce without a chance to find meaningful work. This bulge is statistically associated with social instability, an increased risk of conflict and criminal violence. Countries where young people between the ages of 15 and 24 account for more than 25% of the population over 15 years of age, are considered particularly conflict-laden (Kröhnert, 2006). All countries in Southern Africa fall into this category.

Only when fertility rates decline and the share of the working-age population increases relative to the economically dependent population, consisting of children and the elderly, does the age structure become favourable to economic growth and a demographic window of opportunity opens. This "demographic bonus" can be turned into a "demographic dividend" when sufficient employment is provided for the young workforce. In this case, a dynamic economy unfolds. About one-third of the economic growth of the Asian tiger nations can be attributed to the use of this favourable age structure (Canning, 2015).

All less developed countries in Southern Africa are far away from this situation. They have relatively few workers, which have to support large numbers of children. This high youth dependency ratio constrains the chance of rapid per capita income gains. In Angola, Zambia, Mozambique and Malawi the population under 15 makes up between 41% (Malawi) and 48% (Angola) (UN DESA, 2017). At the same time, youth unemployment is widespread because new jobs are not created at nearly the rate at which the population is growing. Official unemployment figures for these countries are not reliable, as there is no regular registration and the majority of the young workforce is active in the informal sector.

South Africa, where statistics are more reliable, reports a youth unemployment rate (ages 15–24) of more than 50% (World Bank, 2018a). This is dramatic because South Africa is further advanced in the demographic transition, and the ratio between individuals in working age and economically dependent young and older people is already better than in the rest of the region. The country already has an age structure that makes it possible to harness a demographic dividend but unfortunately does not make use of it. There are not enough jobs to make the available workforce productive. As a result, South Africa's economy is not growing fast enough to keep pace with population growth, leaving today's population with a lower real per capita income than in 2014 (Bisseker, 2018).

It is possible to speed up the demographic transition that opens the window for more economic growth and to make use of the demographic dividend. This will only happen if fertility rates decline but this requires an improved livelihood. Therefore, it is necessary to enhance educational attainment and quality,

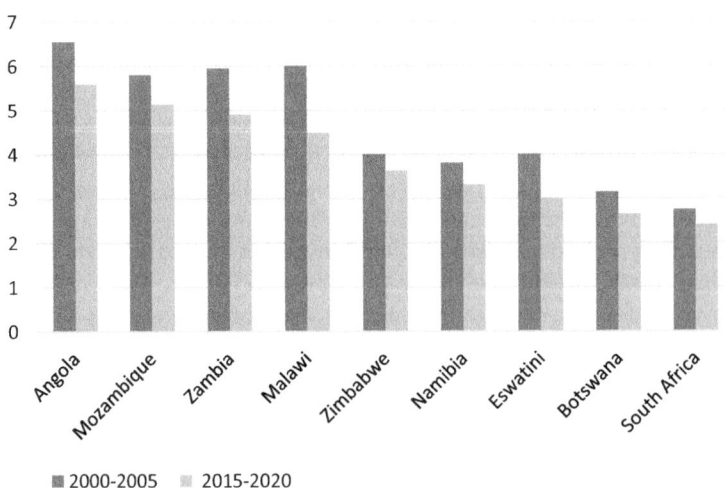

Figure 3.4a Total fertility rate in Southern Africa

Source: UN DESA, 2017

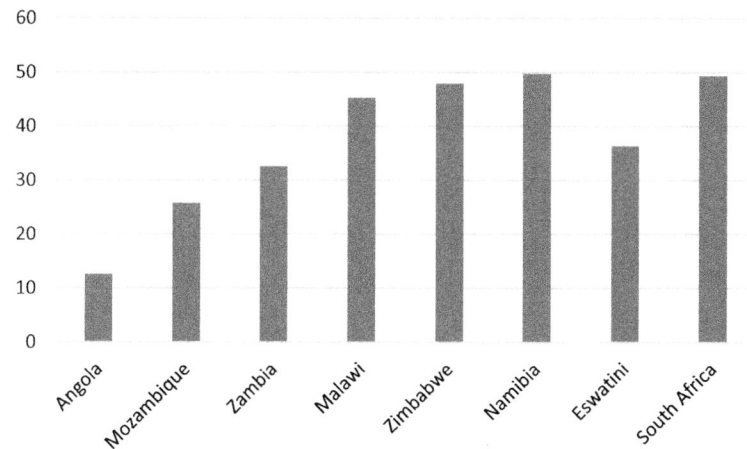

Figure 3.4b Current use of modern contraception by women (in percentage)
Source: The DHS Program. The Demographic and Health Services, 2019.

labour productivity and employment. As a short-term intervention family planning programmes and services, the supply of affordable modern contraception and high-level political commitment for family size limitations can reduce population pressure. Without an active population policy, the current demographic situation in the less developed part of Southern Africa will not improve and this will preclude rapid improvement of livelihood (Cilliers, 2018).

Education is the best contraception

Education is the central instrument for development, as it enhances human capital in terms of skills and health. In addition, education for girls has an enormous effect on fertility rates. The number of children born per woman declines with the number of years young women have spent in school in all developing countries of the world. African women with no education have, on average 5.4 children, whereas women who completed secondary education have 2.7 and those who went to college only 2.2 children (Engelmann, 2016). In Angola, women with upper secondary education have on average of 2.6 children compared to 6.5 children of women that never went to school. In South Africa the numbers are 2.0 and 3.8 respectively (Wittgenstein Centre for Demography and Global Human Capital, 2015).

This fertility gap can be easily explained: a longer time spent in education leads to later pregnancies. If girls in poor countries have the chance of obtaining a higher education we believe they can escape the early marriage market, where they would otherwise be placed at the age of 15 or 16. These girls usually become mothers as teenagers whereas girls with at least secondary education

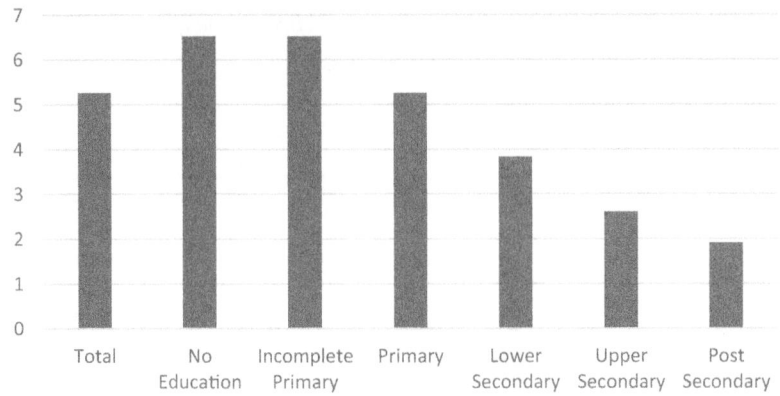

Figure 3.5a Total fertility rate by education level in Angola, 2010–15

Source: Wittgenstein Centre for Demography and Global Human Capital, 2018

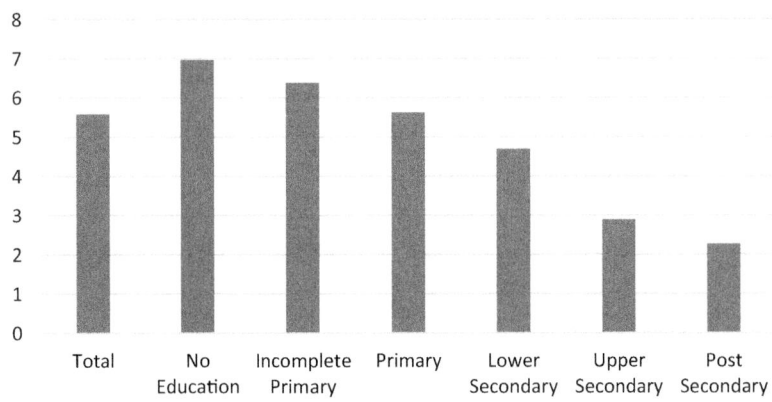

Figure 3.5b Total fertility rate by education level in Malawi, 2010–15

Source: Wittgenstein Centre for Demography and Global Human Capital, 2018

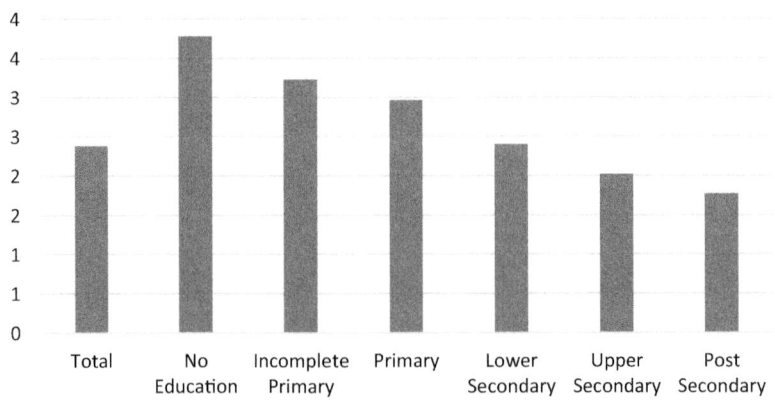

Figure 3.5c Total fertility rate by education level in South Africa, 2010–15

Source: Wittgenstein Centre for Demography and Global Human Capital, 2018

can decide to postpone pregnancy. They have less children in total and can plan for longer intervals between the births, which has a positive effect on the health of the mothers as well as of the newborns. Education for women upgrades a woman's position in the household and affects the bargaining power relative to that of their husbands when it comes to decide on the number of children.

Education for a woman also is an investment in the next generation because educated women tend to invest more into the education of their own children than do educated men. Female education not only reduces population growth but leads to an increase in human capital, elevated female labour force participation and to improved productivity, which makes it easier to unlock economic growth. Finally, less population growth requires less infrastructure investments and allows governments to spend more money per capita on health and education (Canning, 2015; Lutz and Klingholz, 2017).

Crucial development in rural areas

In general, fertility rates in Southern Africa are higher in rural areas compared to urban areas, where education levels are higher, and more jobs are available. In Zambia for example urban fertility stands at 4.4 children per woman vs. 6.4 in rural regions (Republic of Zambia, Central Statistical Office, 2018). This creates even more population pressure in rural areas, where a rising number of young people face unemployment. Many of them are forced to migrate. As a result, population growth in the villages fuels the growth of the cities.

In order to reduce population growth in the countries of Southern Africa, it will be necessary to improve living conditions in rural areas. This requires better access to health and education services in the countryside and more jobs in the agricultural sector, the dominant provider of employment in rural areas.

With the exception of South Africa, Namibia and partly Zambia and Zimbabwe, little has changed to this day in the traditional structure of agriculture. Farming is still carried out mainly on small family farms. Most smallholders are poor, and their output provides mainly for subsistence. Poverty and population growth condemn the majority of these people to remain in subsistence farming. There is a dire need for the transformation of agricultural production systems throughout Southern Africa. Only if productivity increases and households manage to save money, will they have taken an important step towards escaping poverty – a crucial precondition to reduce fertility rates. If that happens, households become less dependent on external support at harvest time. The freed-up workforce can be deployed at new enterprises that emerge in a downstream processing sector that create more value than do raw agricultural products. These enterprises include grain mills, slaughterhouses, dairies, shops selling agricultural supplies, factories and repair shops for agricultural equipment and machinery, among many other things (World Bank, 2018b).

With its vast reservoir of labour simply waiting for jobs to emerge along the agricultural value chain, the less developed countries in Southern Africa possess a valuable resource. This valuable human reservoir should be used to push ahead both economic and demographic transformation. Successful transformation of agriculture can give communities in rural areas the opportunity to earn a living beyond subsistence, they will be exposed to new prospects. If that happens, an economy develops, where child labour is no longer required, fertility rates will decline and the demographic pressure will ease. Family planning begins only when individuals (people) are in a position to plan their future. This was observed in all early industrialized and emerging countries (Suetterlin, 2018). There is no reason to believe that the least developed countries, which can be found primarily in Africa, cannot follow the same course of development.

Note

1 Defined as all countries including, and south of, Angola, Zambia, Malawi and Mozambique

References

Bisseker, C. (2018) SA's population is booming and the economy is struggling to keep up. *Financial Mail*, 27 September.

Canning, D., Raja, S. and Yazbeck, A. S. (2015) *Africa's Demographic Transition: Dividend or Disaster?* Africa Development Forum. Washington, DC.

Cilliers, J. (2018) *Getting to Africa's Demographic Dividend*. Africa Report 13. Institute for Security Studies. Pretoria.

The DHS Program. The Demographic and Health Services (2019) https://dhsprogram.com/Topics/Family-Planning.cfm.

Engelmann, R. (2016) Africa's population will soar dangerously unless women are more empowered. *Scientific American* Vol 314, Issue 2.

Kröhnert, S. (2006) *Warum entstehen Kriege?* Berlin Institute for Population and Development. Berlin.

Lutz, W. and Klingholz, R. (2017) *Education First: From Martin Luther to Sustainable Development*. SUNMeDIA, Stellenbosch.

Population Reference Bureau (2018) *World Population Data Sheet*. Washington, DC.

Republic of Zambia, Central Statistical Office (2018) *Zambia in Figures*. www.zamstats.gov.zm/phocadownload/Dissemination/Zambia%20in%20Figure%202018.pdf.

Suetterlin, S., Reinig, A. and Klingholz, R. (2018) *Food, Jobs and Sustainability: What African Agriculture Needs to Achieve*. Berlin Institute for Population and Development. Berlin.

UN Department of Economic and Social Affairs (2015) *World Population Prospects: The 2015 Revision*. United Nations. New York.

UN Department of Economic and Social Affairs (2017) *World Population Prospects: The 2017 Revision*. United Nations. New York.

Weeks, J. R. (2018) *Population: An Introduction to Concepts and Issues*. Wadsworth, Belmont.

Wittgenstein Centre for Demography and Global Human Capital (2018) Wittgenstein Centre Human Capital Data Explorer. Vienna.

World Bank (2018a) *International Labour Organization*. ILOSTAT Database.

World Bank (2018b) *DataBank. Agriculture in Africa: Telling Myths from Facts*. World Bank. Washington, DC.

4 Climate change and the threat to food production in Southern Africa

Paul L.G. Vlek, Eugene R. Terry
and Richard A. Sikora

Climate change and its impact

Climate change in Southern Africa is having an impact on life and liveli-hoods, and farmers are struggling to cope with its impact. Both, floods and droughts have increased in frequency and intensity and the onset of the rainy season has shifted and has become less predictable. In 2017, the South-ern African Development Community (SADC) published a handbook enti-tled Climate Risk and Vulnerability (Davis-Reddy and Vincent, 2017) with up-to-date climatic analysis and projections for Southern Africa, which is used extensively in the following section.

Average land-surface temperature has increased across Africa over the last decennia, particularly since the 1970s, and is continuing today (Figure 4.1). The subtropical Southern African region is among the most affected with a trend of 0.4 °C per decade. The rate of change in temperature over Africa is more than twice the global estimate (Osborn and Jones, 2014).

Trends in rainfall for Southern Africa are masked by the high variability and the lack of observational data in the region. The data for 1900 to 2014 (Harris et al., 2013), represented in Figure 4.2, show this high variability with some clear drought periods. The 2000s were wetter for most of the region except for the countries along the southwestern coast of Africa and the eastern coastline of Tanzania.

Trends in extreme events sumarized in Figure 4.3, are difficult to discern. However, the number of hot extremes have increased and the number of cold extremes have decreased, as they have globally (Stocker et al., 2013). Evidence suggests that the frequency of dry spells as well as daily rainfall intensity has increased (New et al., 2006). There is some evidence to suggest that droughts have become more intense and widespread over Southern Africa (Masih et al., 2014; New et al., 2006). The projected increase in temperature is expected to lead to more frequent droughts, particularly during periods of reduced rainfall (Engelbrecht et al., 2015; Shongwe et al., 2011; Stocker et al., 2013). The fre-quency of high fire danger days is projected to increase across Southern Africa which is consistent with the increases in heat-wave days (Engelbrecht et al., 2015).

Climate change is expected to alter the magnitude, timing and distribution of storms that produce flood events (Engelbrecht et al., 2013; Stocker et al., 2013).

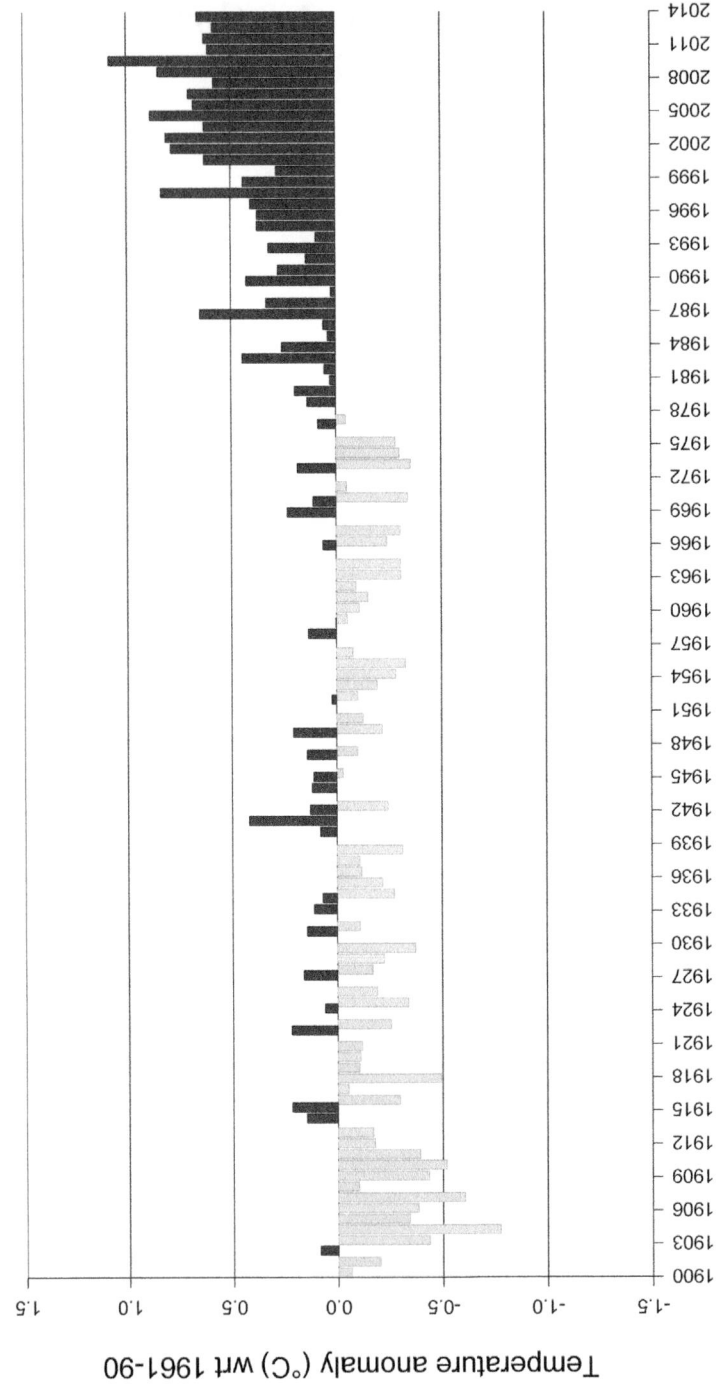

Figure 4.1 Mean annual temperature anomaly (°C) over Southern Africa from 1901 to 2014 with respect to the long-term average climatology 1961–1990; based on the gridded CRUTEMv4 data set; red represents a positive anomaly and yellow a negative temperature anomaly

Source: Davis-Reddy and Vincent, 2017

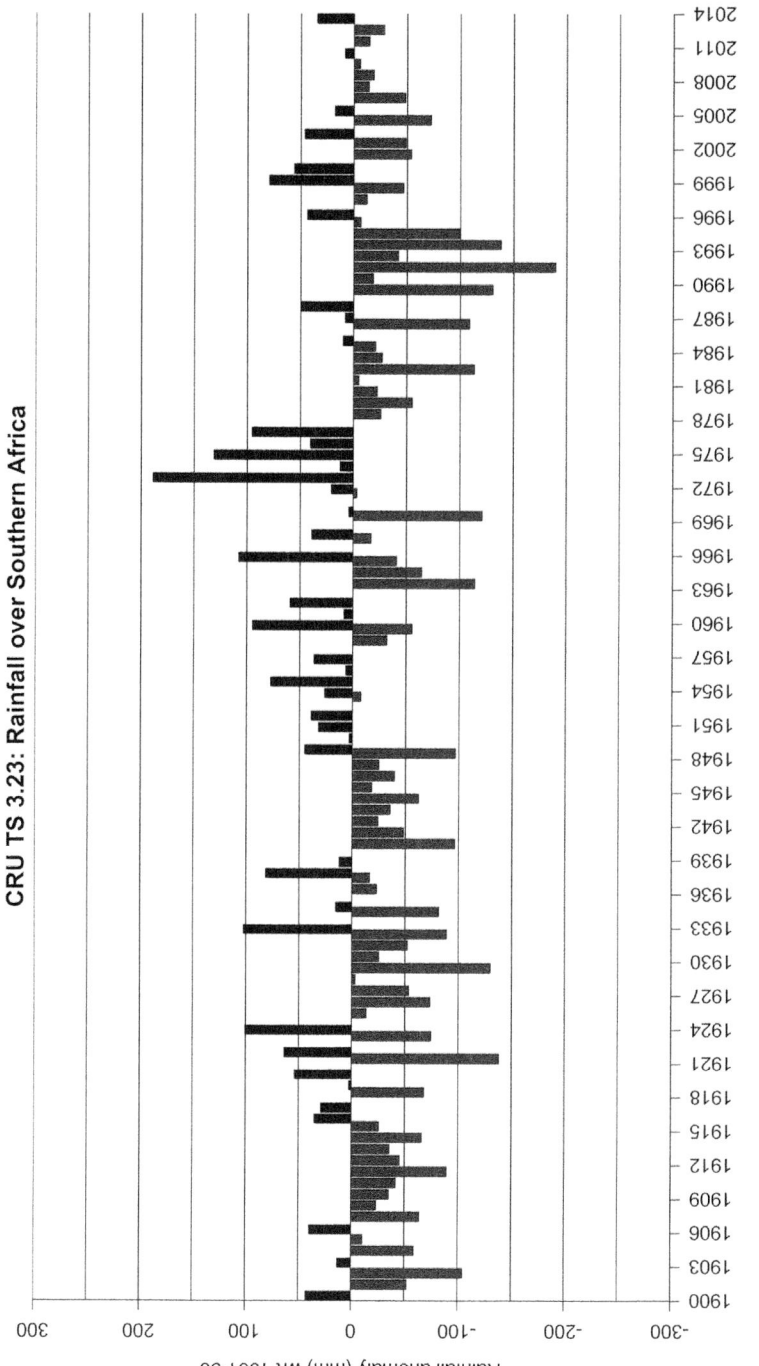

Figure 4.2 Mean annual rainfall anomaly (mm) over Southern Africa from 1901 to 2014 with respect to the long-term average climatology 1961–1990; based on the gridded CRU TS 3.23 data set; red represents positive anomaly and blue a negative anomaly in temperature

Source: Davis-Reddy and Vincent, 2017

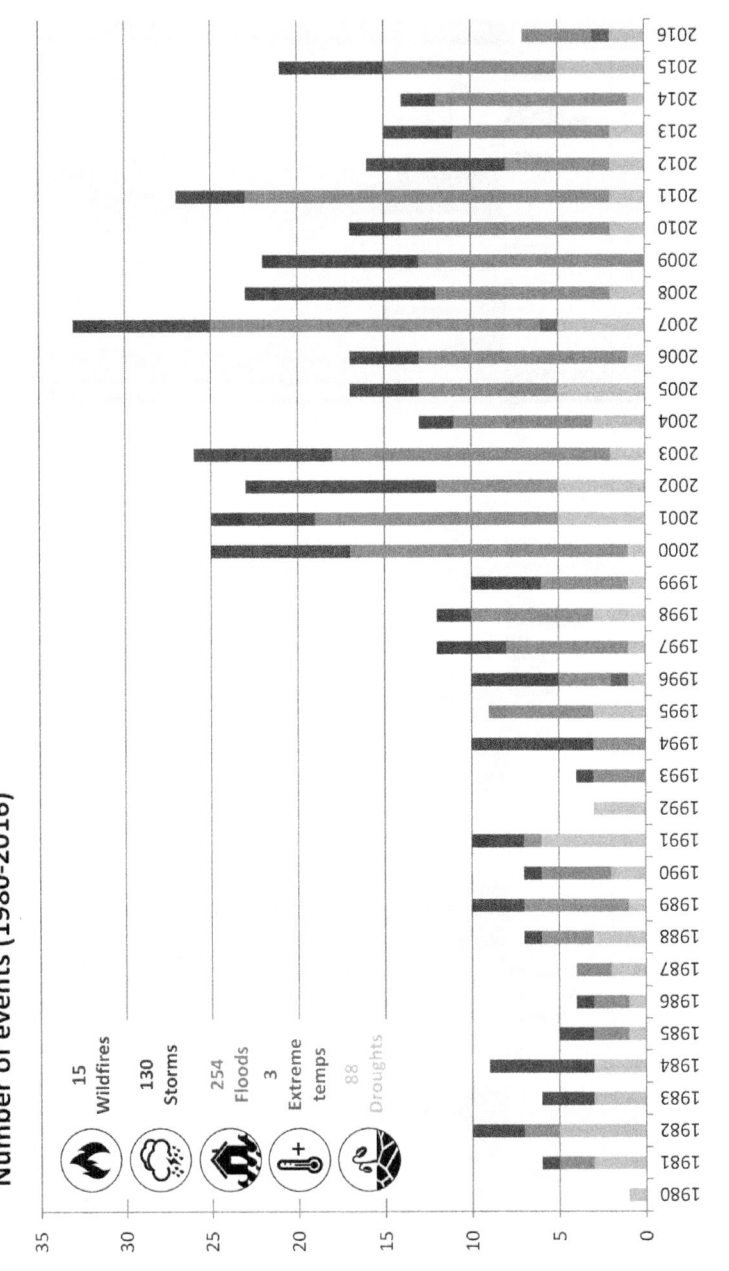

Number of events (1980-2016)

15 Wildfires
130 Storms
254 Floods
3 Extreme temps
88 Droughts

Drought   Extreme temperature   Flood   Storm   Wildfire

Figure 4.3 Number of recorded climate-related events over Southern Africa since 1980 (Source: EM-DAT CRED, 2016); "wild-fires" refers to any uncontrolled and nonprescribed burning of plants in a natural setting; "storms" to tropical, extratropical and convective storm events; "floods" to riverine, flash and coastal flood events

Source: Davis-Reddy and Vincent, 2017

Tropical cyclones used to be rare in Southern Africa and of low intensity (Fitchett, 2018). The ocean temperature of the South Indian Ocean is creeping up, enhancing the intensity of cyclones. Moreover, the cyclone activity has moved southward where they are more likely to make landfall. Since 1994, category 5 storms in the South Indian Ocean have become more frequent and their frequency is likely to keep increasing. Most of this can be ascribed to an increase in sea surface temperature as a result of global warming. The greatest threat is to the northern half of Madagascar, Mozambique and to the islands of Reunion and Mauritius.

The SADC report (Davis-Reddy and Vincent, 2017) summarizes the climate projections for Southern Africa based on the CORDEX effort using over a dozen different Global Circulation Models with different downscaling approaches and two key development trajectories (RCPs). If greenhouse gases emissions continue to rise unmitigated, central Southern Africa will experience drier summers in the future. Parts of Tanzania and northern Mozambique will possibly be wetter. Winter rainfall in the Western Cape of South Africa is likely to decline in the future. Projections on rainfall tend to deviate among GCM projections and thus remain somewhat uncertain. The projections uniformly predict that the 21st century will see continued increases in temperatures, somewhat less in the coastal regions.

The record of extreme events over the past 35 years in Southern Africa are summarized in Figure 4.3. Drought events were a regular occurrence whereas extreme temperatures occurred episodically. The frequency of floods and storms has increased drastically as have wildfires to some extent. The capacity to cope with these events in the region are limited, and the population is indeed highly vulnerable.

Climate change and agriculture

Agricultural in Southern Africa has limited adaptive capacity to the hazards of climate change due to endemic poverty and restricted capital and technology access and poor infrastructure greatly enhancing its vulnerability (Hachigonta et al., 2013; Parry, 2007). Climate change is also having a dramatic impact on food and nutrition security in most of sub-Saharan Africa (NOAA, 2007). The projected increase in rainfall variability, temperature and extreme events for the region will exacerbate the vulnerability of the predominantly rainfed systems on which food supply depends. Knox et al. (2012) conducted a meta-analysis of climate change on the yield of eight major crops in Africa and projected mean yield losses of 17% (wheat), 5% (maize), 15% (sorghum) and 10% (millet) for Africa by mid-century. Dinesh et al. (2015) suggest that the area suitable for maize production could decline by 20 to 40% relative to the period from 1970 to 2000. Maize is particularly sensitive to temperature, losing 1% for each growing day spent at a temperature above 30°C (Lobell et al., 2011). Wheat has an even lower temperature threshold value (Adhikari et al., 2015).

Climate change will affect livestock productivity indirectly as well as directly. Climate change will diminish feed resources for livestock through changes in the primary productivity of crops, forages and rangelands. Changes in species composition in rangelands and some managed grasslands will affect the animal species that can graze them (Thornton et al., 2007). Domesticated species perform best at temperatures between 10 and 30°C and they will produce around 3 to 5% less for each 1°C increase above those levels. These temperatures are already exceeded in several regions.

According to Haywood (2015), the biophysical impacts of climate change can have a ripple effect along the whole agricultural value chain which spans input companies, farmers, distributors, agro-processing companies and retailers. The chapters in this book offer many guidelines on how to cope with the multiple stressors associated with climate change. Policymakers ignoring or denying the hazards associated with climate change do so at the peril of their society.

Climate change and biotic stresses

Crop loss will be complicated by a host of factors such as: 1) a decrease in host plant resistance; 2) a reduction in the efficacy of pesticides; 3) shifts in the effectiveness of ecosystem services that naturally regulate pest densities; 4) shifts in strains of fungi responsible for toxin production in postharvest; 5) arrival of alien pest species; and 6) enhanced weed growth and ultimate competition with crops (Muatinte and Van den Berg, 2018; Oerke and Dehne, 2004; Pratt et al., 2017; Tefera, 2012). Increased temperature associated with climate change will likely have a major impact on crop losses due to increased biotic stress (outlined in chapter 5 in this volume) from weeds, insects, fungal pathogens, viruses, nematodes and rodents. These pests cause yield loss at all stages in the production cycle from planting to postharvest with total harvest losses approaching 60% (see chapter 6 in this volume). Yield losses of major staple crops (i.e., maize, rice, wheat and soybean), due to increased insect pests alone will expand by 10 to 25% for each degree of global mean surface warming (Deutsch et al., 2018; Henson et al., 2011).

These pests usually have enhanced development rates even with slight increases in daily temperatures, producing additional generations per cropping season and increased numbers of offspring. In addition, there will be shifts in pest species composition but also an increased spread of invasive pests into new zones with suitable climatic conditions. Furthermore, increased soil temperature, as affected by direct solar heating, will not only increase evapotranspiration but also the density of soil-borne pests in rhizosphere causing root system degradation that will affect water and nutrient uptake (Sikora, 2018).

Smallholder farmers in Africa will be more vulnerable to these consequences of climate change than large family and commercial farmers who have access to management options. Small farmers lack the necessary knowledge of pest biology and integrated management strategies and also face financial constraints (Biber-Freudenberger et al., 2016). Agriculture in Southern Africa, especially that of smallholders, will need to learn to cope with expected increases in biotic stresses. Therefore, it is of utmost importance that government institutions in the region develop response strategies and improve support to small-scale farmers to avoid crop losses and failures and ensure adequate food production.

Conclusions and recommendations

Climate change is a threat to food security (IPCC, 2019). If farmers are to cope with this threat, it is imperative to better understand the three intertwined food security and climate change challenges, namely 1) ensuring food security; 2) coping with the impact of climate change on agriculture; and 3) mitigating the impact of agriculture on climate change. To properly address these intertwined challenges, the Southern African agri-food system has to become more efficient and resilient at every scale from the farm to the regional food system. (Climate Smart Agriculture Sourcebook, FAO, 2013).

Resilience, the capacity of systems, and communities, to prevent, mitigate or cope with risks or recover from shocks, has various dimensions; biophysical, economic and social, operating at various scales. The manner in which the various dimensions and scales interact is crucial and needs to be better understood.

The current and emerging technologies to enhance agricultural transformation in Southern Africa and create resilient livelihoods are well documented in the various chapters in this book. They were selected for their significant role in traditional farming practices in Southern Africa and farming practices useful for climate change adaptation in agriculture in Southern Africa (IIED, 2011). Priority focus in the scaling up of proven technologies and practices is to reduce the vulnerability of farming in the region. The technologies described in this book as they relate to resilience include crop diversification of cereals and legumes, crop insurance schemes, drought or flood tolerant crop varieties, hazard-proof grain storage facilities, livestock shelters, strategic fodder reserves, bio-security of animal production systems, water reserves to buffer droughts and resilient animal breeding (FAO, 2011). However, traditional knowledge systems should be given due consideration in the technology packages.

Technologies and practices designed to address Disaster Risk Reduction (DRR) and climate change adaptation can be delivered as a coherent package of solutions, scaling up of proven risk reduction practices can be achieved through policies that support their systematic expansion. Climate Smart Disaster Risk Management (CSDRM) is designed to address the need to integrate disaster risk reduction and climate adaptation. There are 12 components of CSDRM comprising three action-oriented pillars (Davis-Reddy and Vincent, 2017) as follows: Pillar 1. Addressing changing disaster risks and uncertainties; Pillar 2. Enhancing adaptive capacity; and Pillar 3. Tackling poverty and vulnerability and their structural causes. Pillar 2 is of particular relevance here, since its key focus is on enhancing adaptive capacity and building resilience. This has policy implications for resource allocation for interventions at institutional and network levels that provide opportunities for learning, knowledge sharing for solving problems. In this context the factors essential for increasing resilience would include community participation, promoting diversity, acknowledging the importance of social values and structures in planning, preparedness and readiness. (Davies-Reddy and Vincent, 2017)

References

Adhikari, U., Pouyan Nejadhashemi, A. and Woznicki, S.A. (2015) Climate change and Eastern Africa: A review of impact on major crops. *Food and Energy Security* 4: 110–113.

Biber-Freudenberger, L., Ziemacki, J., Henri, E. Z., Tonnang, H.E.Z. and Borgemeister, C. (2016) Future risks of pest species under changing climatic conditions. *PLoS.* https://doi.org/10.1371/journal.pone.0153237.

Davis-Reddy, C.L. and Vincent, K. (2017) *Climate Risk and Vulnerability: A Handbook for Southern Africa.* CSIR, Pretoria, South Africa.

Deutsch, C.A., Tewksbur, J.J., Tigchelaar, M., Battisti, D.S., Merril, S.C., Huey, R.B. and Naylor, R.L. (2018) Increase in crop losses to insect pests in a warming climate. *Science.* 361: 916–919. doi:10.1126/science.aat3466.

Dinesh, D., Bett, B., Boone, R., Grace, D., Kinyangi, J., Lindahl, J., Mohan, C.V., Ramirez-Villegas, J., Robinson, R., Rosenstock, T., Smith, J. and Thornton, P. (2015) *Impact of Climate Change on African Agriculture: Focus on Pests and Diseases.* CGIAR Research Program on Climate Change, Agriculture and Food Security (CCAFS), Copenhagen, Denmark. www.ccafs.cgiar.org.

Engelbrecht, C., Engelbrecht, F. and Dyson, L. (2013) High resolution model-projected changes in mid-tropospheric closed-lows and extreme rainfall events over Southern Africa. *International Journal of Climatology* 33: 173–187.

Engelbrecht, F.A., Adegoke, J., Bopape, M.J., Naidoo, M., Garland, R.M., Thatcher, M., McGregor, J.L., Katzfey, J., Werner, M., Ichoku, C., et al. (2015) Projections of rapidly rising surface temperatures over Africa under low mitigation. *Environmental Research Letters* 10.

FAO (2011) *Save and Grow: A Policymaker's Guide to the Sustainable Intensification of Smallholder Crop Production.* FAO, Rome.

FAO (2013) *Climate-Smart Agriculture Sourcebook.* FAO, Rome.

Fitchett, J. M. (2018) Recent emergence of CAT5 tropical cyclones in the South Indian Ocean. *South African Journal of Science* 114(11–12).

Hachigonta, S., Nelson, G.C., Thomas, T.S. and Sibanda, L.M. (2013) *Southern African Agriculture and Climate Change.* International Food Policy Research Institute, Washington DC, 337.

Harris, I., Jones, P., Osborn, T. and Lister, D. (2013) Updated high-resolution grids of monthly climatic observations – the CRU TS3. 10 dataset. *International Journal of Climatology* 34: 623–642.

Haywood, L. (2015) *Climate Change and Its Impact on Agribusiness, Natural Resources and the Environment.* Council for Scientific and Industrial Research, Pretoria.

Henson, R. (2011) *Climate Stabilization Targets: Emissions, Concentrations, and Impacts Over Decades to Millennia.* National Academies Press, Washington, DC. www.nap.edu.

IIED. (2011) *Adapting Agriculture with Traditional Knowledge.* IIED Briefing.

IPCC. (2019). www.ipcc.ch/srccl-report-download-page/?fbclid=IwAR2sZJ_lluCSHJcedq9 Hx9pw2HF5f8iM-vzwYX3sxZOgIPP7-vYkV-ysJQU.

Knox, J., Hess, T., Daccache, A. and Wheeler, T. (2012) Climate change impacts on crop productivity in Africa and South Asia. *Environmental Research Letters* 7(3).

Lobell, David B., Bänziger, M., Magorokosho, C. and Vivek, B. (2011) Nonlinear heat effects on African maize as evidenced by historical yield trials. *Nature Climate Change* 1: 42–45.

Masih, I., Maskey, S., Mussá, F. and Trambauer, P. (2014) A review of droughts on the African continent: A geospatial and long-term perspective. *Hydrology and Earth System Sciences* 18: 3635–3649.

Muatinte, B. and Van den Berg, J. (2018) The use of mass trapping to suppress population numbers of *Prostephanus truncatus* (Horn) (Coleoptera: Bostrichidae) in small scale farmer granaries in Mozambique. *African Entomology* 26: 301–316.

New, M., Kululanga, E., et al. (2006) Evidence of trends in daily climate extremes over Southern and West Africa. *Journal of Geophysical Research* 111(7): D14102.

NOAA (2007) *Report on Climate Change, National Oceanic and Atmospheric Administration of the United States Department of Commerce.* www.nws.noaa.gov/om/brochures/climate/Climatechange.pdf.

Oerke, E.C. and Dehne, H.W. (2004) Safeguarding production – losses in major crops and the role of crop protection. *Crop Protection* 23: 275–285.

Osborn, T. and Jones, P. (2014) The CRUTEM4 land-surface air temperature data set: Construction, previous versions and dissemination via Google Earth. *Earth System Science Data* 6: 61–68.

Pandey, D.N., Gupta, A.K. and Anderson, D.M. (2003) Rainwater harvesting as an adaptation to climate change. *Current Science* 85: 46–59.

Parry, M.L. (2007) *Climate Change 2007: Impacts, Adaptation and Vulnerability: Contribution of Working Group II to the Fourth Assessment Report of the Intergovernmental Panel on Climate Change.* Cambridge University Press, Cambridge.

Pratt, C.F., Constantine, K.L. and Murphy, S.T. (2017) Economic impacts of invasive alien species on African smallholder livelihoods. *Global Food Security* 31–37.

Savary, S., Willocquet, L., Pethybridge, S.J., Esker, P., McRoberts, N. and Nelson, A. (2019) The global burden of pathogens and pests on major food crops. *Nature – Ecology & Evolution.* Published online in Nature Ecology and Evolution.

Shongwe, M.E., van Oldenborgh, G.J., van den Hurk, B. and van Aalst, M. (2011) Projected changes in mean and extreme precipitation in Africa under global warming. Part II: East Africa. *Journal of Climate* 24(14): 3718–3733.

Sikora, R.A., Coyne, D., Hallmann, J. and Timper, P. (Eds) (2018) *Plant Parasitic Nematodes in Subtropical and Tropical Agriculture.* CABI, Wallingford, 3rd ed., 876.

Stocker, T., Qin, D. and Platner, G. (2013) *Climate Change 2013: The Physical Science Basis: Working Group I Contribution to the Fifth Assessment Report of the Intergovernmental Panel on Climate Change. Summary for Policymakers.* Cambridge University Press, Cambridge.

Tefera, T. (2012) Post-harvest losses in African maize in the face of increasing food shortage. *Food Security* 4(2): 267–277.

Thornton, P., Herrero, M., Freeman, H., Okeyo, A., Rege, E., Jones, P. and McDermott, J. (2007) Vulnerability, climate change and livestock – opportunities and challenges for the poor. *Journal of Semi-Arid Tropical Agricultural Research* 4(1).

5 Land rich but water poor

The prospects for agricultural intensification in Southern Africa

Paul L.G. Vlek, Lulseged Tamene and Janos Bogardi

Introduction

Given the prognosis for population growth over the coming decennia elaborated in Chapter 3, the pressure on the natural resources of Southern Africa (Botswana, Lesotho, Malawi, Mozambique, Namibia, South Africa, Swaziland, Zambia and Zimbabwe) is likely to grow. Add to this the keen interest of outside agents in gaining a foothold in the region for export-oriented food production (20 million ha have been appropriated in Africa over the past decade according to the New Partnership for Africa's Development (NEPAD) program of the African Union) and the stage is set for pressure on land to increase dramatically (Chapter 3). Available arable land in the 1960s estimated by the FAO (Rakotoarisoa et al., 2012) in Southern Africa ranged from 1.35 and 0.94 ha/person in Zambia and Namibia, respectively, to around 0.30 ha/person for Swaziland, Malawi and Mozambique. In 2001–04, these values had dropped to 0.5 and 0.4 for Zambia and Namibia and to between 0.3 and 0.2 ha/person for the rest of Southern Africa. With the exception of South Africa and Namibia, there are still prospects of expanding the area of arable land (NEPAD, 2011), but the low quality or protection status of this land are constraints.

This chapter will look at the land and water endowments of the Southern African states and assess their quality and the impact of their use as well as their management. An assessment will be made of the options to intensify their use without depleting them in order to ensure their availability for generations to come.

Land and land cover

According to the World Bank database (2016), the total land area of contiguous Southern Africa covers 4.7 million km^2, 54% of which is currently cultivated or used by (agro) pastoralists, well above the world average of 38%. The fraction of land under agriculture varies widely from 75–80% in South Africa and Lesotho to as little as 32% in Zambia. The remaining countries hover around the regional average of 50%. There is a strong rainfall gradient running from > 1500 mmyr^{-1} NE to less than 50 mm yr^{-1} at the Namibian coast. Rainfall in excess of 1000 mm yr^{-1} is typical for the northern halves of Zambia and Mozambique as well as northern Malawi. Less than 500 mm yr^{-1} is seen in most of Namibia, Botswana, parts of southern Malawi and the western half of South Africa. Most of Namibia, South Africa and Botswana receive less than 250 mm with extremely high potential evapotranspiration rates of over 2500 mm yr^{-1}. The band between these regions experiences between 500- and 1000-mm rain per year (Hijmans et al., 2005). Land cover is largely determined by this gradient with forest dominating the > 1000mm rainfall belt in the north and east of Southern Africa and shrubby and herbaceous vegetation in the larger parts of Namibia, Botswana and South Africa that grows sparser as one moves further to the southwest (Bartholomé and Belward, 2005).

Land use and management

The agro-ecology of the region is largely determined by the rainfall gradient with semi-arid and desert-like conditions in the southwest and humid tropical conditions in the north and east, respectively. Most of South Africa falls in the subtropical zone. The natural vegetation is in accordance with the rainfall, as are the farming systems that have emerged over the centuries in these regions. Where rainfall allows, cereal production is dominant, covering more than 50% of the agricultural land in Southern Africa. Southern Africa is characterized by four ecozone bands oriented East-West, dissected by the maize-producing highlands that reach from Tanzania into Zambia, Malawi and Zimbabwe with a North-South orientation. North of this belt is the root crop and mixed cereal-root crop region that stretches from Angola through Zambia to Mozambique. South of this belt the climate is substantially drier, characterized by agro-pastoralism with millet/sorghum and livestock of equal importance and further south, pastoral zones dominated by livestock. The Namibian/South African tip of Africa is arid and unproductive in southern Namibia and the South African Karoo. The remainder is semi-arid to sub-humid and comprises large holdings and scattered smallholdings dominated by maize in the north and east, while the west is dominated by sorghum and millet. Overall, maize is the dominant crop (nearly 10 million ha) followed by sorghum, millet and wheat approaching 1 million ha each with the latter largely found is South Africa (FAO stat). The prime maize producers are South Africa (2.5 million ha) followed by Zimbabwe, Malawi, Mozambique and Angola with around 1–1.5 million ha each.

Garrity et al. (2012) compiled an in-depth analysis of the African farming systems. In their synthesis they argue that major challenges across the different farming systems are the pressure on land due to population growth, the rapid rate of soil fertility depletion and the poor access to input and output markets, even domestically. Land in Southern Africa is managed with widely different intensity but two farming systems dominate in terms of land coverage. The dominant maize-mixed system of Southern Africa has the potential to ensure food security with ample options for diversification. Yet, to date, this area suffers from more poverty than any other farming system in the region. The agro–pastoral system exhibits low ecosystem productivity and economic risks due to rainfall constraints that can be mitigated through collective resources management and strategic intensification, particularly where water resources can be made accessible. Some of these strategies will depend on government support. The root and tuber-based systems of northern Angola is deemed high potential as it has growth options through a real expansion and mechanization. Also, the modest level of livestock productivity offers room for improvement in this zone.

Land degradation

Worldwide, land degradation presents a serious threat to agricultural productivity and food security and sustainable development goals (Vlek et al., 2017). The degradation processes can be readily observed, but quantifying the degree of degradation of larger regions is very difficult, particularly in Africa, as baseline data are often lacking and panel data from observation points are few and far between. Efforts have been made to use space observations to monitor the state of the land (Vlek et el., 2008; Bai et al., 2008; Le et al., 2016), but considerable discrepancies are seen among the reported states of degradation due to the methods employed in the trend analysis. However, there is consensus that land degradation is seriously affecting the ecosystem functioning of lands in Southern Africa, both due to land conversion, nutrient mining and overgrazing (IPBES, 2018).

Soil resources and productivity

Soils in the Southern African region are characterized by varying fertility levels depending on physiography, land use and land management techniques. The soils in the majority of the countries are generally poor with low soil organic matter (SOM) content and low water retention (Tamene et al., 2019). As a result, they are vulnerable to water and wind erosion, leaching and salinization if put under irrigation. For example, in Namibia, nearly all of the soils have clay contents of less than 5% and thus have very low water holding capacity (Liebenberg, 2005). In Botswana, 70% of the soils are sandy, geologically old and highly leached, poorly structured and infertile (Moroke, 2005). In Lesotho, more than 70% of the soils are acidic, have low organic matter, low pH and are infertile (Ranthamane, 2005). In South Africa, over 30% of the soils are sandy; 60% of the soils have low soil organic matter content and exhibit high levels of degradation and low productivity (Villiers, 2005). In Malawi, soil erosion and nutrient depletion are major causes of degradation and limited productivity. For instance, estimates show that Malawi loses in excess of 30 kg of N and 20 kg of P per hectare per year through erosion on arable land (Henao and Baanante, 1999). According to a study by Folmer et al. (1998), Mozambique is estimated to lose soil nutrients at a rate of about 112 kg/ha of N, 60 kg/ha of P_2O_5, 116 kg/ha of K_2O annually through nutrient mining. Based on the recently published ISRIC 250 m resolution soil property map, the majority of Namibia, Botswana and northwestern parts of South Africa show very low SOM while northern Angola, the majority of Zambia, Malawi and Mozambique have relatively better SOM content (Hengl et al., 2015). In some parts of the region where frequent fires are observed, there is an overall low SOM content in relatively good forest cover areas (Ryan and Williams, 2011). This observed low SOM could result from rapid turnover of SOM due to high temperature or to insufficient moisture limiting decomposition in arid and semi-arid areas (Liddicoat et al., 2010). Drier and warmer conditions predicted to occur in southern Africa in the future are expected to increase rates of soil C mineralization and the associated loss of important soil functions.

Soil management and conservation

Frequent droughts that have occurred in the region for over a century have exacerbated land degradation, and its impact on livelihoods and food security has worsened. This makes land management and conservation a pivotal but formidable task particularly during the critical moisture deficient periods. Governments, nongovernmental Organizations (NGOs) and in some instances, the private sector have made efforts to address the issue of land degradation and identified suitable remedial options. In Southern Africa, land tenure determines the willingness to invest in land and address sustainability in agriculture and rural livelihoods (Bond et al., 2004). SADC as a regional organization has also put in place specific operational units to address the land tenure, land degradation and land management issues in the region. Generally, fertilizer application is the primary soil health replenishing option in the region, and some countries have promoted this intervention with subsidy programmes such as in Malawi and Zambia. Although fertilizer input in the region is below the recommended levels, substantial gains have been achieved in terms of maize yield (Jama et al., 2017). Complemented with organic inputs, conservation agriculture and good agronomic practices, overall agricultural productivity can be enhanced. However, additional land management interventions such as terraces, buffers and agroforestry will be needed at the landscape scale, as intensification cannot be achieved sustainably at the plot/farm level if degradation continues to occur at the watershed level. Integrated watershed management interventions have shown positive effects, specifically in tackling soil erosion and its associated on- and off-site effects (Henry, 2015), but such initiatives remain rare.

Despite some isolated successes, numerous interventions targeted at reducing poverty through improvement in land resource management have not achieved their targets due to lack of coordination as well as rigidity of implementation which failed to recognize and incorporate indigenous knowledge and peoples' preferences (Msangi, 2006). In addition, weak land tenure systems combined with high poverty and low literacy levels common among the rural population in the region complicate improving land management (Msangi, 2006). Low technological capacity, poor governance and poorly conceived management policies and their implementation further undermine the potential role of land management practices in tackling land degradation and enhancing food security (Henry, 2015). Some of the introduced technologies do not address the problems facing a specific area and, hence, lack relevance and applicability to different micro-environments. As a result, technology adoption and proper land management is limited in many parts of the region (Msangi, 2004). Despite the fact that most countries in the region have made notable efforts in promoting land management through policies that provide the basic foundation for managing land degradation, it is observed that their implementation remains a considerable challenge in achieving sustainability in the management of land and other natural resources in those countries (Mango et al., 2017).

Water resources and distribution

The nine SADC countries of Southern with more than 15% of the land area receive less than 7% of the continental annually renewable water resources (ARWR), estimated at 274 km^3/year (UNECA et al., 2000, Thamae, 1997). The Zambezi basin in the northern part of the region claims over 40% of this resource (Dai and Trenberth, 2002; The World Bank, 2010). This precarious water situation is likely to deteriorate due to the change of climate, land degradation and the increasing water demand due to a growing population. Kusangaya et al. (2014) estimate the expected decrease of streamflow in most of the river basins to be between 18–75% by 2050, while a few basins may show increasing streamflow trends of between 5–38%. The Zambezi is predicted to lose 26–40% of its discharge. These predictions will have to be taken into account when planning the strategic and efficient use of water.

Approximately 7% of the ARWR in Southern Africa is withdrawn to satisfy societal needs, including agriculture (UNECA et al., 2000). However, increasing this fraction is complicated as water availability and water demand seldom match in space and time and will thus require investments in massive reservoir capacities and water transfer systems. South Africa consumes approximately 24% of its ARWR with the help of more than 500 reservoirs (Wikipedia, 2014), providing an aggregate storage space equaling two-thirds of the entire ARWR of the country for an average year.

In southern Africa at present about 85% of the water withdrawn from nature (rivers, streams, wetlands and aquifers) is consumed in agriculture (WWF, 2013), well above the global average (70%). The share of irrigation in Southern Africa remains well below 20%, which is the average share of land irrigated in developing countries (FAO, 2003). Almost three-quarters of irrigation in Southern Africa takes place in South Africa with its formidable and distributed storage coupled with an advanced network of inter-basin water transfer facilities providing the backbone to the irrigation of approximately 1,500,000 ha in the country (SADC, 2012). However, the water available for irrigation in South Africa is insufficient to irrigate the entire equipped area unless deficit irrigation is a common practice. Even though the share of irrigated cropland is only slightly above 10%, it provides 30% of the crops (WWF, 2013)

Increasing the share of irrigated crops in agricultural production in the region would require the rehabilitation of many dilapidated irrigation perimeters (Liang, 2008), the construction of additional dams, transfer facilities and in-field infrastructure. With the most favourable dam sites already developed, such investment would be riskier. Relying on groundwater as the source of irrigation is not a widespread option in Southern Africa (Pavelic et al., 2013; Villholth, 2013). Recharge rates are erratic even in areas endowed with easily accessible and abundant aquifers. In most cases, increasing groundwater withdrawals would diminish surface water resources, which might threaten animal husbandry by eliminating drinking water supplies.

Theoretically, Southern Africa has room to increase withdrawals by 3% of the regional ARWR without violating recommended threshold of 10% (Rockström et al., 2009). With most other rivers already fully exploited, the best prospects are in the relatively water rich Zambezi basin, mainly in Mozambique and Zambia and some of Zimbabwe (NEPAD, 2013a, 2013b; The World Bank, 2010). Two mega dams on the Zambezi together can store the entire average annual flow of the Zambezi River. However, both reservoirs are operated for hydroelectric energy output (Magadza, 2006). Any new dam investment may have to consider and optimize the competing uses for the energy and agricultural sectors.

The natural resource constraints of Southern Africa

Southern Africa is well endowed with land resources, but with the ever-increasing demand for food and the slow adoption of sustainable intensification practices in agriculture, land degradation is a continuous threat to feeding Southern Africa in the future. Soil erosion, nutrient mining and climate change are all affecting agricultural productivity of the region. Sustainable intensification could alleviate this problem through the widespread use of fertilizers, which has been stuck at around 10kg ha^{-1} for nearly half a century but should be combined with the restitution of agricultural residues to overcome nutrient mining. Mulching and green manuring are important components of this strategy. Erosion control would be improved through these practices as well but would benefit greatly from community initiatives such as land use planning and terracing. Doubling the rate of fertilizer application alone would add 100 kg of grain per ha which for the 10 million ha under maize would amount to one million tons of grain.

The limited role of irrigation in increasing food production in Southern Africa can be illustrated by the following rough estimations. Assuming that the average regional withdrawal rate can go above the recommended 10% to 20% of the ARWR, approximating the current situation in South Africa and that 85% of the withdrawn volume of 54.8 km^3/annum is readily available for irrigation, with an irrigation efficiency of 40% prevailing in developing countries (FAO, 2003). Finally, assuming staple crops require only 6,000 m³ water per hectare, the 18.6 km³/year net available irrigation water would suffice to irrigate 3.1 million hectares of land in the region. At present, there is already 2 million hectares of irrigated land in the nine countries considered. Thus, at a regional scale even this assumed, but unrealistic water withdrawal would enable the conversion of 1.1 million hectares into irrigated agriculture, roughly 44% of the irrigable land of the Southern African part (excluding Angola) of the Zambezi basin (FAO, 1997). Clearly, water rather than land is the limiting factor and more so, once we impose realistic assumptions of water extraction. In reality, the available water resources may dwindle because of climate change. Irrigation expansion will help but not solve the food security issues of Southern Africa. Even improvements of the quite moderate 40% water use efficiency would not revise this conclusion.

If food security in Southern Africa is to be realized from the land that is cultivated, it will be paramount that the quality of land is preserved in order to deliver the ecosystem services on which its population depends. To that end, both, land conversion and land degradation should be held in check. Land sparing of pristine ecosystems and regeneration of fragile lands through intensification of agriculture on resilient cultivated land should be the guiding principle of policies, investments and incentives for future development. These should aim to fully capture the benefits of locally available resources such as water and biodiversity but also ensure that the fertility of the soil is eliminated as a constraint to agricultural productivity. Policies to these ends should be put in place and enforced in order to ensure a stable and sustainable supply of food for the region.

References

Bai, Z.G., Dent, D.L., Olsson, L. and Schaepman, M.E. (2008). Proxy global assessment of land degradation. *Soil Use and Management*, 24: 223–234.

Bartholomé, E. and Belward, A.S. (2005). GLC 2000: A New Approach to Global Land Cover Mapping from Earth Observation Data. *International Journal of Remote Sensing*, 26(9): 1959–1977.

Bond, I., Child, B., de la Harpe, D., Jones, B., Barnes, J. and Anderson, H. (2004). Private land contribution to conservation in South Africa. In: Child, B. (ed.). *Parks in Transition*. London, UK: Earthscan: 29–61.

Dai, A. and Trenberth, K.E. (2002) Estimates of Freshwater Discharge from Continents: latitudinal and Seasonal Variations. *Journal of Hydrometeorology*, 3: 660–687.

FAO (1997) *Irrigation Potential in Africa – A Basin Approach*. FAO Land and Water Bulletin No.4. Rome.

FAO (2003) Agriculture, Food and Water. Chapter 3. The Use of Water in Agriculture a Contribution to the World Water Development Report 2003 Water for People, Water for Life. www.fao.org/docrep/006/y4683e/y4683e07/htm.

Folmer, L.C.R., Guerts, P.M.H. and Francisco, J.R. (1998) Assessment of Soil Fertility Depletion in Mozambique. *Agriculture, Ecosystems and Environment*, 71: 159–167.

Garrity, D., Dixon, J. and Boffa, J.M. (2012) *Understanding African Farming Systems*. Conference Paper "Food Security in Africa", Sydney November 29–3–2012. ACIAR.gov.au/aifsc.

Henao, J. and Baanante, C. (1999) *Estimating Rates of Nutrient Depletion in Soils of Agricultural Lands of Africa*. Muscle Shoals: IFDC.

Hengl, T., Heuvelink, G.B.M., Kempen, B., Leenaars, J.G.B., Walsh, M.G., Shepherd, K.D., Sila, A., MacMillan, R.A., den Jesus, J.M., Tamene, L. and Tondoh, J.E. (2015) Mapping Soil Properties of Africa at 250 m Resolution: Random Forests Significantly Improve Current Predictions. *PLoS One*. doi:10.1371/journal.pone.0125814.

Henry, S. (2015) The Role of Governance in Sustainable Land Management Malawian Experiences. *Journal of Environment and Earth Science*, 5(7): 7–11.

Hijmans, R.J., Cameron, S.E., Parra, J.L., Jones, P.G. and Jarvis, A. (2005) Very High Resolution Interpolated Climate Surfaces for Global Land Areas. *International Journal of Climatology*, 25: 1965–1978. doi:10.1002/joc.1276. (20) (PDF) Very high resolution interpolated climate surfaces of global land areas. https://www.researchgate.net/publication/224839897_Very_high_resolution_interpolated_climate_surfaces_of_global_land_areas [accessed Oct 4 2019].

IPBES (2018) Summary for Policymakers of the Assessment Report on Land Degradation and Restoration of the Intergovernmental Science Policy Platform on Biodiversity and Ecosystem Services. In: Scholes, R., Montanarella, L., Brainich, A., Barger, N., ten Brink, B., Cantele, M., Erasmus, B., Fisher, J., Gardner, T., Holland, T.G., Kohler, F., Kotiaho, J.S., Von Maltitz, G., Nangendo, G., Pandit, R., Parrotta, J., Potts, M.D., Prince, S., Sankaran, M. and Willemen, L. (eds.). *IPBES Secretariat*. Bonn, Germany: 44.

Jama, B., Kimani, D., Harawa, R., Mavuthu, A.K. and Sileshi, G.W. (2017) Maize Yield Response, Nitrogen Use Efficiency and Financial Returns to Fertilizer on Smallholder Farms in southern Africa. *Food Security*, 9: 577–593.

Kusangaya, S., Warburton, M., Archer van Garderen, E. and Jewitt, G.P.W. (2014) Impacts of Climate Change on Water Resources in Southern Africa: A Review. *Elsevier Physics and Chemistry of the Earth*, 67–69: 47–54.

Le, Q.B., Nkonya, E. and Mirzabaev, A. (2016) Biomass Productivity-Based Mapping of Global Land Degradation Hotspots. In: Nkonya, E. Mirzabaev, A. and von Braun, J. (eds.). *Economics of Land Degradation and Improvement – A Global Assessment for Sustainable Development*. New York: Springer: 55–84. doi:10.1007/978-3-319-19168-3.

Liang Zhi You (2008) *Africa Infrastructure Country Diagnostic Irrigation Investment Needs in Sub-Saharan Africa*. Summary of Background Paper 9, IFPRI/World Bank, p. 9.

Liddicoat, C., Schepel, A., Davenport, D. and Dwyer E. (2010) *Soil Organic Carbon and Climate Change*. A Discussion Paper. Rural Solution, South Australia.

Liebenberg, P. (2005) Land and Water Management: Namibia Situation Analysis. In: Nhira, C. and Mapik, A. (eds.). *Proceedings of the Workshop on Regional Situation Analysis: Land and Water Management in the SADC Region*. Southern Africa Development Community.

Magadza, C.H.D. (2006, December) Kariba Reservoir: Experience and Lessons Learned. *Wiley Journal Lakes and Reservoirs*, 11(4). https://doi.org/10.1111/j.1440-1770.2006.00308.x.

Mango, N., Makate, C., Tamene, L., Mponela, P. and Ndengu, G. (2017) Awareness and Adoption of Land, Soil and Water Conservation Practices in the Chinyanja Triangle, Southern Africa. *International Soil and Water Conservation Research*, 5: 122–129.

Moroke, T.S. (2005) Land and Water Management: Botswana Situation Analysis. In: Nhira, C. and Mapik, A. (eds.). *Proceedings of the Workshop on Regional Situation Analysis: Land and Water Management in the SADC Region*. Southern Africa Development Community.

Msangi, J.P. (2004) Drought Hazard and Desertification Management in the Drylands of Southern Africa. *Environmental Monitoring and Assessment*, 99: 75–87.

Msangi, J.P. (2006) Land Degradation Management in Southern Africa. In: Sivakumar, M.V.K. and Ndiang'ui, N. (eds.). *Climate and Land Degradation: Environmental Science and Engineering Subseries*. Berlin, Heidelberg and New York: Springer. ISBN 10 3-540-72437-0.

NEPAD (2011) Agriculture in Africa − − Transformation and outlook.

NEPAD (2013a) Country Water Resource Profile Mozambique, 49p.

NEPAD (2013b) Country Water Resource Profile Zambia, 24p.

Pavelic, P., Villholth, K.G., Shu, Y., Rebelo, L-M. and Smakhtin, V. (2013) Smallholder Groundwater Irrigation in Sub-Saharan Africa: Country-Level Estimates of Development Potential, Routledge. *Water International,* 38(4): 392–407. http://dx.doi.org/10.1080/02508060.2013.819601.

Rakotoarisoa, M. A., Iafrate, M. and Paschali, M. (2012) *Why Has Africa Become a Net Food Importer? Explaining Africa Agricultural and Food Trade Deficits*. Rome Italy: FAO.

Ranthamane, M. (2005) Land and Water Management: Lesotho Situation Analysis. In: Nhira, C. and Mapik, A. (eds.). *Proceedings of the Workshop on Regional Situation Analysis: Land and Water Management in the SADC Region*. Southern Africa Development Community.

Rockström, J., et al. (2009) A Safe Operating Space for Humanity. *Nature* 461: 472–475.

Ryan, C. M. and Williams, M. (2011) How Does Fire Intensity and Frequency Affect Miombo Woodland Tree Populations and Biomass? *Ecological Applications,* 21: 48–60.

SADC (2012) Regional Infrastructure Development Master Plan Water Sector Plan, 72p.

Tamene, L., Sileshi, G.W., Ndengu, G., Mponela, P., Kihara, J., Sila, A. and Tondoh, J. (2019) Soil Structural Degradation, and Nutrient Limitations Across Land Use Categories and Climate Zones in Southern Africa. *Land Degradation & Development*, doi:10.1002/ldr.3302.

Thamae, L. (1997) Institutional Arrangements for the Effective Development of Southern Africa's Water Resources. In: Haddad, L. (ed.). *Achieving Food Security in Southern Africa*: 226–234, Chapter 8.

UN Economic Commission for Africa, AU, AfDB. (2000) Africa Water Vision 2025, 28p.

Villholth, K.G. (2013) Groundwater Irrigation for Smallholders in Sub-Saharan Africa − A Synthesis of Current Knowledge to Guide Sustainable Outcomes. *Routledge Water International,* 38(4): 369–391. http://dx.doi.org/10.1080/o2508060.2013.821644.

Villiers, M.C. (2005) Land and Water Management: South Africa Situation Analysis. In: Nhira, C. and Mapik, A. (eds.). *Proceedings of the Workshop on Regional Situation Analysis: Land and Water Management in the SADC Region*. Southern Africa Development Community.

Vlek, P.L.G., Le, Q.B. and Tamene. L. (2008) *Land Decline in Land-Rich Africa: A Creeping Disaster in the Making*. CGIAR Science Council. Rome, Italy: c/o FAO.

Vlek, P.L.G., Khamzina, A., and Tamene, L. (eds.). (2017) *Land Degradation and the Sustainable Development Goals: Threats and Potential Remedies*. CIAT Publication No. 440. Nairobi, Kenya: International Center for Tropical Agriculture (CIAT): 67.

Wikipedia (2014) *List of Dams in South Africa*, October 16. https://en.wikipedia.org/wiki/List_of_dams_in_South_Africa.

The World Bank (2010) The Zambezi River Basin a Multi-Sectoral Investment Opportunities Analysis, Volume 3 State of the Basin. *The World Bank Water Resources Management Africa Region,* 202.

The World Bank (2016) *ESA CCI Land Cover - Sentinel-2A Prototype Land Cover 20 meter Map of Africa 2016*. https://datacatalog.worldbank.org.

WWF (2013) Agriculture: Facts & Trends South Africa, 27.

6 The big giveaway

Farmers and biological constraints

Richard A. Sikora, Johnnie van den Berg and Erich-Christian Oerke

Introduction

It is estimated that the countries of Southern Africa need to increase food production by 2% or more per year to feed present and future populations (FAO, 2011). One of the most effective means of achieving this goal is through the reduction of losses due to pests and diseases.

The amount of food lost during production and postharvest due to the activity of pests and diseases has been estimated to be as high as 60% (Oerke et al., 1994; Oerke and Dehne, 2004; Savary et al., 2019).

Biotic constraints in Southern African agriculture are numerous and include insects, mites, fungal pathogens, viruses, bacteria, nematodes, rodents, birds and weeds. Please note that the term "pest" will be used in this chapter as a catch-all term for all these detrimental organisms.

These pests are major limiting factors to food production in Southern Africa, where small family farms of < 2 ha produce crops under largely rainfed conditions, with poor access to basic agronomic inputs, and often under climatic stress conditions. Even in the Southern African countries where larger size family and commercial farms predominate, and where sound agronomic practices and pest management systems are in place, crop loss due to biotic constraints is ever present.

According to the authors – "there is no such thing as a healthy plant in nature". All plants are under constant attack by one or more pests, both in the field from planting up until harvest and then in storage.

Pest damage to crop plants in the field and to stored food produce are to a major degree responsible for periodic hunger and for dependency on food imports, and in some cases food aid. Small family farms in rural communities are not producing at levels that ensure sufficient short- and long-term food availability. This is especially true at specific times of the year, when drought conditions exist or when there are major pest outbreaks. These small farms, however, do have the potential to produce healthier and thereby higher yielding crops. In this chapter we will demonstrate: 1) the massive losses farmers incur due to pests; 2) the impact of inadequate pest management on crop yield; 3) the losses caused by postharvest pests; and 4) the threat of invasive pests.

We will conclude by underscoring the need for new or improved agricultural policy and associated processes to make plant protection technologies available to small- and medium-size farmers.

The magnitude of crop loss

A number of key studies have demonstrated the enormous amounts of yield lost to pests in Africa and around the world on a wide range of crops (Oerke et al., 1994; Tefera, 2012; Sikora et al., 2018; Savary et al., 2019).

To demonstrate the devastation these biotic stressors have on crop production in Southern Africa, we compared maize and tomato yields in Southern African countries with that of the world average (Table 6.1).

The differences in yield between the world average and Southern Africa, with only a few exceptions, are enormous. The major Southern African maize producing countries harvest only 37.4% of the world average. For tomatoes, a similar discrepancy was found with a yield of 45.3%. This means the countries of Southern Africa are greatly underproducing many food crops when compared to other regions of the world. These differences in yield are even more pronounced if the crop yields in the Republic of South Africa, which is subjected to similar climatic conditions, are used for comparison. Yield reductions of at least 50% for maize and 80% for tomatoes are evident in most of the other countries in the region. These results illustrate the large potential existing in

Table 6.1 Production of maize and tomato in ten countries of Southern Africa (average 2015–2017)

Region/ Country	Maize			Tomatoes		
	Area [1000 ha]	Yield [kg/ha]	Production [1000 t]	Area [1000 ha]	Yield [kg/ha]	Production [1000 t]
World	194,328.3	5,637	1,095,689.8	4,836.4	37,169	179,770.3
Africa	38,987.3	1,975	77,080.6	1,292.6	16,787	21,685.2
Southern Africa, total	10,663.5	2,106	22,459.3	69.0	22,895	1,580.6
– relative to world	5.5	37.4	2.0	1.4	61.6	0.9
– relative to Africa	27.4	106.6	29.1	5.3	136.4	7.3
Angola	2,098.7	1,087	2,271.1	6.0	**2,703**	16.2
Botswana	43.3	**223**	9.2	0.1	53,997	5.7
Lesotho	102.2	722	85.8	–	–	–
Madagascar	180.3	1,714	308.9	4.9	8,527	41.8
Malawi	1,691.9	1,693	2,870.0	24.4	19,863	485.6
Mozambique	1,676.2	883	1,484.4	0.7	13,386	9.8
Namibia	31.9	1,958	62.4	1.5	6,090	9.4
South Africa	2,409.4	**4,716**	11,517.8	7.8	**75,353**	584.7
Zambia	1,151.8	2,676	3,032.6	2.6	9,782	25.7
Zimbabwe	1,275.8	626	802.5	3.7	7,020	25.9

Source: FAO (2019)

the region to produce higher yields when appropriate agronomic practices and pest management are adopted.

Crop production with and without pest management

In Figure 6.1, we present what is estimated to be maximum production potential for maize (large circle), maize production with pest management (middle circle) and production without integrated pest management (small circle) on a global scale.

The estimates provided in Figure 6.1 were extrapolated from data collected from large farms as well as university and industrial experimental fields (Oerke et al., 1994). Regardless of the source of the data, it is clear that under good agronomic conditions the lack of adequate pest management leads to significantly high crop loss as seen by the large differences between the middle and small circles.

With the exception of a few countries in Southern Africa, the small circle probably best represents the current situation with regard to the level of production and the degree of pest damage to maize on most small farms. In fact, the level of losses is probably even higher. These are farms characterized by inadequate agronomic inputs and ineffective pest management. Similar negative production estimates for other crops grown in the region under similar agronomic conditions and without pest management are to be expected.

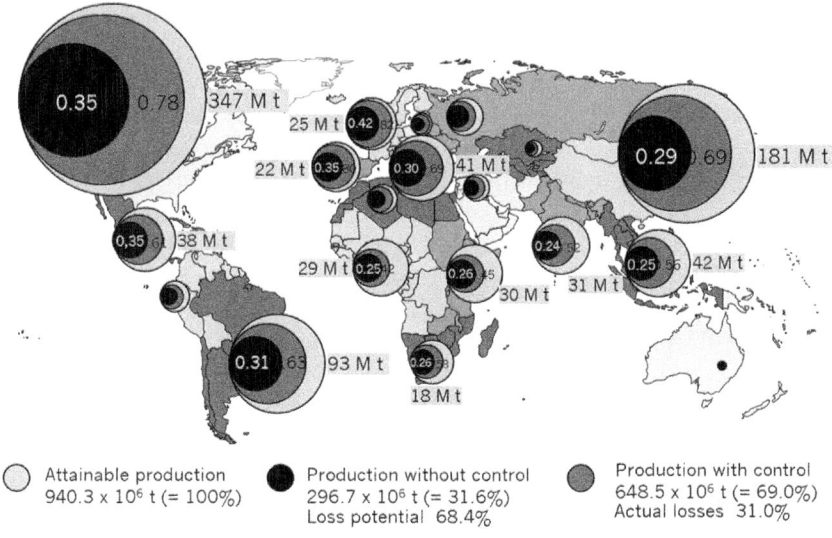

Figure 6.1 Estimated production potential for maize (large circle); production with good agronomic practices and pest management (centre circle) and production with good agronomy but without plant protection (smallest circle)

Source: Data for production worldwide and by region for the period 2002–04; data from Oerke (2006)

The giveaway

The differences in production presented in the previous table and figure are what we call the farmer's *giveaway*, with pests being the primary recipients. In the end, farmers are attempting to produce food for human consumption, but a large proportion of their yield ends up being given away for consumption by pests. This unwanted *giveaway* is the result of 1) lack of knowledge of the presence and/or importance of the pests and how to manage them, 2) insufficient access to pest management inputs and 3) poor access to extension information on pest management.

The fact that the majority of family farmers in Southern Africa are giving away vast amounts of yield to pests is catastrophic. This is especially true when one considers the serious food shortages and poor nutritional status that affects people in the region.

Conversely, larger family farms and commercial farming enterprises in some countries in Southern Africa – those who have a stronger financial basis and access to modern agronomic and pest management inputs – seem to be less affected by pests.

Crop-specific losses

The overall loss potential due to major pests in different crops around the world was evaluated by Oerke et al. (1994) and Oerke and Dehne (2004). The actual amount of loss, i.e., losses despite the use of modern integrated crop protection practices were estimated for wheat, rice, maize, barley, potatoes, soybeans, sugar beet and cotton for the period from 1996–1998 on a regional basis for 17 regions. Actual losses were estimated at 35%, 39% and 40% for maize, potatoes and rice, respectively.

In a recent study, yield losses for 137 pest species on wheat, rice, maize, potato and soybean worldwide were estimated at different hotspots around the world including sub-Saharan Africa (Savary et al., 2019). Their survey showed that in Southern Africa average loss and the range of losses for different crops were wheat 21.5% (10.1–28.1%), rice 30.0% (24.6–40.9%), maize 22.5% (19.5–41.1%), potato 17.2% (8.1–21.0%) and soybean 21.4% (11.0–32.4%). The greatest losses were detected in food-deficit regions with fast-growing populations.

On a worldwide basis, weeds had the most severe impact on yield (32%) followed by animal pests and then pathogens at 18 and 15%, respectively. Added together the total loss due to these major pests is over 60%. Weeds are a farmer's worst enemy, with smallholder farmers, mainly women, spending 50–70% of their total labour time hand weeding (Gianessi and Williams, 2011).

Pest damage chain

The amount of yield lost to pests is complex in that farmers are constantly confronted with a series of losses caused by the simultaneous occurrence of pests along the food chain from sowing until harvest and thereafter in storage (Figure 6.2). Pests seldom if ever come packaged alone.

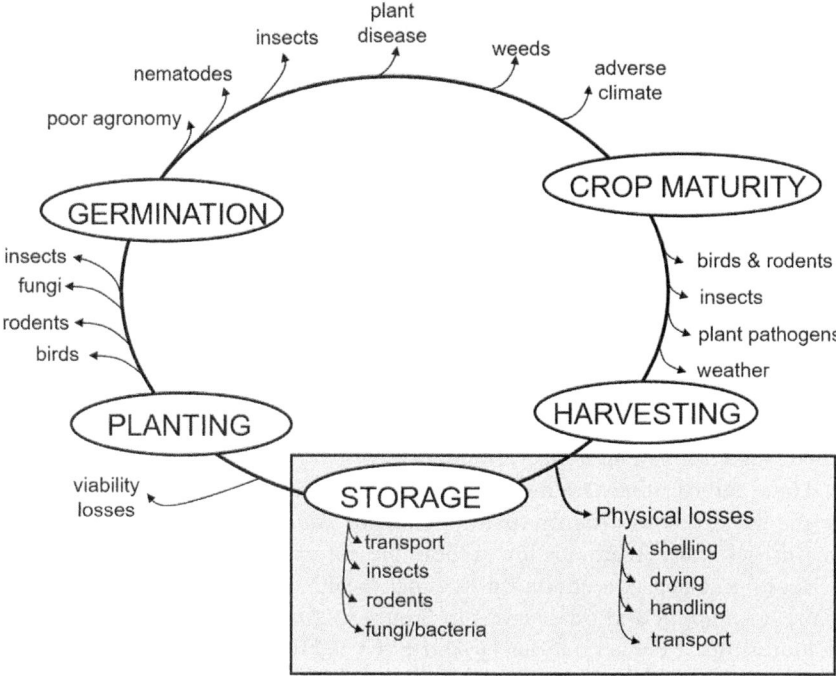

Figure 6.2 The yield loss chain, illustrating the biotic and abiotic constraints that may affect crops during the crop cycle from before seedling emergence up to storage

Source: Adapted from Harris and Lindblad, 1976

At every stage of crop production, from the field to the marketplace, a certain percentage of yield is slashed off by one or more pests acting singly or in consortium, resulting in cumulative damage that exceeds that caused by any single pest acting alone. A simplified chain of events affecting plant health could include: 1) fungal and viral contamination of seed in storage for planting; 2) ever present soil-borne weeds, fungal pathogens and nematodes; 3) fungal spores distributed by wind; 4) highly mobile below- and above-ground insects; 5) insect vectors of disease causing organisms; and 6) birds and rodents.

Depending on 1) initial pest density; 2) pest complex; 3) susceptibility of the crop variety; 4) climatic conditions; 5) the nutritional status of the plant; and 6) quality of storage silos, the pest damage chain can and does result in very high accumulated losses, as shown in Figure 6.1.

Postharvest losses

Pest damage does not stop at harvest but continues during postharvest storage when the grain is fed upon by insects and rodents and is further degraded through the activity of fungal molds.

Insect damage during postharvest storage can be extensive and is usually the results of poor methods of exclusion from storage bins (see Chapter 13) and when no pest control is applied. Two major examples are the larger grain borer (*Prostephanus truncatus*) and the maize weevil (*Sitophilus zeamais*), which can cause losses of up to 65 and 80% respectively (Boxall, 2002; Muatinte and Van den Berg, 2018).

In addition, postharvest losses can occur at different stages, from harvest to the marketing process itself (Tefera, 2012) (Figure 6.3). At this time, for example, pests reduce the quality of horticultural products such as fruits and vegetables in storage and even during transport to markets.

A major unseen enemy that directly affects consumer health is the molds that cause indirect damage and quality loss due to contamination of grain in storage bins (Bandyopadhyay et al., 2016; Savary et al., 2019; Akello et al., Chapter 21 in this volume). A number of fungi infect maize and other grains as well as vegetables and contribute to both quantitative as well as qualitative loss in food value and a decrease in market value.

These fungal molds are responsible for the formation of toxins (Fandohan et al., 2005) that are highly toxic to humans and livestock (see Chapter 21 in this volume). Aflatoxins are responsible for stunting in children, cancer in adults and negative effects on livestock health. These fungal molds infect maize, groundnut and other crops in Southern Africa and produce aflatoxins in food stuffs at concentrations far above the WHO recommended standards. The presence of these toxins in many cases would lead to 100% loss in many countries of the world where aflatoxin levels are monitored and thresholds enforced.

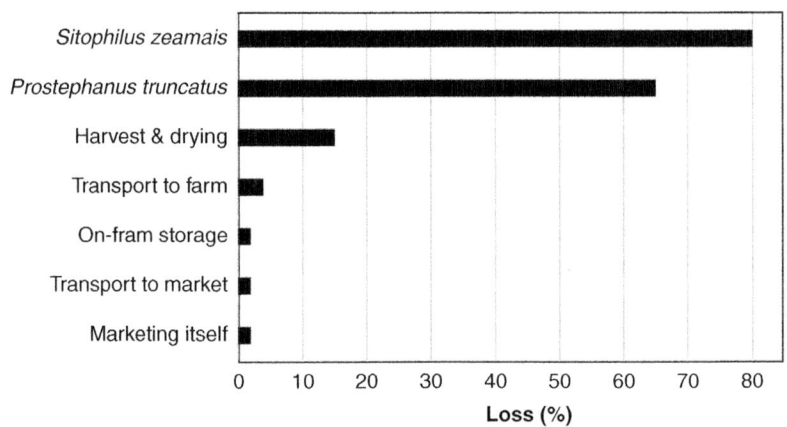

Figure 6.3 Graphic presentation of the degree of losses caused by storage pests which take place during the harvest and other grain handling processes

Source: Own representation, modified from Boxall, 2002; Muatinte and Van den Berg, 2018; Tefera, 2012

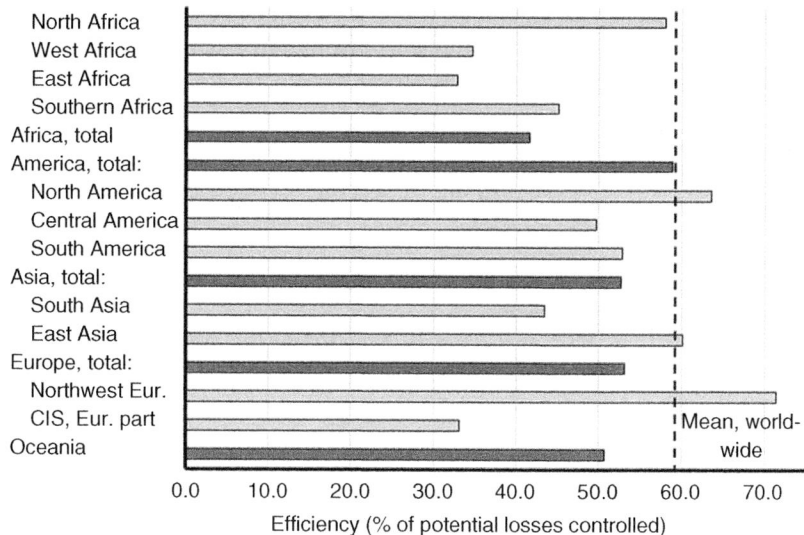

Figure 6.4 Regional differences in the overall efficacy of crop protection practices in 11 crops (barley, cotton, maize, oilseed rape, peanuts, potatoes, rice, soybean, sugar beet, tomatoes and wheat); calculations are based on loss estimates between 2002–2004

Source: Own representation

Distortions in the effectiveness of pest management

To make matters worse, there are strong regional differences in the efficacy of plant protection practices applied in the region (Figure 6.4). The effectiveness of pest management in West, East and Southern Africa, for example, is significantly below the worldwide average measured over 11 crops (Oerke and Dehne, 2004).

This discrepancy in control efficacy between regions is probably due to 1) a lack of awareness of the cause of damage; 2) improper selection and/or access to plant protection products; 3) ineffective application procedures; and 4) poor access to extension service knowledge support systems.

The unwanted guests

The threat of introduction of invasive pests represents another major challenge to food production. A recent review by Sileshi et al. (2019) reported that 16 current alien invasive insect and mite pests have invaded new areas in Africa. These pests will, over time, spread to countries in Southern Africa, where they will cause significant crop damage (Muatinte and Van den Berg, 2018). Unless suitable measures of quarantine are in place to prevent the introduction and spread of alien invasive species, crop damage cannot be avoided (CABI, 2019). Unfortunately, the often understaffed and poorly funded phytosanitary programmes in Southern African countries leads to enhanced spread of invasive pests in the region.

Five destructive invasive species which threaten agricultural production and food security in East and Southern Africa were discussed by Pratt et al. (2017). The estimated economic impacts of these five invasive species on maize cultivated in production systems with beans and tomato caused combined annual

Table 6.2 List of important invasive pest species, their host crops and estimated economic impact

Invasive pest	Major hosts	Regions affected	Estimated losses
Spotted stem borer *Chilo partellus*	Maize and sorghum	East Africa, 1920s. (Now East and Southern Africa)	US$450 million/yr Smallholder losses
Maize Lethal Necrosis Disease (MLND)	Maize, sorghum, millet and some grasses	Kenya in 2011. (Now Tanzania, Uganda and Rwanda)	US$350 million/yr Smallholder losses
Weed species *Parthenium hysterophorus*	Grasses in pastures, maize	Southern Africa	US$82 million/yr
Leaf miner *Liriomyza trifolii*	Horticultural crops, vegetable, affects marketability	Most of East and Southern Africa	Total fruit losses Farm closures
South American tomato leaf miner *Tuta absoluta*	Vegetables, affects marketability	Tunisia, 2007. Most of East and Southern Africa	US$791.5 million
Fall Armyworm *Spodoptera frugiperda*	Attacks 350 plant species incl. maize, sorghum, millet, rice, groundnut, vegetables	West Africa, 2016. All of SSA	Maize 20.6 million t/yr in 12 countries 53% of production, US$6.2 billion
Panama fungal wilt disease *Fusarium oxysporum* Race 4	Banana – Cavendish	Mozambique, 2017	Total crop loss
Fruit fly *Bactrocera invadens*	Horticultural crops, Mango, avocado	Kenya, Mozambique, Tanzania, Uganda	60% fresh fruit loss Negative local and export impact
Liberibacter asiaticus (Asian greening disease) Transmitted by the psyllid *Diaphorina citri*	Citrus	Ethiopia, Tanzania, Kenya, Réunion and Mauritius.	Millions of trees destroyed in the Americas

Sources: CABI, 2019; Mahuku et al., 2015; Rubaba et al., 2017; Kiruwa et al., 2016; Goergen et al., 2011; Viljoen, 2019

losses of US\$0.9–1.1 billion. They estimated future annual losses over the next five–ten years at US\$1.0–1.2 billion and indicated that it could be much higher.

Conclusions

Biotic stresses in the form of a wide array of pest organisms are a major constraint to food production. Losses in yield of food crops in the region are far more serious than that of the world average. These pests impact yield from seeding to postharvest storage and indicate a major lack of pest management at farm level. These pests are responsible for food insecurity in many countries during years affected by low levels of production (e.g., drought), between crops and harvests (hunger-season) but also over longer periods of time due to chronic underproduction.

These losses are manageable with proper pest management as discussed in a number of chapters in this volume. In many cases small family farmers are not familiar with pests and diseases, cannot access this type of information or the inputs to control these pest organisms are too expensive, which leads to immense losses.

By reversing the food *giveaway*, Southern Africa could produce enough food to reduce food insecurity until at least 2050 and beyond by using currently available pest management technologies to reduce pest-induced losses in all crops through improved: 1) integrated pest management programmes (IPM) supported by extension programmes; 2) crop breeding for resistance to biotic and abiotic stresses and seed distribution programmes; 3) improved postharvest storage technology; 4) food toxin management; and 5) improved phytosanitary services (all outlined in the chapters in this volume).

It is imperative that government policies are put in place that improve small- and medium-size farmer access to pest management systems as well as the other inputs of modern agriculture. A significant improvement in extension services and their information technological knowledge dissemination systems also is needed.

Intensification of crop production without addressing pest management will have disastrous effects in terms of food security and is probably economically irrational and ecologically irresponsible.

References

Abassa, A.B., Ndungurua, G., Mamirob, P., Alenkhec, B., Mlingia, N. and Bekunda, M. (2014) Post-harvest food losses in a maize-based farming system of semi-aridsavannah area of Tanzania. *Journal of Stored Products Research* 57: 49–57.

Bandyopadhyay, R., Ortega-Beltran, A., Akande, A., Mutegi, C., Atehnkeng, J., Kaptoge, L., Senghor, A.L., Adhikari, B.N. and Cotty, P.J. (2016) Biological control of aflatoxins in Africa: current status and potential challenges in the face of climate change. *World Mycotoxin Journal* 9: 771–789.

Boxall, R.A. (2002) Damage and loss caused by the larger grain borer *Prostephanus truncatus*. *Integrated Pest Management Reviews* 7: 105–121.

CABI. (2019) Invasive species: the hidden threat to sustainable development. Taken from: www.invasive-species.org/wp-content/uploads/sites/2/2019/02/Invasive-Species-The-hidden-threat-to-sustainable-development.pdf.

Day, R., Abrahams, P., Bateman, M., Beale, T., Clottey, V., Cock, M., Colmenarez, Y., Corniani, N., Early, R., Godwin, J., Gomez, J., Gonzalez Moreno, P., Murphy, S.T., Oppong-Mensah, B., Phiri, N., Pratt, C., Silvestri, S. and Witt, A. (2017) Fall armyworm: impacts and implications for Africa. *Outlooks on Pest Management* 28(5): 196–201.

Fandohan, B., Gnonlonfin, B., Hell, K., Marasas, W.F.O. and Wingfield, M.J. (2005) Impact of indigenous storage systems and insect infestation on the contamination of maize with fumonisins. *African Journal of Biotechnology* 5: 546–552.

FAO. (2011) *Global Food Losses and Food Waste-Extent, Causes and Prevention.* FAO, Rome, Italy.

FAOStats. (2019) Taken from: www.fao.org/faostat/en/#country/.

Gianessi, L. and Williams, A. (2011) Overlooking the obvious: The opportunity for herbicides in Africa. *Outlooks on Pest Management* 22: 211–205.

Goergen, G., Ois Vayssie'Res, J-F., Gnanvossou, D. and Tindo, M. (2011) *Bactrocera invadens* (Diptera: Tephritidae), a new invasive fruit fly pest for the afrotropical region: Host plant range and distribution in west and central Africa. *Environmental Entomology* 40(4): 844–854.

Harris, K.L. and Lindblad, C.J. (1976) *Postharvest Grain Loss Assessment Methods: A Manual of Methods for the Evaluation of Postharvest Losses.* American Association of Cereal Chemists, Washington, DC, 193p.

Kiruwa, F.H., Feyissa, T. and Ndakidemi, P.A. (2016) Insights of maize lethal necrotic disease: a major constraint to maize production in East Africa. *African Journal of Microbiology Research* 10: 271–279.

Mahuku, G., Lockhart, B.E., Wanjala, B., Jones, M.W., Kimunye, J.N., Stewart, L.R., Cassone, B.J., Sevgan, S., Nyasani, J.O., Kusia, E., Kumar, P.L., Niblett, C.L., Kiggundu, A., Asea, G., Pappu, H.R., Wangai, A., Prasanna, B.M. and Redinbaugh, M.G. (2015) Maize lethal necrosis (MLN), an emerging threat to maize-based food security in sub-Saharan Africa. *Phytopathology* 105: 956–965.

Muatinte, B. and Van den Berg, J. (2018) The use of mass trapping to suppress population numbers of *Prostephanus truncatus* (Horn) (Coleoptera: Bostrichidae) in small scale farmer granaries in Mozambique. *African Entomology* 26: 301–316.

Oerke, E.C. (2006) Crop losses to pests. *Journal of Agricultural Science* 144: 31–43.

Oerke, E.C. and Dehne, H.W. (2004) Safeguarding production – losses in major crops and the role of crop protection. *Crop Protection* 23: 275–285.

Oerke, E.C., Dehne, H.W., Schönbeck, F. and Weber, A. (1994) *Crop Production and Crop Protection: Estimated Losses in Major Food and Cash Crops.* Elsevier Science, Amsterdam.

Pratt, C.F., Constantine, K.L. and Murphy, S.T. (2017) Economic impacts of invasive alien species on African smallholder livelihoods. *Global Food Security:* 31–37.

Rubaba, O., Chimbari, M. and Mukaratirwa, S. (2017) Scope of research on *Parthenium hysterophorus* in Africa. *South African Journal of Plant and Soil* 34: 323–332.

Savary, S., Willocquet, L., Pethybridge, S.J., Esker, P., McRoberts, N. and Nelson, A. (2019) The global burden of pathogens and pests on major food crops. *Nature – Ecology & Evolution.* published online Nature Ecology and Evolution.

Sikora, R.A., Coyne, D., Hallmann, J. and Timper, P. (2018) *Plant Parasitic Nematodes in Subtropical and Tropical Agriculture.* CABI, 3rd ed., 876p.

Sileshi, G.W., Gebeyehu, S. and Mafongoya, P.L. (2019) The threat of alien invasive insect and mite species to food security in Africa and the need for a continent-wide response. *Food Security.* doi:10.1007/s12571-019-00930-1.

Tefera, T. (2012) Post-harvest losses in African maize in the face of increasing food shortage. *Food Security* 4(2): 267–277.

Thurow, R. (2012) *The Last Hunger Season*. Public Affairs and Perseus Books Group, Philadelphia, PA, Southern Africa, 295.

Viljoen, A., Ma, L-J. and Molina, A.B. (2019) Fusarium wilt (Panama disease) and monoculture banana production: resurgence of a century-old disease. In: *Emerging Plant Diseases and Global Food Security*. Chapter 8. APS Press.

7 The impact of global and regional markets on agricultural transformation in Southern Africa

Ferdi Meyer, Tracy Davids and Nick Vink

Introduction

Over the past two decades, agricultural markets, trade and food systems in Southern Africa have experienced dramatic transformation. Most of the region's economies were characterized by periods of relatively fast economic growth (around 3 to 6%), increasing population and strong patterns of urbanization that have triggered a diversification in diets. Consumers have progressively added more perishable and processed foods to diets previously dominated by grains and other staples, unleashing a wave of structural transformation of the food system. Consequently, markets responded and a number of countries in the Southern African region have expanded agricultural production beyond what local markets can consume, boosting the value and volume of exports of a wide number of commodities.

In high value commodities such as fruit and wine, the region typically has a resource-based competitive advantage in global exports markets, mainly due to suitable climatic conditions, water for irrigation, relatively low wage rates and, in the case of South Africa, competitive infrastructure. However, countries like Zambia and Angola remain net importers of high-valued and processed agricultural products, yet these countries are barely scratching the surface of their natural resource potential for producing high-value crops. In the case of Zimbabwe, production of high-value crops has dwindled following the introduction of the land grab in the early 2000s. Thus, there is significant opportunity for expansion well beyond what local markets can absorb, yet these countries are faced by poor infrastructure, very little investment in cold chains, complex bureaucratic systems and an inconsistent policy environment that increases the perceived risk for long-term investments in high value crops.

In contrast to high value crops, many countries in the region have made significant progress in increasing production of field crops (e.g., maize, soybeans, sugar, etc.), with the sharp rise in commodity prices and some policy reforms with respect to access to land being the main drivers of change. Predictably, though, most of the increase in production north of South Africa's borders was driven by expansion of the area under production and not by higher yields.

Figure 7.1 presents the net trade position of total agricultural products for the Southern African Development Community (SADC) region. It illustrates that the region is typically a net exporter of unprocessed products. Whereas imports of processed products increased until 2012, significant investments in midstream and downstream industries, such as a major expansion of soybean crushing facilities, fruit juice concentrate plants, etc., have resulted in a reversal of this trend in recent years. While the region as a whole remains a net importer of processed products, the real net import value of these in 2017 was less than a third of the 2012 value.

Figure 7.2 presents a different view on the net trade for the SADC region by aggregating all agricultural products into 15 commodity groups and illustrating the average net trade position for each commodity group between 2010 and 2017. For specific commodity groups where the region has limited production capacity (such as wheat and rice), or where the competitiveness of domestic value chains is under pressure (such as chicken meat), the region continues to import significant volumes.

At aggregate levels, the changes observed in agricultural trade in recent years reflect growth opportunities arising from the rapid transformation of agricultural markets, major private sector investments throughout the value chain and the rapid adoption of world-class technology. At the same time, a number of critical challenges remain and are in fact becoming more important. Many of these relate to more complex bilateral and multilateral trade negotiations, sanitary and phytosanitary protocols (SPS) and Non-Tariff Trade Measures (NTM's), rules of origin and many more, which will influence the future growth trajectory of the sector. One underlying element of all these challenges relates to effective management, the capacity and the required skills by governments

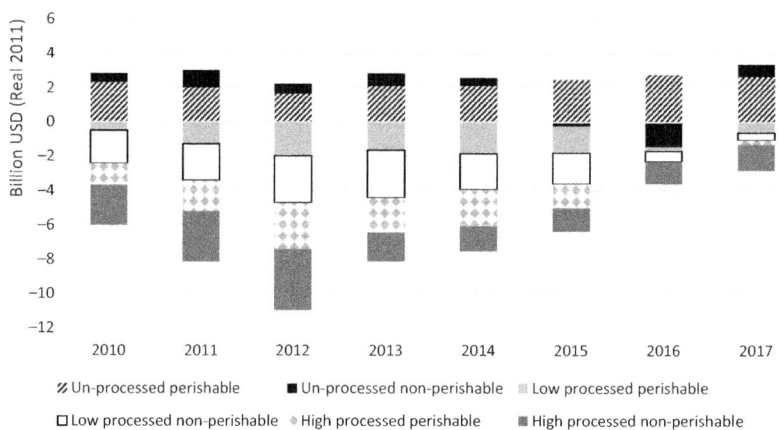

Figure 7.1 Net trade in agricultural products for the SADC region

Source: Compiled from World Integrated Trade Solutions & World Bank, 2019

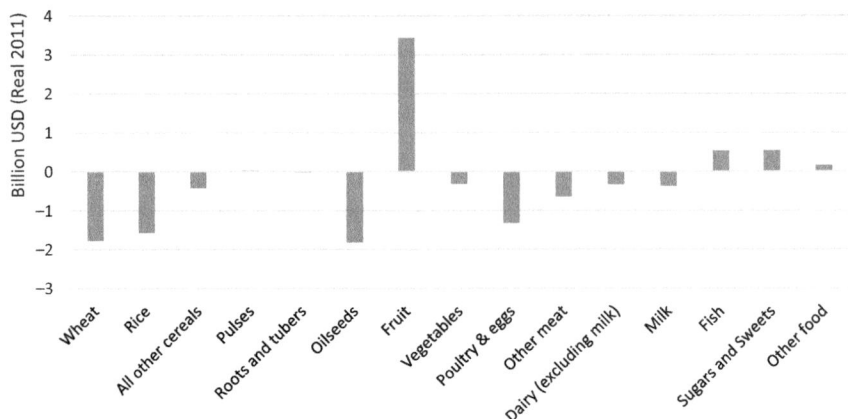

Figure 7.2 Average net trade position of all agricultural products, 2010–2017
Source: Compiled from World Integrated Trade Solutions & World Bank, 2019

and industry to address these issues effectively. Countries that do not have the capacity to deal with these issues effectively will over the long run lose out on growing export markets and will face increasing competition from imports.

These challenges and opportunities are best illustrated through case studies of three key commodities where rapid transformation has been evident over the past decade.

Commodity specific case studies

These case studies are based on three commodity groups that reflect a spectrum of imported and exported products. The first is vegetable oil, where the SADC region remains a net importer, despite significant investment in oilseed processing facilities in a number of individual countries. The second is citrus, the largest contributing industry to the net export position of fruit products and a key sector driving agricultural growth in South Africa in recent years. The third is beef, where a number of countries reflect remarkable export driven growth, but substantial opportunities remain unutilized due to a lack of market access.

Vegetable oil

Since 2007, the value of vegetable oil imports into the SADC region has increased by almost 3% per year. In many countries, investments have occurred in response to the development of domestic value chains and growing supply. In South Africa, for example, dedicated soybean crushing capacity increased from 340,000tonnes in 2012 to 1.8 million tonnes in 2017. In Zambia, oilseed crushing capacity expanded from 125,000 tonnes in 2010, to 375,000 tonnes in 2016, and in Malawi, from 95,000 tonnes in 2010 to 465,000 tonnes in 2016 (Meyer et al., 2018).

Despite the expansion in processing facilities, crush volumes in most of these countries remain well below capacity. In South Africa, soybean production growth of almost 15% per annum has still been insufficient to ensure optimal utilization of the installed crush capacity, but in 2017, the domestic crop of 1.5 million tonnes came close. The sharp increase in local crushing has reversed the increasing trend in imported soybean meal for the feed industry and, compared to import levels around 950,000 tons of soybean meal six years ago, imports have declined to around 450,000 tons. In its latest baseline, the Bureau for Food and Agricultural Policy (BFAP) projects that South Africa can become self-sufficient in soybean meal within the next five years (BFAP, 2018).

In Zambia and Malawi however, rapid soybean production growth has resulted in growing soybean exports, despite surplus crush capacity. Both Malawi and Zambia remain net importers of vegetable oil, but the demand for animal feed from the intensive livestock sectors (poultry, eggs, dairy, pig meat) has not yet expanded at the same rate as soybean production – yielding a surplus of soybean oilcake (Table 7.1). The reason for this is that both countries have a small intensive livestock industry relative to the amount of feed they can produce. For instance, a substantial share of Zambia's poultry production has also transformed into a highly intensive operation, total production amounts to approximately 100,000 tons, which is sufficient to supply the local demand for chicken meat. Therefore, if the government of Zambia will not actively drive potential export markets for chicken meat into the region and boneless portions into the EU markets, the potential for value addition of soybeans and meal will be limited.

Table 7.1 Soybean production and trade in South Africa, Zambia and Malawi

		South Africa		Zambia		Malawi	
		Average 2015–17	*10-year growth p.a.*	*Average 2015–17*	*10-year growth p.a.*	*Average 2015–17*	*10-year growth p.a.*
Soybean production	1000 t	1042.67	14.99%	281.74	16.77%	116.51	−1.27%
Soybean area	1000 ha	588.02	13.20%	168.97	17.78%	131.27	0.27%
Soybean yield	t/ha	1.78	1.79%	1.70	−1.01%	0.89	−1.55%
Imports							
Soybeans	1000 t	161.44	24.8%	0.77	−19.5%	0.36	−28.3%
Soybean meal	1000 t	545.64	−6.8%	1.11	−7.9%	0.70	−60.4%
Soybean oil	1000 t	187.93	−2.9%	16.94	4.2%	12.42	−2.7%
Exports							
Soybeans	1000 t	5.25	−3.9%	34.01	25.5%	14.00	23.2%
Soybean meal	1000 t	82.522	26.0%	54.686	106.1%	15.266	56.0%
Soybean oil	1000 t	71.15	41.5%	0.70	−12.1%	0.00	NA

Source: Compiled from SAGIS, 2019; ReNAPRI, 2018; ITC Trademap, 2019

Currently, the surplus is typically exported to South Africa, resulting in growing intraregional trade, but high transport costs (freight rates, waiting time at the border, etc.) in the region implies that price levels, and consequently also crush margins, decline sharply when surpluses have to be exported. As a result, production growth has stagnated in recent years.

Going forward, rapid growth in South African soybean production is projected to take the country to self-sufficiency in soybean meal, implying that current regional exporters such as Zambia and Malawi will likely need to find new markets for their products, either domestically or elsewhere in the region. Domestic consumption would, of course, require substantial growth in intensive livestock production, which is possible if current trends in agricultural transformation continue. However, exports into the rest of the Southern African region would benefit from investments that reduce the cost of trade.

While the region arguably has the potential to produce and process enough oilseeds to also become self-sufficient in vegetable oil, it will continue to be challenged by highly competitive palm oil imports from other parts of the world (e.g., Indonesia and Malaysia). This is particularly relevant to high oil yielding seeds such as sunflower and canola.

Citrus

In its National Development Plan, compiled in 2012 (NDP, 2012), South Africa's National Planning Commission identified a number of high value, export orientated commodities with significant growth potential. Citrus represents the largest industry among these and is currently the leading contributor to South Africa's agricultural exports. Table 7.2 indicates that, over the past decade, exports of oranges have increased by nearly 2% per year, while the exports of other citrus products increased by almost 5% per year. The increase in export

Table 7.2 Growth in the South African citrus industry

		Oranges		Other citrus[1]	
		Average 2015–2017	10-year avg. growth p.a.	Average 2015–2017	10-year avg. growth p.a.
Production	1000 tons	1427.67	1.05%	919.21	3.34%
Area	1000 ha	43.79	1.54%	29.09	5.68%
Yield	t/ha	32.65	−0.49%	31.91	−2.35%
Domestic fresh consumption	1000 tons	95.88	−3.75%	48.18	8.38%
Domestic processing	1000 tons	262.93	−0.10%	233.11	−0.67%
Exports	1000 tons	1068.86	1.73%	637.91	4.77%

1 Soft Citrus, Grapefruit, Lemons and Limes

Source: Compiled from BFAP, 2018

value was even greater, supporting significant area expansion. Many parts of South Africa possess the climatic conditions and natural resource base to produce citrus successfully, and at different times of the year, provide a comprehensive portfolio of products through most of the year. The labour intensive nature of production provides South Africa with a competitive advantage, allowing the country to become the second largest exporter of fresh citrus fruit in the world after Spain.

While the citrus industry is an undoubted success story within South African agriculture, it remains challenged by a number of factors. The prevalence of Citrus Black Spot (CBS), a fungal disease that predominantly affects the rind of fruit, has led to the implementation of a number of costly compliance measures to retain access to South Africa's largest export markets in the EU. As an alternative and diversification strategy, producers have targeted Asian markets for export growth. However, many of South Africa's competitors in the Southern Hemisphere, such as Chile, Peru and Australia, have been more successful in negotiating preferential access into these Asian markets, leaving South Africa's producers at a competitive disadvantage.

Other challenges that the industry will have to address is the effective use of water. In a recent agricultural census undertaken in the Western Cape Province, flyover and remote sensing data shows clear shifts of orchards out of wine production into citrus over the past five years. The water demand per hectare is much higher for citrus trees than for wine grapes, and through closer analysis of the data, it is evident that the total area under irrigation has in fact decreased over the past five years due to this increased demand for water per hectare and changing cropping patterns. Apart from a handful of new water infrastructure investments and the revitalization of existing idle irrigation schemes, South Africa will mainly have to achieve further expansion in irrigation of high valued crops by effective maintenance of existing irrigation infrastructure and the implementation of water saving technology.

Beef

At aggregate level, the SADC region remains a net importer of beef products, but selected countries have been very successful in driving export growth. The most prolific of these is South Africa, which increased beef product exports by an annual average of 27% between 2007 and 2017 (BFAP, 2018). Most of this beef is produced in intensive feedlot operations that are responsible for approximately 80% of all beef that is produced in South Africa. In contrast, beef production in Botswana and Namibia is mainly characterized by extensive pasture-based systems. Both of these countries have also been successful in building export-orientated beef industries, though adverse weather conditions over the past five years have limited export growth. Nevertheless, they have established good market access and continue to export premium products very successfully into the EU in particular. Furthermore, it is also important to note that there is significant intraregional trade in live animals. For example, South

Table 7.3 Beef export growth in South Africa, Namibia, Botswana and Zambia

	Average 2012–2017 ('000 tons)	10-year avg. growth p.a. (%)
South Africa	31.55	27
Namibia	18.84	−6
Botswana	25.00	1
Zambia	0.08	19

Source: Compiled from ITC Trademap, 2019

African feedlots are importing in the order of 200,000 weaner calves annually because feed costs are significantly lower in South Africa than in Namibia.

The beef industry in Zambia has also transformed significantly over the past decade, mainly due to higher maize and soybean production, which has boosted the production of beef in feedlots. Exports are still small in absolute terms but have grown impressively at just under 20% per annum over the past decade (Table 7.3).

One of the greatest challenges facing beef exports from the Southern African region is the prevalence of Foot and Mouth Disease (FMD), which poses challenges in complying with the SPS protocols of many importing countries. The often close proximity of cattle to buffalo, which carry the disease, makes it difficult to control. In Botswana and Namibia, the challenges associated with FMD have been overcome by compartmentalization and zoning, combined with strong traceability systems. This has allowed both countries access to the EU market, where they obtain a premium for grass fed beef exported from the FMD free zones. In both countries, government entities have played a key role in beef exports.

South Africa also utilizes a zoning system. FMD remain endemic to the Kruger National Park, but a protection zone has been established around the park and the World Organisation for Animal Health (OIE) declared the rest of the country free of FMD without vaccination in 2014 (DAFF, 2015). Consequently, exports expanded rapidly, first into high value markets in the Middle East and later into China. This growth in exports has supported prices, despite challenging weather conditions and herd liquidation in 2015 and 2016. Unlike Namibia however, South Africa does not have a centralized traceability system and, hence, is not able to access premium markets such as the EU and the USA. Had these markets been accessible to South African producers, both exports and production would have performed even better.

Conclusion

The agricultural sector in South Africa is changing rapidly, driven by changing patterns of demand that have come about as a result of demographic change, especially rapid urbanization and the rise of a middle class with sufficient disposable income. This has set in motion a number of fundamental changes in the food retail sector, giving rise to a "supermarket revolution" that started more than a decade ago and had already spread into secondary cities and the larger towns (Weatherspoon and Reardon, 2003) because of the liberalization of foreign direct investment rules and the rise of the middle class.

In this chapter, we have provided three case studies (vegetable oils, citrus and beef) which illustrate how production and trade patterns are changing in reaction to the changes in the structure of demand. The changes are not uniformly positive and are often retarded by poor infrastructure and excessive bureaucracy but generally point to potential positive changes.

References

BFAP. (2018) *BFAP Baseline: Agricultural Outlook 2018–2027.* Bureau for Food and Agricultural Policy, Pretoria.

International Trade Council (ITC). (2019) *Trademap Database.* [Online] Available at: www.trademap.org.

Meyer, F.H., Traub, L.N., Davids, T., Chisanga, B., Kachule, R., Tostao, E., Vilanculos, O., Popat, M., Binfield, J. and Boulanger, P. (2018) *Modelling Soybean Markets in Eastern and Southern Africa.* Regional Network of Agricultural Policy Research Institutes (ReNAPRI), EUR 28978 EN. doi:10.2760/20598.

National Planning Commission: South African Presidency. (2012) *National Development Plan 2030 Our Future – Make It Work.* Government of South Africa. ISBN: 978-0-621-41180-5.

Regional Network of Agricultural Policy Research Institutes (ReNAPRI). (2018) Unpublished Database.

South African Department of Agriculture, Forestry and Fisheries (DAFF). (2015) Media Briefing on the Status of Foot and Mouth Disease in South Africa – 13 January. [Online] Available at: www.nda.agric.za/docs/media/Media%20brief%20on%20FMD.pdf.

South African Grain Information Service (SAGIS). (2019) Historic Production Database. [Online] Available at: www.sagis.org.za/historic%20hectares%20&%20production%20info.html.

Weatherspoon, D.D. and Reardon, T. (2003) The Rise of Supermarkets in Africa: Implications for Agrifood Systems and the Rural Poor. *Development Policy Review* 21(3).

World Bank. (2019) World Development Indicators Database. [Online] Available at: https://databank.worldbank.org/data/reports.aspx?source=world-development-indicators.

World Integrated Trade Solutions (WITS). (2019) World Integrated Trade Solutions Database. [Online] Available at: https://wits.worldbank.org/.

Part III

Current technologies

This Part describes a number of current and key technologies that are part of good farming practices used in many places in the world for crop improvement. Many of these technologies also are available in Southern Africa but are often only readily available to financially secure larger farm operations. These are technologies that could be readily and quickly adopted by farmers to offset the biological and physical constraints listed in Part II.

8   Soil fertility maintenance and nutrient management for agricultural transformation

Bernard Vanlauwe, Pauline Chivenge and Shamie Zingore

Introduction

Sub-Saharan Africa (SSA) continues to experience growing food insecurity underpinned by low crop productivity and rapid population growth (FAO et al., 2018). The problems of low crop productivity and malnutrition are most severe in Southern Africa (Misselhorn, 2005) due to widespread problems of soil fertility depletion, land degradation and unfavourable climatic conditions (Nyamangara et al., 2000; Msangi, 2007). Yields for cereal crops in smallholder farming systems in Southern Africa are less than 30% of attainable yields, and low use of fertilizer and other nutrient resources are recognized as one of the major limiting factors. That said, overall crop production has quadrupled over the past seven decades, with the relative contribution of area expansion and productivity increase varying between various countries. For example, between 1961 and 2017 maize productivity increase supported by increased fertilizer use contributed substantially to maize production increase in Zambia and Malawi (Table 8.1). In contrast, maize yields and fertilizer use have remained low in Mozambique, and area expansion has accounted for the largest maize production increase (Table 8.1).

Large areas where crops are produced with little or no fertilizer and organic nutrient inputs are characterized by severe loss in carbon stocks, biodiversity and ecosystem service provision in natural vegetation and soils (Zingore et al., 2005; Vlek et al., 2008). The majority of soils in Southern Africa are inherently infertile due to a bedrock that consists of mostly granites and gneiss (Deckers et al., 2000). Although average fertilizer use has remained low in Africa at about 16 kg nutrients ha^{-1}, national fertilizer use data show a growing number of countries in Southern Africa achieving at least 30 kg nutrients ha^{-1}. Further significant increase in fertilizer and other organic resources will be required to offset negative nutrient balances regionally (Fixen et al., 2015).

To be effective, technologies for increasing productivity must be adapted to the complex and highly variable biophysical and socioeconomic conditions in smallholder farming systems. At the regional scale, agro-ecological and soil conditions have led to diverse farming systems with different crops, cropping patterns, soil management considerations and access to inputs and commodity

Table 8.1 Changes in maize yield, production areas and recent fertilizer use for Malawi, Mozambique and Zambia in Southern Africa

Country	Base yield 1961 (t/ha)	Yield 2017 (t/ha)	Yield Increase (%)	Base Area 1961 (million ha)	Area 2017 (million ha)	Area Increase (%)	Fertilizer NPK nutrient use 2016 (kg/ha)
Malawi	1.02	2.01	97	0.80	1,73	116	21.6
Mozambique	0.87	0.93	7	0.43	1,83	331	3.7
Zambia	0.88	2.52	186	0.75	1,43	91	58.5

Source: FAOSTAT: www.fao.org.

markets. Within farming communities, distinctive features that characterize smallholder farming systems in Southern Africa as in most of sub-Saharan Africa (SSA) is the wide diversity of farming households and marked heterogeneity for both biophysical and socioeconomic conditions (Zingore et al., 2011; Kamanga et al., 2009). The intensity of nutrient use varies between farms of different resource endowment and production orientation, leading to variation in soil fertility status and crop productivity at the farm level. In smallholder farming systems resources are general preferentially allocated to fields closer to the homestead largely due to shortage of inputs and labour (Mtambanengwe and Mapfumo, 2005; Tittonell et al., 2010). This has created soil fertility gradients on the farms, with lower fertility on the outfields, and these also vary depending on farm typology and soil types.

Soil fertility decline: a slow variable that can result in non-responsiveness

Traditionally, soil fertility in African smallholder farming systems was regenerated through shifting cultivation where land was cleared and cropped for a few years followed by multiyear fallow periods. However, increasing population pressure has eroded fallows in Southern Africa and resulted in continuous cropping, usually without crop rotations, with inadequate inputs due to limited resources, resulting in nutrient mining. Soil fertility decline has led to corresponding declines in crop productivity (Mtambanengwe and Mapfumo, 2005; Zingore et al., 2011). Consequently, this has resulted in declining above- and below-ground biomass that can be returned to the soil as source of C and nutrients. A study in the Eastern Cape Province of South Africa showed that soil fertility was low for cultivated land in smallholder fields largely due to inadequate nutrient inputs compared to their commercial counterparts (Mandiringana et al., 2005). These soils were associated with a critically low pH, suggesting that crop response to fertilizer addition may be low if soils are not limed. The combination of low soil organic matter, lack of surface cover and lack of crop rotations makes the soils in smallholder farming systems susceptible to erosion, further degrading the soils, with effects being greater on coarse rather than fine textured soils.

Soil fertility degradation on sandy soils tends to have greater repercussions on crop productivity than clayey soils (Rusinamhodzi et al., 2013; Vanlauwe et al., 2015). Clayey soils, on the other hand, tend to have greater soil organic matter due to greater physical protection (Chivenge et al., 2007), giving them greater buffering capacity and less sensitivity to degradation. Sandy soils generally have lower water and nutrient holding capacity, low soil organic matter and low soil pH. Exchangeable basic cations are low in sandy soils and are depleted on degraded sandy soils, contributing to low crop productivity and soil acidification, especially when continuously cropped without addition of organic fertilizers (Juo et al., 1995). Application of mineral fertilizers in such soils often gives small yield responses with low returns to investment and low resource use efficiencies. These soils are known as non-responsive due to other limitations affecting crop response including micronutrients and soil pH (Vanlauwe et al., 2010). That study further outlined that addition of organic resources or other amendments in non-responsive soils is necessary if any return to fertilizer investment is to be realized.

The ability of soil management practices to restore crop productivity of degraded soil depends on the extent and the path of the degradation process; a phenomenon analogous to hysteresis (Tittonell et al., 2008). The restoration of soil fertility and crop productivity, with the combined application of manure with mineral fertilizer, has been observed to be faster on a degraded clayey than degraded sandy soil (Rusinamhodzi et al., 2013). Moreover, it was possible to restore soil fertility and crop productivity with the application of mineral fertilizers on the clayey soil but not on the sandy soil. This emphasizes the need for repeated application of organic resources on degraded non-responsive soils before benefit from applied mineral fertilizers can be realized.

Restoration of soil fertility in degraded sandy soils often requires the addition of organic materials due to multiple nutrient deficiencies, coupled with acidity in some cases (Zingore et al., 2011). Similarly, fast-growing trees have been used in Southern Africa to enhance soil fertility and improve crop yields (Sileshi et al., 2012) since these can access nutrients and water from deeper, often less sandy soil horizons (Pierret et al., 2016). Organic materials contain macronutrients and micronutrients to alleviate the multi-nutrient deficiencies. In addition, the slow release of nutrients from organic materials offers greater synchrony between demand and supply of nutrients, particularly for sandy soils where nutrients tend to leach easily. In a nine year study in Murewa, Zimbabwe, Rusinamhodzi et al. (2013) observed that mineral fertilizer alone was not adequate to restore fertility of degraded sandy soils. However, even with addition of large quantities of manure, crop productivity and soil organic carbon build-up was delayed in degraded sandy soils compared to degraded clayey soils. They attributed this to multiple nutrient deficiencies associated with inherent low fertility and nutrient depletion due to previous management. Availability of manure is limited to farmers who own livestock, while those who do not have livestock often have their fields grazed by livestock from their neighbours, further exporting nutrients from poor farmers' fields. The quantity of manure on smallholder farming systems is often inadequate to apply to the whole farm even for farmers who own livestock (Zingore et al., 2011). Consequently, the restoration of degraded sandy soils remains a challenge, especially for resource poor smallholder farmers.

Need for sustainable intensification and improved soil fertility status

Smallholder agriculture in sub-Saharan Africa (SSA) needs to either intensify expanding agricultural land is no longer an option for densely populated areas, or ensure that natural ecosystems, such as the forest in the Congo Basin, are preserved. Even in areas where land expansion still occurs, intensification of agricultural production is needed to keep pace with an ever-growing population. The discourse on intensification is currently framed as "Sustainable Intensification" (SI) and commonly encompasses three dimensions: 1) increased productivity; 2) maintenance of ecosystem services; and 3) increased resilience to shocks (e.g., Pretty et al., 2011; Vanlauwe et al., 2014), accompanied by a set of dimensions that enable the generation of previous outputs in the human, social and economic realm (Snapp et al., 2018). Sustainable Intensification thus aims at generating the required food for a growing population while operating within the planetary boundaries (Rockström et al., 2009) and addressing important drivers of change affecting crop yields such as climate change.

Soil fertility interacts with each of the previous dimensions and is thus key to the delivery of SI. First, increased crop yields require improved soil fertility conditions, and more so in SSA where smallholder agriculture has been based largely on the mining of soil nutrient stocks. This concerns mainly those soil fertility components that interact directly with crop growth including the supply of nutrients, the control of soil moisture dynamics, the provisioning of appropriate soil physical conditions for root growth and the regulation of soil biota that affect plant growth. Second, soil fertility provides a number of important ecosystem services that are not directly related to crop growth. These include the regulation of nutrient and water use efficiencies, the accumulation of atmospheric CO_2 and the control of erosion. Considering that the soil organic matter status is probably one of the most important indicators for soil fertility status, Figure 8.1 sketches conceptually how soil fertility status can interact with the provision of crop productivity and other soil-based ecosystem services. Lastly, fertile soils can reduce the effects of weather- or pest and disease-related stresses by providing improved rooting and soil moisture conditions or supporting healthier plants, respectively. In a sense, one could argue that SI will not happen under poor soil fertility conditions.

Integrated soil fertility management – a path towards SI?

If SI is unlikely under poor soil fertility conditions, and since most soils under smallholder farming are degraded to a certain extent because of nutrient mining and associated degradation processes, it follows that improvement of soil

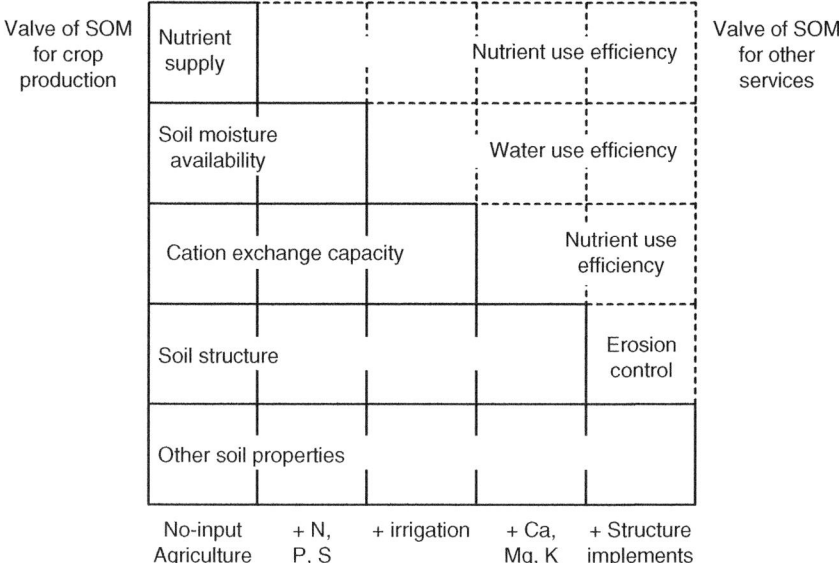

Figure 8.1 Conceptually depicted functions of the soil organic matter (SOM) pool in relation to crop production and other environmental services; note that the sketch does not include a relative valuation of each of the functions

fertility is a prerequisite on the path towards SI. Several technologies and combinations thereof have been conceptualized, promoted and evaluated over time, including those aiming at seeking alternatives to fertilizer (Figure 8.2). Over the years, however, it was found that technologies that do not generate immediate improvements in crop yield are unlikely to be adopted by large numbers of smallholder farmers since they face large risks for food insecurity and compete with labour for other production units (Vanlauwe et al., 2017). While in the 1970s, a lot of work was initiated on fertilizer use and appropriate land preparation methods, during the 1980s and 1990s the balance was swung away from fertilizer towards more organic matter-based systems. Since the latter did not provide those immediate benefits to the farmers during the last ten to 15 years, soil fertility management R&D is now often placed in the context of Integrated Soil Fertility Management (Vanlauwe et al., 2010; Vanlauwe et al., 2015).

Integrated Soil Fertility Management has been defined as:

> A set of soil fertility management practices that necessarily include the use of fertilizer, organic inputs and improved germplasm, combined with the knowledge of how to adapt these practices to local conditions, aimed at maximizing agronomic use efficiency of the applied nutrients and improving crop productivity. All inputs need to be managed following sound agronomic principles.
>
> (Vanlauwe et al., 2010)

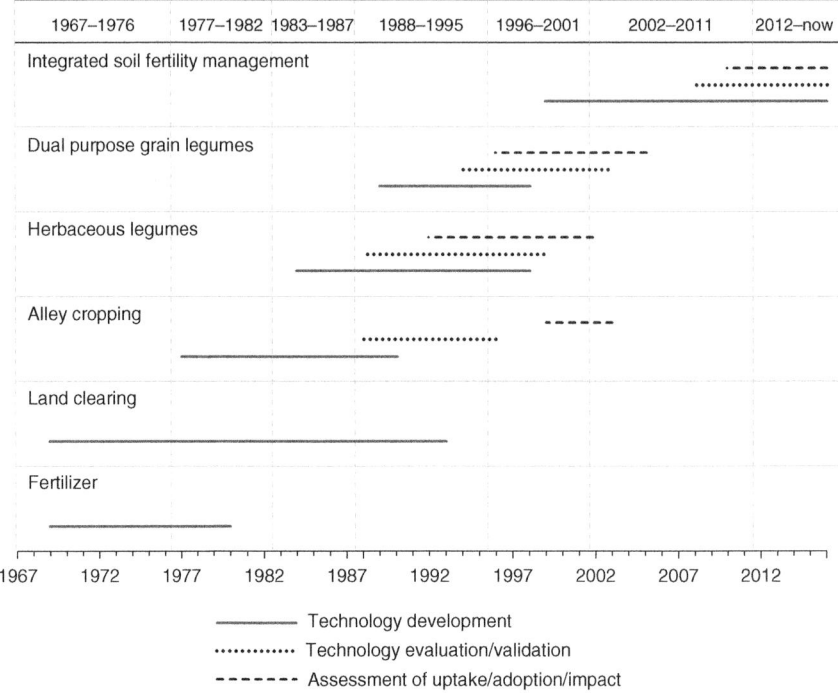

Figure 8.2 Evolution of the technologies and interventions prioritized by soil and soil fertility research initiatives at the International Institute of Tropical Agriculture, with
an indication of the technology development, evaluation/validation and uptake/
adoption/impact phases from 1967 to the present

Source: Vanlauwe et al., 2017

Although conclusive evidence is limited, data from certain long-term trials
confirm that ISFM practices increase crop yields, enhance soil C and are more
stable over time (e.g.,Vanlauwe et al., 2005).

ISFM applied to Southern Africa
and implications for policy

Soils in smallholder farming systems have high spatial heterogeneity within
short distances either due to parent material or management, associated with
resource availability and farmer typology. Nutrient management in those situations needs to be tailored to suit different site-specific conditions, instead of the
blanket recommendations that have been promoted by extension. While ISFM
has been shown to improve crop yields compared to no input control or sole
applied organic or mineral fertilizer, yield gains tend to be influenced by soil
type and agro-climatic conditions. On sandy soils, addition of organic resources

tends to improve water productivity, particularly in drier areas, thereby influencing the utilization efficiency of added mineral fertilizers. ISFM has been shown to improve maize yields and restore soil fertility compared to sole mineral fertilizer, especially on degraded sandy soils (Rusinamhodzi et al., 2013). This is because the organic resources contribute secondary nutrients and micronutrients that are deficient in degraded soils, thereby enhancing the utilization efficiency of the mineral fertilizers. In a study in Eastern Zimbabwe, ISFM was shown to be more effective at restoring soil fertility and improving maize yields when herbaceous legumes were included in the cropping sequence (Nezomba et al., 2015). This is important for smallholder farmers without adequate mineral fertilizers and organic manures to restore degraded soils.

Climate change predictions in Southern Africa estimate declining precipitation associated with high interannual variability and increasing temperature (Conway et al., 2015). This has been predicted to cause reductions in maize productivity. Using simulation modelling using future climate scenarios, Rurinda et al. (2015) predicted that fertilizer addition would increase crop yields in current and future climates but yield gain with fertilizer decreased over time, suggesting that nutrient management will remain important in future climates in order to avoid loss of crop productivity. Thus, breeding efforts to improve drought tolerance would need to be considered, together with shifts towards crops that are more drought tolerant such as sorghum and millets.

The need for appropriate policy initiatives to guide agricultural transformation

The uptake of ISFM towards the transformation of smallholder agriculture in Southern Africa is highly dependent on institutional and government policies. Investments on ISFM are going to be driven by land security, with farmers likely to invest when there is land tenure security or ownership. Addressing the land tenure issue in smallholder farming systems in Southern Africa will also be important to leverage farmers' access to financial services, since most of the farmers operate at marginal economic levels. This will also require linking farmers to input and output markets, to enable farmers to sell their produce at favourable prices and avoid being short-changed by middlemen and ensure access to inputs (Koppmair et al., 2017). Access to inputs needs to be improved, since one of the key elements of ISFM is the use of fertilizers, which are generally not available or expensive for smallholder farmers. Consequently, the successful uptake of ISFM would be high if governments ensure timely availability of fertilizers and, where possible offer smart subsidies for both fertilizers and good quality seeds. Such incentives can be linked to ICT, for faster and wider information delivery through mobile phones, taking advantage of high cellphone use in smallholder farming systems (Aker and Mbiti, 2010). ICT is useful across the whole value chain including development of tools for deployment of ISFM recommendations to farmers, translated from highly technical and knowledge-intensive format into readily understandable and actionable formats. This will also require investments in training extension agents who can deliver ISFM information to farmers and ensure local adaptation of ISFM depending on ecological and socioeconomic conditions.

References

Aker, J.C. and Mbiti, I.M. (2010) 'Mobile phones and economic development in Africa', *Journal of Economic Perspectives*, vol 24, pp. 207–232.

Chivenge, P.P., Murwira, H.K., Giller, K.E., Mapfumo, P. and Six, J. (2007) 'Long-term impact of reduced tillage and residue management on soil carbon stabilization: Implications for conservation agriculture on contrasting soil', *Soil and Tillage Research*, vol 94, pp. 328–337.

Conway, D., Van Garderen, E.A., Deryng, D., Dorling, S., Krueger, T., Landman, W., Lankford, B., Lebek, K., Osborn, T. and Ringler, C. (2015) 'Climate and southern Africa's water – energy – food nexus', *Nature Climate Change*, vol 5, p. 837.

Deckers, J., Laker, M., Vanherreweghe, S., Vanclooster, M., Swennen, R. and Cappuyns, V. (2000) 'State of the art on soil-related geo-medical issues in the world', in J. Låg (ed) *Geomedical Problems in Developing Countries*. Norwegian Academy of Science and Letters, Oslo.

FAO, IFAD, UNICEF, WFP and WHO. (2018) *The State of Food Security and Nutrition in the World 2018. Building Climate Resilience for Food Security and Nutrition.* FAO, Rome.

Fixen, P., Brentrup, F., Bruulsema, T., Garcia, F., Norton, R. and Zingore, S. (2015) 'Nutrient/fertilizer use efficiency: Measurement, current situation and trends', in P. Drechsel, P. Heffer, H. Magen, R. Mikkelsen and D. Wichelns (eds) *Managing Water and Fertilizer for Sustainable Agricultural Intensification*. International Fertilizer Industry Association (IFA), International Water Management Institute (IWMI), International Plant Nutrition Institute (IPNI), and International Potash Institute (IPI), Paris, France.

Juo, A.S.R., Dabiri, A. and Franzluebbers, K. (1995) 'Acidification of a kaolinitic Alfisol under continuous cropping with nitrogen fertilization in West Africa', *Plant and Soil*, vol 171, pp. 245–253.

Kamanga, B.C.G., Waddington, S.R., Robertson, M. and Giller, K.E. (2009) 'Risk analysis in maize-legume cropping systems with smallholder farmer resource groups in central Malawi', *Experimental Agriculture*, vol 46, pp. 1–21.

Koppmair, S., Kassie, M. and Qaim, M. (2017) 'Farm production, market access and dietary diversity in Malawi', *Public Health Nutrition*, vol 20, pp. 325–335.

Mandiringana, O., Mnkeni, P., Mkile, Z., Van Averbeke, W., Van Ranst, E. and Verplancke, H. (2005) 'Mineralogy and fertility status of selected soils of the Eastern cape province, South Africa', *Communications in Soil Science and Plant Analysis*, vol 36, pp. 2431–2446.

Misselhorn, A.A. (2005) 'What drives food insecurity in southern Africa? A meta-analysis of household economy studies', *Global Environmental Change*, vol 15, pp. 33–43.

Msangi, J.P. (2007) 'Land degradation management in Southern Africa', in M.V.K. Sivakumar and N. Ndiang'ui (eds) *Climate and Land Degradation*. Springer, Berlin.

Mtambanengwe, F. and Mapfumo, P. (2005) 'Organic matter management as an underlying cause for soil fertility gradients on smallholder farms in Zimbabwe', *Nutrient Cycling in Agroecosystems*, vol 73, pp. 227–243.

Nezomba, H., Mtambanengwe, F., Tittonell, P. and Mapfumo, P. (2015) 'Point of no return? Rehabilitating degraded soils for increased crop productivity on smallholder farms in eastern Zimbabwe', *Geoderma*, vol 239, pp. 143–155.

Nyamangara, J., Mugwira, L.M. and Mpofu, S.E. (2000) 'Soil fertility status in communal areas of Zimbabwe in relation to sustainable crop production', *Journal of Sustainable Agriculture*, vol 16, pp. 15–29.

Pierret, A., Maeght, J-L., Clément, C., Montoroi, J-P., Hartmann, C. and Gonkhamdee, S. (2016) 'Understanding deep roots and their functions in ecosystems: An advocacy for more unconventional research', *Annals of botany*, vol 118, pp. 621–635.

Pretty, J., Toulmin, C. and Williams, S. (2011). 'Sustainable intensification in African agriculture', *International Journal of Agricultural Sustainability*, vol 9, pp. 5–24.

Rockström, J., Steffen, W., Noone, K., Persson, A., Chapin, III, F.S., Lambin, E., Lenton, T.M., Scheffer, M., Folke, C., Schellnhuber, H., Nykvist, B., De Wit, C.A., Hughes, T., van der Leeuw, S., Rodhe, H., Sörlin, S., Snyder, P.K., Costanza, R., Svedin, U., Falkenmark, M., Karlberg, L., Corell, R.W., Fabry, V.J., Hansen, J., Walker, B., Liverman, D., Richardson, K., Crutzen, P. and Foley, J. (2009) 'Planetary boundaries: Exploring the safe operating space for humanity', *Ecology and Society*, vol 14, no 2, p. 32.

Rurinda, J., Van Wijk, M.T., Mapfumo, P., Descheemaeker, K., Supit, I. and Giller, K.E. (2015) 'Climate change and maize yield in southern Africa: What can farm management do?' *Global Change Biology*, vol 21, pp. 4588–4601.

Rusinamhodzi, L., Corbeels, M., Zingore, S., Nyamangara, J. and Giller, K.E. (2013) 'Pushing the envelope? Maize production intensification and the role of cattle manure in recovery of degraded soils in smallholder farming areas of Zimbabwe', *Field Crops Research*, vol 147, pp. 40–53.

Sileshi, G.W., Debusho, L.K. and Akinnifesi, F.K. (2012) 'Can integration of legume trees increase yield stability in rainfed maize cropping systems in Southern Africa?' *Agronomy Journal*, vol 104, pp. 1392–1398.

Snapp, S.S., Grabowski, P., Chikowo, R., Smith, A., Anders, E., Sirrine, D., Chimonyo, V. and Bekunda, M. (2018) 'Maize yield and profitability tradeoffs with social, human and environmental performance: Is sustainable intensification feasible?' *Agricultural Systems*, vol 162, pp. 77–88.

Tittonell, P., Corbeels, M., Van Wijk, M.T., Vanlauwe, B. and Giller, K.E. (2008) 'Combining organic and mineral fertilizers for integrated soil fertility management in smallholder farming systems of Kenya: Explorations using the crop-soil model FIELD', *Agronomy Journal*, vol 100, pp. 1511–1526.

Tittonell, P., Muriuki, A., Shepherd, K.D., Mugendi, D., Kaizzi, K.C., Okeyo, J., Verchot, L., Coe, R. and Vanlauwe, B. (2010) 'The diversity of rural livelihoods and their influence on soil fertility in agricultural systems of East Africa – A typology of smallholder farms', *Agricutural Systems*, vol 103, pp. 83–97.

Vanlauwe, B., AbdelGadir, A.H., Adewopo, J., Adjei-Nsiah, S., Ampadu-Boakye, T., Asare, R., Baijukya, F., Baars, E., Bekunda, M., Coyne, E., Dianda, M., Dontsop-Nguezet, P.M., Ebanyat, P., Hauser, S., Huising, J., Jalloh, A., Jassogne, L., Kamai, N., Kamara, A., Kanampiu, F., Kehbila, A., Kintche, L., Kreye, C., Larbi, A., Masso, C., Matungulu, P., Mohammed, I., Nabahungu, L., Nielsen, F., Nziguheba, G., Pypers, P., Roobroeck, D., Schut, M., Taulya, G., Thuita, M., Uzokwe, V.N.E., van Asten, P., Wairegi, L., Yemefack, M. and Mutsaers, H.J.W. (2017) 'Looking back and moving forward: 50 years of soil and soil fertility management research in sub-Saharan Africa', *International Journal of Agricultural Sustainability*, vol 15, pp. 613–631.

Vanlauwe, B., Diels, J., Sanginga, N. and Merckx, R. (2005) 'Long-term integrated soil fertility management in South-western Nigeria: Crop performance and impact on the soil fertility status', *Plant and Soil*, vol 273, pp. 337–354.

Vanlauwe, B., Bationo, A., Chianu, J., Giller, K.E., Merckx, R., Mokwunye, U., Ohiokpehai, O., Pypers, P., Tabo, R., Shepherd, K., Smaling, E., Woomer, P.L. and Sanginga, N. (2010) 'Integrated soil fertility management: Operational definition and consequences for implementation and dissemination', *Outlook on Agriculture*, vol 39, pp. 17–24.

Vanlauwe, B., Coyne, D., Gockowski, J., Hauser, S., Huising, J., Masso, C., Nziguheba, G. and Van Asten, P. (2014) 'Sustainable intensification and the smallholder African farmer', *Current Opinion in Environmental Sustainability*, vol 8, pp. 15–22.

Vanlauwe, B., Descheemaeker, K., Giller, K.E., Huising, J., Merckx, R., Nziguheba, G., Wendt, J. and Zingore, S. (2015) 'Integrated soil fertility management in sub-Saharan Africa: Unravelling local adaptation', *SOIL*, vol 1, pp. 491–508.

Vlek, P.L.G., Le, Q.B. and Tamene, L. (2008) *Land Decline in Land-Rich Africa – A Creeping Disaster in the Making*. CGIAR Science Council Secretariat, Rome.

Zingore, S., Manyame, C., Nyamugafata, P. and Giller, K.E. (2005) 'Long-term changes in organic matter of woodland soils cleared for arable cropping in Zimbabwe', *European Journal of Soil Science*, vol 57, pp. 727–736.

Zingore, S., Tittonell, P., Corbeels, M., Van Wijk, M.T. and Giller, K.E. (2011) 'Managing soil fertility diversity to enhance resource use efficiencies in smallholder farming systems: A case from Murewa District, Zimbabwe', *Nutrient Cycling in Agroecosystems*, vol 90, pp. 87–103.

9 The role of seed systems development in African agricultural transformation

Joseph D. DeVries

Introduction

Subsistence-level farming in Africa is unsustainable. As rural populations grow and spread, agricultural lands and other resources are steadily depleted and rendered incapable of providing a stable, decent existence. Poverty, hunger and malnutrition become the dominant themes in rural communities, and people lose faith in farming as a livelihood. Climate change, with its attendant droughts, floods and other extreme weather events, exacerbates this trend. The result is high rates of rural–urban migration, leading to an overflow of nonproductive people living in Africa's cities. Many, especially the youth, make desperate attempts to migrate to Europe and other developed regions. Others turn to radical religious and political factions which threaten the stability of whole regions of the continent and other parts of the world.

For decades, this has been the dominant trend across much of rural Africa. For lack of better options, smallholder farmers have continued to depend on the same, subsistence-style farming practices as generations of farmers before them. Grain crop yields in many countries have fluctuated around approximately 1 MT/ha, and rural economies, with a few exceptions, have stagnated. Meanwhile, population growth rates averaging 2.7% drive a steady increase in demand for food, education and health services which weak economies simply cannot supply. Much of the fallout from failing agricultural systems is absorbed by women, who care for children and also supply a large portion of the labour on Africa's farms (Palacios-Lopez et al., 2015).

A straightforward solution

There is a solution to the trap of subsistence agriculture. Throughout history and around the world, sustained increases in agricultural productivity and rural economic growth have been catalyzed by the introduction of seed of improved, locally adapted crop varieties which make more efficient use of sunlight, water and soil nutrients, resist pests and diseases and mature more quickly. Broad experience throughout the world – and now, in Africa as well – has shown that few other attempts to increase farmers' yields have proven as successful or as sustainable.

In recent years, the decades-long trend of static or declining crop yields in Africa has been reversed in several countries following the broad introduction and promotion of seed of improved, adapted crop varieties. In the past 13 years, yield increases in Africa have outstripped the progress of the previous 40 years. Food production on a per capita basis has likewise increased by 12% since 2000 (The Bill and Melinda Gates Foundation, 2018, unpublished). The biggest gains have been made in East Africa, where per capita cereal production has risen 50% since 2000. Not surprisingly, East Africa is also where the adoption of improved seed, especially hybrid maize seed, has flourished (The Economist, 2016).

Southern Africa, too, has benefited significantly from modern crop breeding and seed supply, especially in South Africa and Zimbabwe, and increasingly in Zambia and Malawi as well. However, the farmers of a number of countries in Southern Africa are yet to gain access to higher-yielding seed, with the situations in Angola, Botswana, the Democratic Republic of Congo and Madagascar being of particular concern. As all of these are maize-producing countries, one relatively quick solution is likely to be the introduction and promotion of higher-yielding maize varieties, including hybrids.

From 2007 to 2017, average maize yields in Uganda increased by 70%, from less than 1.5 MT/ha to 2.5 MT/ha. Meanwhile, Uganda's maize harvest increased from 1.26 million metric tons (MT) to over 3 million MT, an increase of 138%. Over this same period, Ethiopia's average maize yields nearly doubled, from less than 2 MT/ha to 3.7 MT/ha (Figure 9.1). Ethiopia's maize harvest grew 143%.

The dramatic gains in maize productivity in Uganda and Ethiopia were achieved during a period of rapid increases in the supply and adoption of improved seed. Data collected from companies during this period show supply increasing from 8.3 MT per year to 26,700 MT per year in Uganda (not shown) and from 260 MT per year to 57,350 MT per year in Ethiopia. This data, however, does not capture the actual baselines for seed supply in either country.

Thanks to recent investments in the breeding of African crops by the Bill and Melinda Gates Foundation, USAID, The Rockefeller Foundation and several

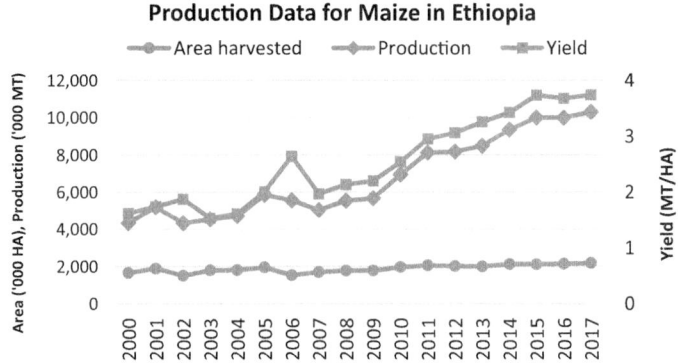

Figure 9.1 Maize production data for Ethiopia, 2000–2017

Source: FAOSTAT, 2019

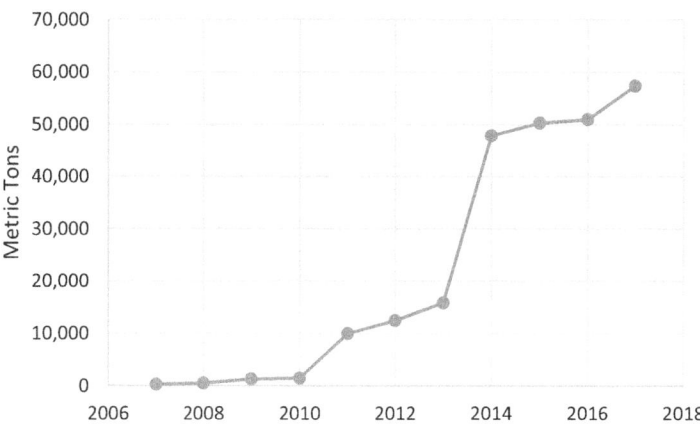

Figure 9.2 Improved seed supply in Uganda and Ethiopia, 2007–2017

Source: AGRA/PASS Database, 2018

other international donor agencies, improved varieties have now been bred for most of Africa's staple crops, which are higher-yielding, drought-tolerant, disease-resistant and earlier-maturing. A recent publication has identified 650 varieties of 14 staple food crops which have been officially released in the past 10 years (AGRA, 2018).

Currently, AGRA, the CGIAR and several other national and international development agencies are receiving financial support to establish seed delivery systems based on investments in private seed companies, agro-dealers, seed awareness building (extension) and better seed policies. The evidence from official data and from observations of farmers' fields is that it is working (Figure 9.2). However, these efforts are concentrated in approximately 12 countries, leaving farmers in many other African countries with major agricultural potential without access to the new crop varieties or delivery systems which could allow them to intensify their crop production and increase yields.

Why is seed so critical?

Crops grown using seed of varieties developed through modern breeding mature quicker, resist pests and diseases, are more drought tolerant and partition a greater portion of their biomass into the harvested portion of the crop. Simply by substituting seed of a genetically improved variety for a traditional variety, farmers can increase their crop yields by 20 to 30%. But the changes don't end there. As farmers observe the greater vigour of the new crop, they begin to apply more manure, buy more fertilizer and improve their overall crop management practices, including planting in rows and weeding the crop more regularly. Thus, the *combined* effects of seed of new varieties, greater nutrient supply and improved crop management allow farmers to *double, triple* or even *quadruple* their yields.

Around the world the introduction and adoption of seed of improved varieties has been central to this challenge. Hybrid maize seed transformed the American Midwest in the 1930s and 40s. Improved, "Green Revolution" wheat varieties bred by Norman Borlaug transformed Mexico and then the Indian subcontinent in the 1950s and 60s. Modern rice breeding and seed supply led to the transformation of the rice crop – and eventually, the economies – of Southeast Asia in the 1970s and 80s (Pingali, 2012). As climate change accelerates bringing more frequent and more intense periods of crop moisture stress, the greater resource-use-efficiency of modern crop genetic improvement has become essential to increasing farmer productivity and the resilience of crop production systems worldwide, especially in Africa (Webber et al., 2018).

Africa's emerging green revolution

Africa's green revolution, in contrast, was delayed by several factors. First, the far greater diversity of Africa's agro-ecologies, food crops and production systems created a longer lag phase during which new varieties were still being bred. Second, the existence of many relatively small countries in SSA has tended to reduce the flow of successful technologies across large areas. Third, whereas the key technologies that drove green revolutions in Latin America and Asia were delivered to farmers through large, publicly managed distribution schemes, supply in Africa has been largely driven by private sector, which historically suffered from less access to capital and skilled human resources. Finally, Africa's governments lacked appropriate seed policies and other measures to enable the private sector to operate freely and take seed to farmers.

In the early years of the 21st century, a small team of programme officers at The Rockefeller Foundation began combining investments in agro-ecology-based crop breeding with support to private, local seed companies and village-based agro-dealers as a means of providing smallholder farmers with regular, dependable access to seed of improved, locally adapted crop varieties (DeVries and Toenniessen, 2002). Local crop yields increased, fueling greater demand for seed and the growth of national seed and fertilizer businesses. The Rockefeller Foundation model for seed systems development led to the establishment in 2006 of the Programme for Africa's Seeds (PASS), which spearheaded the development of seed supply in 13 countries across West, East and Southern Africa.

Over this period, nearly 700 improved crop varieties developed by national agricultural research systems (NARS) and Consultative Group on International Agricultural Research (CGIAR) breeders have been approved by Africa's seed regulatory agencies, and over 100 private, independent seed companies have been established to multiply, package and sell seed to smallholder farmers. By 2016, annual seed production and sales from these companies had risen to 127,000 MT (AGRA, 2017), sufficient to plant an estimated 7 million hectares.

Hence, the food security scenario in the 13 countries targeted for seed systems development is broadly hopeful (Sanchez, 2015).

However, numerous African countries remain where seed supply *can* catalyze higher yields, reduce food shortages and jump-start rural economies, but only where the model described previously is yet to be introduced. These include 16 key agrarian countries with a total population of over 300 million people and an estimated 38 million farm families. The average level of chronic child malnutrition (stunting) in these countries is 38%.

For the vast majority of the farmers in these countries, the only sustainable means of improving their living conditions is through increased agricultural productivity. However, due to national boundaries and a lack of investment in seed systems development at a national level, tens of millions of farmers in these "left-behind" countries remain trapped behind a "low-yield seed barrier". As a result, crop yields in these countries have largely stagnated, whereas crop yields in countries where seed systems have been actively developed have increased significantly, as illustrated in the chart that follows (Figure 9.3).

A model for yield increases proven in Africa

Public-private seed systems are modular in structure, consisting of four main, interconnected parts: crop improvement, seed enterprise development, private sector-led extension and agro-dealer development (Figure 9.4).

These can be briefly described as follows.

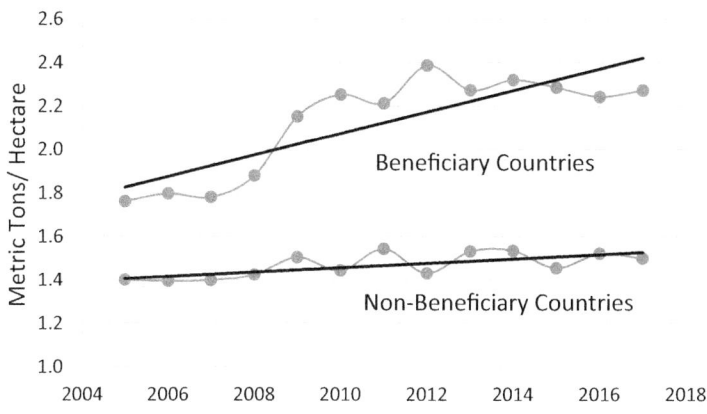

Figure 9.3 Maize and rice crop yield trends in countries benefiting from seed systems assistance vs. those nonbeneficiary countries, 2005–2017

Note: Trendlines for PASS countries including Burkina Faso, Ethiopia, Ghana, Kenya, Malawi, Mali, Mozambique, Niger, Nigeria, Rwanda, Senegal, Sierra Leone, Tanzania, Uganda and Zambia; proposed countries include Angola, Benin, Botswana, Burundi, Chad, Cote d'Ivoire, Democratic Republic of Congo, Eritrea, Guinea, Madagascar and Togo.

Source: FAOSTAT, 2019

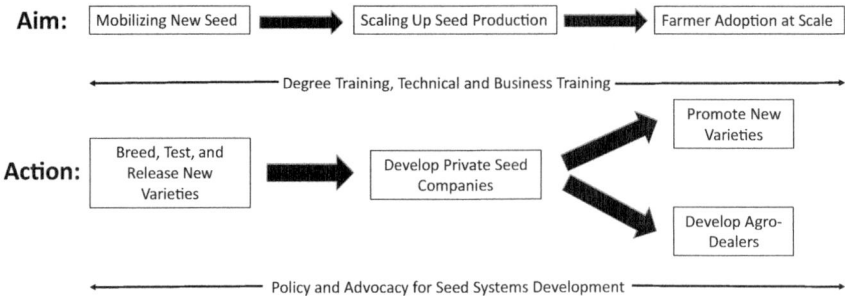

Figure 9.4 Seed systems development schematic, showing main points of entry and their functions

Breeding and release of new varieties

The journey towards adoption of high-yielding crop varieties among smallholder farmers begins with training and supporting a cadre of crop scientists to engage in the introduction, testing and selection of improved varieties of the countries' main food crops. This work is performed by scientists and technicians working in national agricultural research institutes. The large number of new, Africa-adapted crop varieties developed over the past decade represents an enormous advantage in developing the seed systems of countries heretofore left behind. Once released by the national agricultural research system, these varieties can be licensed to private, independent seed companies for commercialization. These licenses often include payment of royalties back to the institution, representing a new source of revenue for these often-neglected public entities.

Seed enterprise development

The identification and release of higher-yielding varieties needs to be paralleled by the establishment of a critical number private, independent seed companies which compete for sales of seed to farmers through an open market system. Experience has shown that entrepreneurs, including seed entrepreneurs, exist throughout Africa but often lack the specialized knowledge or capital to establish formal companies capable of producing, packaging and distributing improved seed. Women seed entrepreneurs have figured strongly in the emergence of the seed business in Africa.

Applying business principles to seed supply in countries historically dependent on government or NGO-led supply models is a positively disruptive intervention, which tends to attract attention and constructive debate at several levels of the society. Seed companies receive start-up capital and intensive training in business practices, which are specifically effective in reaching smallholder farmers (such as selling seed in small, 1 or 2 kg packages, marketing seed in open-air markets, placing demonstrations in areas frequented by farmers, broadcasting information in local languages via radio, etc.). Many of these seed companies grow rapidly and sustainably, eventually replacing the bulk of public and voluntary organizations' seed supply with a more professional, self-improving model. Competition between seed companies seeking to add to their market share leads to added innovations, such as seed treatments which protect the crop against fungal pathogens, nematodes and insects during early stages of growth.

Private sector-led extension

Improved seed can achieve little if farmers are unaware of its value. In parallel with the establishment of seed supply, extension activities focus on the demonstration and promotion of seed of new varieties at farmer level. Self-employed "Village-Based Advisors" (VBAs) are recruited, trained to teach farmers how to cultivate the new seed using fertilizer, row-spacing, weeding and other modern practices and facilitated to rapidly distribute hundreds of thousands of small (50 gram) packs of new seed plus small (200 gram) packs of fertilizer to fellow farmers who then engage in a mass experiential learning exercise by growing the new crop on small portions of their land. The better results obtained on these small plots is usually sufficient to convince most farmers to purchase a larger quantity of seed and fertilizer in the following season and provides a ready source of demand for the new seed companies and fertilizer suppliers.

The adoption and spread of mobile phones among farmers has dramatically increased the impact that can be achieved through private sector-led extension and VBAs, more specifically. VBA's can be facilitated to communicate a wide range of messages regarding seed performance, seed availability, farmer field day meetings, as well as opportunities for accessing fertilizer and other inputs, and even grain marketing opportunities.

Agro-dealer development

The final link in the seed value chain is village-level supply of seed through agro-dealers. This step involves providing start-up capital and technical assistance to village-based entrepreneurs to open their own seed and input supply shops (known widely as agro-dealers). The establishment of private, locally owned input shops ensures the steady supply of seed, fertilizer and other technology at a local level and removes, once and for all, the physical access barrier to these and other technologies. Agro-dealerships also offer new sources for women and youth employment in rural areas. Young, village-based entrepreneurs are often key to introducing new seed through these small businesses.

The role of government

It is important to point out that seed systems development cannot be achieved through voluntary and entrepreneurial action alone. In addition to action-oriented investments in seed supply, placing improved seed at the top of the agenda for agricultural development requires direct, consistent engagement with high-level government officials. The role of public, national agricultural research institutions and seed regulatory agencies is, of course, central to much of the activity described earlier.

As governments begin to recognize the value placed on seed (and fertilizer) among smallholder farmers, they often capitalize on the demand for improved seed by setting up subsidy schemes to encourage broad adoption. These schemes can have a positive or negative impact on seed markets, depending on how well they are administered but are ubiquitous wherever improved seed becomes a matter of importance to large numbers of farmers (Timmer, 1992). Key for farmers is preserving the element of choice in the seed they plant and timely delivery of the seed to local agro-dealers.

The emergence of modernized grain markets

Although grain traders are not usually an integral part of seed systems, they play a crucial role in shaping them. Their growth and development into modernized suppliers of tradable surpluses often parallels the development of seed systems. This happens in several ways. As farmers begin to produce reliable surpluses of grains as a result of the higher yields obtained from improved seed, grain traders are quick to see the new opportunities represented in cleaning, processing, packaging and exporting these surpluses. Modernized grain markets and standardized, packaged grain products, however, require consistency with regard to grain colour, texture, taste and quality. These requirements can only be met through supply of the corresponding seed. As a result, grain traders begin to pay close attention to seed supply, often supplying seed in advance to farmers in their production networks.

The opportunity and the payoff

Much of Africa's future hinges on how the continent as a whole deals with its lingering food and nutritional challenge. Recent experience with the adoption of seeds of modern, high-yielding varieties among farmers in a number of countries has proven that they are capable of producing much higher yields than previously believed possible. This has provided crucial evidence that a broader African Green Revolution is an achievable goal.

For Africa as a continent to develop normally and contribute to feeding the rest of the world, agricultural advances – especially higher-yielding, locally adapted seeds of the continent's major food crops – cannot be limited to a select handful of countries. They must be extended to all countries and to as many farmers as possible.

The experience and lessons learned over the past decade in 15 countries has produced a reliable model for seed systems development in Africa and significantly reduced the risks associated with extending this intervention to a new group of countries which are critical to peace, stability and food security on the continent. Indeed, the juxtaposition of the haves and have-nots with regard to improved seed in Africa has created a new sense of urgency around getting seed to the rest of Africa's farmers and begs the question, "If not now, then when, and if not us, then who?"

References

AGRA. (2017). *Seeding an African Green Revolution*. The Alliance for a Green Revolution in Africa. Nairobi, Kenya. Eds. J.D. DeVries, and Z. Masiga, p. 71

AGRA. (2018) *Catalogue of Varieties from AGRA-Supported Breeding Programs*. Eds. J.D. DeVries, L. Gichuru, J. Ininda, M. Muthama, P. Tongoona. Nairobi, Kenya: AGRA. 143pp.

Bill and Melinda Gates Foundation. (2018) Internal study based on data from FAOSTAT and UN Population Division, shared by Tom Kehoe.

DeVries, J.D. and G.H. Toenniessen (2002) *Biotechnology, Breeding and Seed Systems for African Crops I*. Book of Abstracts for 1st General Meeting of Collaborators of The Rockefeller Foundation held in Entebbe, Uganda, November 2–5.

The Economist. (2016) A Green Revolution: African Agriculture. *The Economist*, March 12, www.economist.com/briefing/2016/03/12/a-green-evolution

FAOSTAT. (2019). Data downloaded from online database of the United Nations Food and Agricultural Organization (http://www.fao.org/faostat/en/#data/QC).

Palacios-Lopez, A., L. Christiaensen, and T. Kilic. (2015) *How Much of the Labor in African Agriculture Is Provided by Women?* World Bank Policy Research Working Paper No. 7282, World Bank Group, Washington, DC, http://documents.worldbank.org/curated/en/9796 71468189858347/pdf/WPS7282.pdf

Pingali, P.L. (2012). Green Revolution: Impacts, limits, and the path ahead. *Proc Natl Acad Sci*, 109(31):12302-8.

Sanchez, P. (2015). En route to plentiful food production in Africa. *Nature Plants*. January 8, 2015.

Timmer, C.P. (1992) *Agriculture and the State: Growth, Employment, and Poverty in Developing Countries*. Ed. C.P. Timmer. Ithaca, NY: Cornell University Press, 321p.

Webber, T., A. Haensler, D. Rechid, S. Pfeifer, B. Eggert, and T. Jacob (2018) Analyzing Regional Climate Change in Africa in a 1.5, 2, and 3°C Global Warming World. *Earth's Future*, Vol. 6, Issue 4, pp. 643–655, https://agupubs.onlinelibrary.wiley.com/doi/full/10.1002/2017EF000714.

10 Scaling climate-smart agriculture for agricultural transformation in Southern Africa

Caroline Mwongera, Christine Lamanna,
Hannah N. Kamau and Evan Girvetz

Climate change and agriculture in Southern Africa

Agriculture accounts for more than 20% of the Gross Domestic Products (GDP) of most African countries (Awokuse and Xie, 2015), supporting over 80% of the rural population (Davis et al., 2017). Sub-Saharan Africa is the region most affected by undernourishment and with the greatest food security risk because of the uncertainties of climate change, land degradation and market fluctuations (Gebbers and Adamchuk, 2010; Ramirez-Villegas and Thornton, 2015; van Ittersum et al., 2013) and a doubling of the population by 2050. Average temperatures have warmed by nearly 0.5°C over the last century and are projected to increase a further 1.4–5.5°C. by 2100 (Adhikari et al., 2015). More troubling for the large proportion of the population that depends on rain-fed agriculture, is the projected increase in variability of precipitation cycles across the continent. Although median precipitation is projected to change by only −2% to 20% on average, this change will be accompanied by increased variability in the onset and duration of rainy seasons and increased intensity of both rainfall events and dry spells. Local climate projections in southern Africa predict an increase in mean annual temperatures from about 1.5°C to 4°C (Zinyengere and Crespo, 2017) and a general drying over the region (IPCC, 2014; Zinyengere et al., 2013). Thus, there is a dire need for agriculture in Southern Africa to transform from its current climate-vulnerable state to a state that is sustainably food secure in the face of climate change.

One proposed solution to the intertwined problems of food insecurity and climate change is Climate-Smart Agriculture (CSA). CSA is an approach to agricultural transformation seeking to 1) sustainably increase agricultural productivity; 2) adapt and build resilience to climate change; and 3) reduce or remove greenhouse gas (GHG) emissions where possible (FAO, 2013). CSA is not a prescriptive set of agricultural practices or technologies but rather endorses locally relevant innovations that meet the three goals of productivity, resilience and mitigation. The popularity of CSA in Africa is evident. In 2014, the African Union endorsed the "Vision 25 x 25", of having at least 25 million smallholder farming households practicing CSA by 2025. More so, 42 African countries prioritize adaptation in the agriculture sector in their Nationally Determined Contributions (NDCs) to the 2015 Paris Climate Agreement (AfDB, 2019), while 29 specifically mention CSA in their NDCs (Richards et al., 2015).

Given the political will and resources behind scaling CSA in Southern Africa, does the approach have the potential to transform agriculture for sustainable food security in the region? Here we give a brief overview of CSA initiatives in the region, make the case for CSA contributing to agricultural transformation and give recommendations for policymakers for up-scaling CSA in Southern Africa based on experiences in the region and elsewhere.

Methodology

To assess the potential for CSA to transform agriculture in Southern Africa, we draw on three key resources for the region: the Evidence for Resilient Agriculture (ERA) database, the CSA Country Profiles and the CCARDESA SAAIKS Knowledge Hub. The ERA, formerly referred to as CSA compendium, is a platform designed to pinpoint what agricultural technologies work and in which locations. It is a systematic review of the English language peer-reviewed literature using search terms related to more than 70 potential CSA practices and 20 outcomes related to the three pillars of CSA (Rosenstock et al., 2015, World Agroforestry, 2019). For inclusion in the resulting meta-analysis, a study had to include at least one potential CSA practice, have a relevant non-CSA control, provide quantitative data on at least one CSA outcome and have taken place within Africa.

The CSA Country Profiles are participatory appraisals of the agricultural challenges in countries and how CSA can help to adapt and mitigate climate change. At their core is an empirical assessment of potential CSA technologies through consultation with 40–50 country-level experts. Each CSA country profile is framed around six key analytical stages: 1) relevance of agriculture in the country; 2) challenges for the agricultural sector; 3) climate change impacts on agriculture; 4) CSA technologies and practices; 5) institutions and policies for CSA; and 6) financing CSA. The paper uses data from the profiles of five Southern Africa countries developed so far: Lesotho, Malawi, Mozambique, Zambia and Zimbabwe. CSA country profiles are a project of the International Center for Tropical Agriculture (CIAT) in collaboration with the CGIAR Research Program on Climate Change, Agriculture and Food Security (CCAFS), the World Bank and the UK Government's Department for International Development (DFID).

The CCARDESA SSAIKS Knowledge Hub is a repository of agricultural research and extension information for the Southern Africa region, maintained by the Center for Coordination of Agricultural Research and Development in Southern Africa (CCARDESA). It currently contains nearly 400 publications on CSA in the region, including research outputs, policy briefs, extension materials and decision-support tools.

What is the evidence base for CSA in Southern Africa?

In Southern Africa, there is a wealth of evidence available on CSA. Based on the Evidence for Resilient Agriculture systematic review, there are approximately 500 studies of potential CSA technologies in the Southern Africa region. While the region includes diverse production systems, nearly 50% of the available data focuses on maize, while 10% deals with legumes (Figure 10.1). Thus, much is known about the impact of CSA in maize systems, but very little is known about the impact of CSA on other economically and nutritionally important crops such as sorghum, dairy or vegetables. More than 70 individual CSA practices have been documented in the region, with addition of inorganic fertilizers being the most common practice studied, followed by mulching and reduced tillage, both of which are components of conservation agriculture, reflecting the long history of the practice in the region.

The participatory CSA Country Profiles identified 61 climate-smart technologies ranging across crop, agroforestry, livestock, soil and water management and energy systems for the 22 production systems selected across the five Southern African countries. Thirty eight percent of the total value chains selected were livestock, followed by pulses and cereals (19%), roots and tubers (14%) and oilseed (10%). In each of the production systems, experts identified the most appropriate technologies and scored them on a point scale of −10 to +10 for climate-smartness based on their contribution to productivity, adaptation and mitigation. The long list of practices for each production system had an average of ten technologies. Based on the average climate smartness scores, Figure 10.2 presents the top three CSA technologies identified in the 22 production systems across five countries. This reveals the expansive array of already tested CSA technologies suited to Africa's diverse production systems and regions (Sova et al., 2018) and available for scaling to reach a wide number of farmers. Yet, the current overall rate of CSA adoption by farmers remains low despite their benefits (Mwongera et al., 2015; Senyolo et al., 2018). For instance, although conservation agriculture, water conservation, drought tolerant and early maturing varieties are most suited climate-smart agriculture technologies to cope with water stress in Southern Africa, high initial investment costs and additional labour requirements limit their adoption (Senyolo et al., 2018). However, farmers in sub-Saharan Africa are already putting CSA into action and increasingly adopting some of these technologies.

CSA as agricultural transformation: examples from the field

Many improved agricultural practices have the potential to achieve both the three goals of CSA (improved productivity, increased resilience to climate change and mitigation of greenhouse gases) as well as agricultural transformation to sustainable food security. Conservation agriculture (CA), or the practice of minimizing tillage, maintaining soil cover and diversifying crop production,

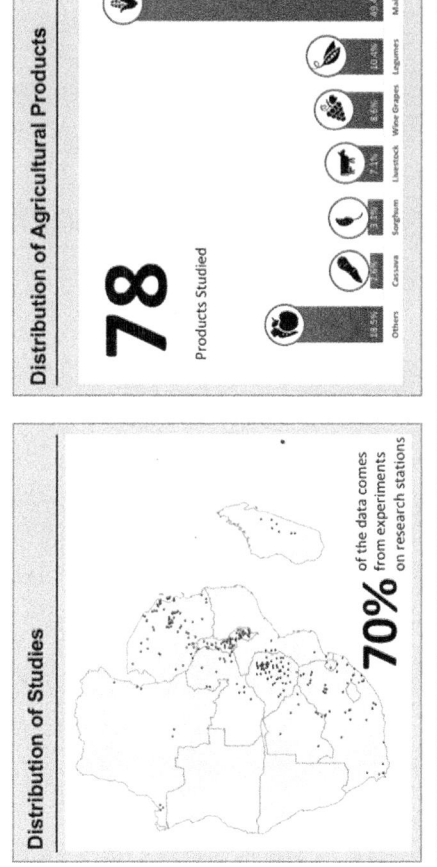

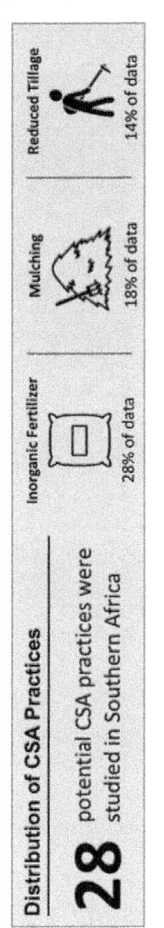

Figure 10.1 Systematic map of the evidence base for CSA in Southern Africa

Source: Own representation

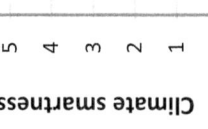

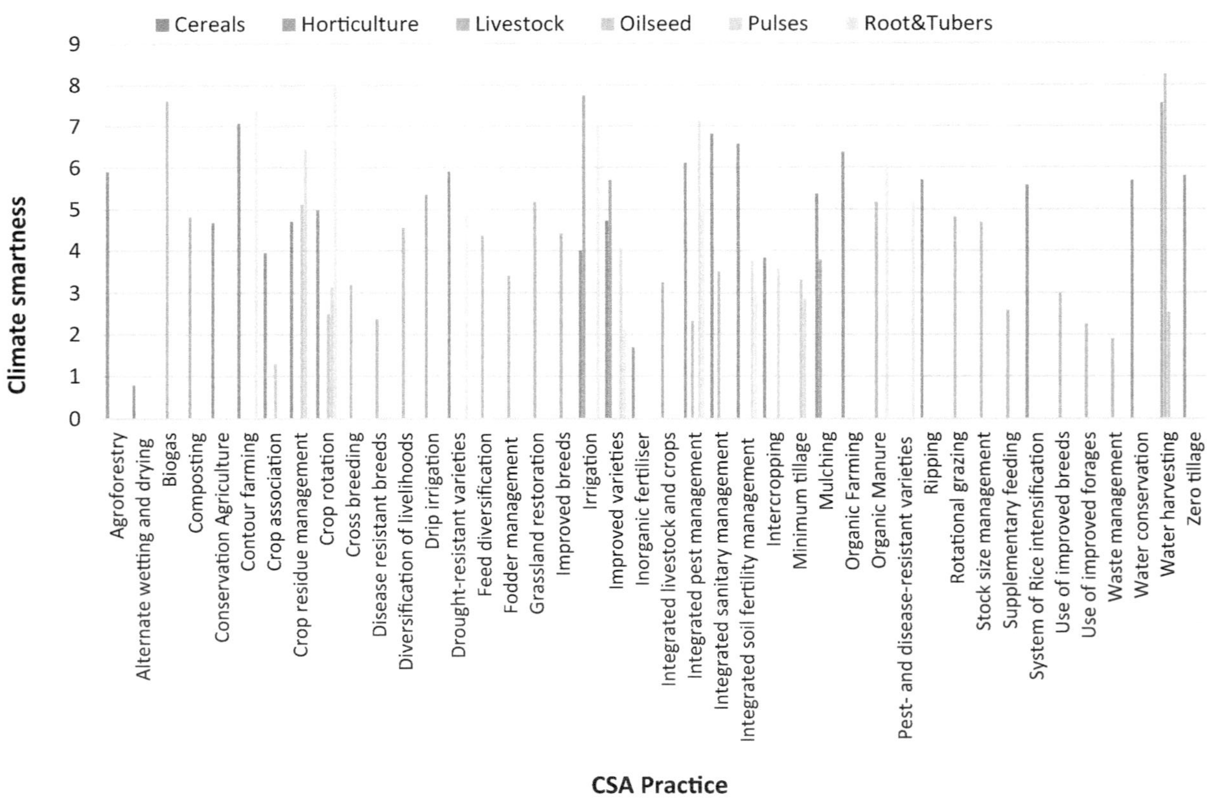

Figure 10.2 The climate-smartness of technologies for selected production systems in Southern Africa

Source: Own representation

Legend (top): ■ Cereals   ■ Horticulture   ■ Livestock   ■ Oilseed   ■ Pulses   ■ Root&Tubers

Y-axis: Climate smartness (0–9)

X-axis (CSA Practice):
Agroforestry
Alternate wetting and drying
Biogas
Composting
Conservation Agriculture
Contour farming
Crop association
Crop residue management
Crop rotation
Cross breeding
Disease resistant breeds
Diversification of livelihoods
Drip irrigation
Drought-resistant varieties
Feed diversification
Fodder management
Grassland restoration
Improved breeds
Irrigation
Improved varieties
Inorganic fertiliser
Integrated livestock and crops
Integrated pest management
Integrated sanitary management
Integrated soil fertility management
Intercropping
Minimum tillage
Mulching
Organic Farming
Organic Manure
Pest- and disease-resistant varieties
Ripping
Rotational grazing
Stock size management
Supplementary feeding
System of Rice intensification
Use of improved breeds
Use of improved forages
Waste management
Water conservation
Water harvesting
Zero tillage

has been shown to increase productivity of crops such as maize, particularly in drier areas such as Southern Africa (Pittelkow et al., 2015). In Malawi, conservation agriculture increased the resilience of maize systems to drought events by maintaining a higher yield during an El Niño event than conventional agricultural systems (Steward et al., 2019). While mitigation benefits from CA through increases in soil carbon may be modest (Theirfelder et al., 2017), in Tanzania, increased productivity of maize under conservation agriculture reduced the emissions intensity of maize production (GHGs produced per unit of yield), resulting in net mitigation benefits from CA (Kimaro et al., 2015). Thus, conservation agriculture has the potential increase sustainable food security, particularly in drier cereal production systems in the region.

Agroforestry, or incorporating trees into crop and livestock systems, is another commonly promoted approach in the region with the potential to deliver on CSA goals and agricultural transformation. For example in semi-arid Tanzania, intercropping maize with *Cajanus cajan* resulted in higher total yield than maize monocropping alone (LER = 1.46), particularly in lower productivity regions (Kimaro et al., 2019). The inclusion of trees into cropping systems can modify the microclimate through shading and hydraulic lift, as well as increase soil health through nitrogen fixation (in the case of leguminous "fertilizer trees") and deposition of organic material, resulting in lower moisture stress for crops. For example, in Zambia, intercropping *Leucaena leucocephala* with maize resulted in higher rain-use efficiency of the maize crop and buffered against yield losses in dry years (Sileshi et al., 2011). Finally, agroforestry has direct mitigation benefits, through the capture and storage of carbon on the landscape. Indeed, it is estimated that trees in agricultural lands accounts for more than 75% of the carbon stored in this land-use globally (Zomer et al., 2016).

Scaling-up CSA in Southern Africa: recommendations for policymakers

In order to realize the benefits of CSA for achieving agricultural transformation in Southern Africa, it must be scaled-up from case-studies to a common approach to agriculture in the region. In order to achieve scaling for CSA, the following are seven key recommendations for policymakers.

1 Prioritizing climate-smart agriculture investments

   Economic, technical, social considerations and priority goals of stakeholders vary and require to be considered in the selection of the portfolio of CSA technologies (Andrieu et al., 2017; Mwongera et al., 2015). Given the broad range of CSA technologies available and the multiple dimensions of CSA, identifying appropriate interventions requires prioritizing what is appropriate for given contexts. Site-specific CSA options that are gender responsive are likely to provide lasting benefits, especially for women, youth, ethnic minorities and the very poor farmers at the community.

2   Moving beyond a single farm to landscapes

A recent global review notes a shift from a technology-oriented approach to a more system-oriented approach that considers the complexity of farming systems (Totin et al., 2018). Climate change is part of the wider set of interacting social, economic, political and cultural dimensions within and beyond a single farm, implying that a more integrated landscape approach is needed to move farmers towards a sustainable and resilient livelihood. This comprehensive approach stimulates potential synergies and sustainable transformation in complex systems (Hounkonnou et al., 2012), such as those found in Africa. Developing responses to climate change through climate-smart agriculture thus requires, for example, integrating market, political and other institutional aspects that shape the context in which farming takes place. Across larger landscapes, impacts need to be assessed for their sustainability.

3   Ensuring CSA is mainstreamed into agricultural programmes

For CSA to be scaled out to establish transformational change within agriculture systems, large-scale systematic investment is needed. There are bottlenecks in mainstreaming climate-smart agriculture in Africa which include weak governance, institutional arrangements and linkages between relevant sectors, e.g., energy, environment, climate, water, agriculture and forestry. Often this contributes to lack of a coordinated policy mix and response to climate change at the subnational and national levels (Ampaire et al., 2017). Linking national priority needs and investments with subnational level experiences and ensuring CSA is incorporated into broader policy and strategy processes provides enabling conditions for CSA implementation. Most importantly, local institutions should be strengthened to improve CSA policy coherence and effective implementation at subnational levels (FAO, 2010).

4   Capacity development for up scaling CSA

The gap between awareness and use of CSA technologies is one of the challenges of shifting from CSA theory to practice. Second, CSA evidence focuses disproportionately on the main agricultural production systems in the Southern African region. Access to training and information, technology transfer and technological capabilities (Martinez-Baron et al., 2018) are some of the key elements required. Strengthening the capacity of key stakeholders (farmers, CSA implementers, researchers, policymakers) through training and information dissemination, identifying and testing, in a participatory manner, the best CSA practices promotes learning. The use of information and communications technology such as local radios and short message service are supporting the dissemination of the technical information and raising awareness on CSA.

5   Institutions and partnerships

An enabling environment for CSA adoption requires defining and developing partnerships between institutions and organizations supporting

CSA implementation. Potential synergies can be realized by strategically capitalizing on key institutional structures and networks, participatory processes, social learning and multi-stakeholder negotiation (O'Donnell and Garrick, 2017) for collaboration, access to resources and the dissemination of the technologies. In Africa, private sector support for CSA technologies is still limited, with Government representing the single largest CSA-related institutional category (Sova et al., 2018). The private sector plays an important role in supporting the uptake of CSA technologies for example, market, credit and insurance services.

6   Monitoring and Evaluation

To test and track the changes resulting from CSA implementation, the evidence of how such successes are measured and achieved is of critical importance (Neate, 2013). Gleaning clear empirical messages to inform farmers and policymakers and support any scaling up will depend on continuous two-way feedback mechanisms between multiple stakeholders, e.g., researchers and practitioners, farmers, extension agents and policymakers on the relevance of particular CSA technologies in a given context. The monitoring should effectively track CSA outcomes through meaningful indicators and provide timely information for adaptive management (Eitzinger et al., 2017).

7   CSA financing

It is evident that the investment needed to adapt to climate change and accelerate adoption of CSA needs to be scaled up considerably. Adoption of climate-smart agricultural innovations may also be associated with new challenges in terms of resources to acquiring new inputs, equipment, knowledge, infrastructure, human resources and institutional support. Actions are required from a broad range of stakeholders from government and the public sector, private sector, academia research, nongovernmental organizations and community based organizations. Farmers have limited assets and lack access to affordable financial services to allow them to invest in adopting CSA technologies at scale. There is need to strengthen financial opportunities, agricultural insurance, credits and organize resource flows to create synergies (Williams et al., 2015).

Conclusion

Considering the urgency of action needed to achieve a 1.5°C future, it is essential to promote a type of agriculture that can help small-scale farmers to adapt to climate change while reducing impact on the environment. Climate-smart agriculture practices exist, the focus should be on the dissemination of proven and successful technologies. Further research on the impacts of CSA should include a broader range of economically and nutritionally important production systems such as the small grain cereals (millet, sorghum) roots, tubers, livestock and aquaculture.

References

Adhikari, U., Nejadhashemi, A.P. and Woznicki, S.A. (2015) Climate change and eastern Africa: A review of impact on major crops. *Food and Energy Security*. 4, 110–132.

AfDB. (2019) *Analysis of Adaptation Components of Africa's Nationally Determined Contributions (NDCs)*. African Development Bank, Abidjan, Côte d'Ivoire. 86 p.

Ampaire, E., Acosta, M., Mwongera, C., Läderach, P., Eitzinger, A., Lamanna, C., Mwungu, C., Shikuku, K., Twyman, J. and Winowiecki, L. (2017) *Formulate Equitable Climate-Smart Agricultural Policies*. International Center for Tropical Agriculture (CIAT), Cali, Colombia. 8 p.

Andrieu, N., Sogoba, B., Zougmore, R., Howland, F., Samake, O., Bonilla-Findji, O., Lizarazo, M., Nowak, A., Dembele, C. and Corner-Dolloff, C. (2017) Prioritizing investments for climate-smart agriculture: Lessons learned from Mali. *Agricultural Systems*. 154, 13–24. https://doi.org/10.1016/j.agsy.2017.02.008.

Awokuse, T.O. and Xie, R. (2015) Does agriculture really matter for economic growth in developing countries? *Canadian Agricultural Economics Society*. 63, 77–99. https://doi.org/10.1111/cjag.12038.

Davis, B., Di Giuseppe, S. and Zezza, A. (2017) Are African households (not) leaving agriculture? Patterns of households' income sources in rural Sub-Saharan Africa. *Food Policy*. 67, 153-174. https://doi.org/10.1016/j.foodpol.2016.09.018.

Eitzinger, A., Mwongera, C., Läderach, P., Acosta, M., Ampaire, E.L., Lamanna, C., Mwungu, C.M., Shikuku, K.M., Twyman, J. and Winowiecki, L.A. (2017) Monitor climate-smart agricultural interventions with a real-time participatory tool. International Center for Tropical Agriculture (CIAT), Cali, Colombia. 6 p.

FAO. (2013) *Climate-Smart Agriculture Sourcebook*. Food and Agriculture Organization of the United Nations, Rome, Italy. 570 p.

FAO. (2010) *"Climate Smart" Agriculture: Policies, Practices and Financing for Food Security*. Food and Agriculture Organization of the United Nation, Rome, Italy. 41 p.

Gebbers, R. and Adamchuk, V.I. (2010) Precision agriculture and food security. *Science*. 327, 828–831. https://doi.org/10.1126/science.1183899.

Hounkonnou, D., Kossou, D., Kuyper, T.W., Leeuwis, C., Nederlof, E.S., Röling, N., Sakyi-Dawson, O., Traoré, M. and van Huis, A. (2012) An innovation systems approach to institutional change: Smallholder development in West Africa. *Agricultural Systems*. 108, 74–83. https://doi.org/10.1016/j.agsy.2012.01.007.

IPCC. (2014) Climate change 2014: Impacts, adaptation, and vulnerability. In: Field, C.B., Barros, V.R., Dokken, D.J., Mach, K.J., Mastrandrea, M.D., Bilir, T.E., Chatterjee, M., Ebi, K.L., Estrada, Y.O., Genova, R.C., Girma, B., Kissel, E.S., Levy, A.N., MacCracken, S., Mastrandrea, P.R. and White, L.L. (Ed.), *Contribution of Working Group II to the Fifth Assessment Report of the Intergovernmental Panel on Climate Change*. Cambridge, UK and New York, USA: Cambridge University Press, pp. 1–32.

Kimaro, A.A., Mpanda, M., Rioux, J., Aynekulu, E., Shaba, S., Thiong'o, M., Mutuo, P., Abwanda, S., Shepherd, K., Neufeldt, H. and Rosenstock, T.S. (2015) Is conservation agriculture 'climate-smart' for maize farmers in the highlands of Tanzania? *Nutrient Cycling in Agroecosystems*. 105, 217–228.

Kimaro, A.A., Sererya, O.G., Matata, P., Uckert, G., Hafner, J., Graef, F., Sieber, S. and Rosenstock, T.S. (2019) Understanding the multidimensionality of climate-smartness: Examples from agroforestry in Tanzania. In: Rosenstock, T.S., Nowak, A. and Girvetz, E. (Eds.), *The Climate-Smart Agriculture Papers*. Springer International Publishing, pp. 153–162.

Martinez-Baron, D., Orjuela, G., Renzoni, G., Loboguerrero Rodríguez, A.M. and Prager, S.D. (2018) Small-scale farmers in a 1.5°C future: The importance of local social dynamics as an enabling factor for implementation and scaling of climate-smart agriculture. *Current Opinion in Environmental Sustainability*. 31, 112–119. https://doi.org/10.1016/j.cosust.2018.02.013.

Mwongera, C., Shikuku, K.M., Winowiecki, L., Twyman, J., Läderach, P., Ampaire, E., van Asten, P. and Twomlow, S. (2015) *Climate-Smart Agriculture Rapid Appraisal (CSA-RA): A Prioritization Tool for Outscaling CSA. Step-by-step Guidelines (No. 409)*. International Center for Tropical Agriculture (CIAT). Cali, Colombia, 44.

Neate, P. (2013) *Climate-Smart Agriculture Success Stories from Farming Communities Around the World*. Wageningen, Netherlands: CGIAR Research Program on Climate Change, Agriculture and Food Security (CCAFS) and the Technical Centre for Agricultural and Rural Co-operation (CTA), 44.

O'Donnell, E.L. and Garrick, D.E. (2017) Chapter 26 – Defining success: A multicriteria approach to guide evaluation and investment. In: Horne, A.C., Webb, J.A., Stewardson, M.J., Richter, B. and Acreman, M. (Eds.), *Water for the Environment*. Academic Press, pp. 625–645. https://doi.org/10.1016/B978-0-12-803907-6.00026-7.

Pittelkow, C.M., Liang, X., Linquist, B.A., van Groenigen, K.J., Lee, J., Lundy, M.E., van Gestel, N., Six, J., Venterea, R.T. and van Kessel, C. (2015) Productivity limits and potentials of the principles of conservation agriculture. *Nature*. 517, 365–368.

Ramirez-Villegas, J. and Thornton, P.K. (2015) *Climate Change Impacts on African Crop Production*. CGIAR Research Program on Climate Change, Agriculture and Food Security (CCAFS) Working Paper no. 119, Copenhagen, Denmark.

Richards, M., Bruun, T.B., Campbell, B.M., Gregersen, L.E., Huyer, S., Kuntze, V., Madsen, S.T.N., Oldvig, M.B. and Vasileiou, I. (2015) *How Countries Plan to Address Agricultural Adaptation and Mitigation: An Analysis of Intended Nationally Determined Contributions*. CGIAR Research Program on Climate Change, Agriculture and Food Security (CCAFS) InfoNote.

Rosenstock, T.S., C. Lamanna, S. Chesterman, et al. (2015) *The Scientific Basis for Climate-Smart Agriculture: A Systematic Review Protocol*. CGIAR Research Program on Climate Change, Agriculture and Food Security (CCAFS) Working Paper no. 138, Copenhagen, Denmark.

Senyolo, M.P., Long, T.B., Blok, V. and Omta, O. (2018) How the characteristics of innovations impact their adoption: An exploration of climate-smart agricultural innovations in South Africa. *Journal of Cleaner Production*. 172, 3825–3840. https://doi.org/https://doi.org/10.1016/j.jclepro.2017.06.019.

Sileshi, G.W., Akinnifesi, F.K., Ajayi, O.C. and Muys, B. (2011) Integration of legume trees in maize-based cropping systems improves rain use efficiency and yield stability under rain-fed agriculture. *Agricultural Water Management*. 98, 1364–1372.

Sova, C.A., Grosjean, G., Baedeker, T., Nguyen, T.N., Wallner, M., Jarvis, A., Nowak, A., Corner-Dolloff, C., Girvetz, E. and Laderach, P.M.L. (2018) Bringing the concept of climate-smart agriculture to life: Insights from CSA country profiles across Africa, Asia, and Latin America. https://doi.org/doi:10.1596/31064.

Steward, P.R., Thierfelder, C., Dougill, A.J. and Ligowe, I. (2019) Conservation agriculture enhances resistance of maize to climate stress in a Malawian medium-term trial. *Agriculture, Ecosystems & Environment*. 277, 95–104.

Theirfelder, C., Chivenge, P., Mupangwa, W., Rosenstock, T.S., Lamanna, C. and Eyre, J.X. (2017) How climate-smart is conservation agriculture (CA)? It's potential to deliver on adaptation, mitigation, and productivity on smallholder farms in southern Africa. *Food Security*. 9, 537–560.

Totin, E., Segnon, C.A., Schut, M., Affognon, H., Zougmoré, B.R., Rosenstock, T. and Thornton, K.P. (2018) Institutional perspectives of climate-smart agriculture: A systematic literature review. *Sustainability*. https://doi.org/10.3390/su10061990.

van Ittersum, M.K., Cassman, K.G., Grassini, P., Wolf, J., Tittonell, P. and Hochman, Z. (2013) Yield gap analysis with local to global relevance – A review. *Food Crops Research*. 143, 4–17. https://doi.org/10.1016/j.fcr.2012.09.009.

Williams, T.O., Mul, M., Cofie, O., Kinyangi, J., Zougmore, R., Wamukoya, G., Nyasimi, M., Mapfumo, P., Speranza, C.I., Amwata, D., Frid-Nielsen, S., Partey, S., Girvetz, E., Rosenstock, T. and Campbell, B.M. (2015) *Climate Smart Agriculture in the African Context*. Feeding Africa Conference, 21–23 October.

World Agroforestry. (2019) *Evidence for resilient agriculture. Database*. https://era.ccafs.cgiar.org.

Zinyengere, N. and Crespo, O. (2017) Chapter 2 – Assessing local impacts of climate change on crop production in Southern Africa: Critiquing an approach. In: Zinyengere, N., Theodory, T.F., Gebreyes, M. and Speranza, C.I. (Eds.), *Beyond Agricultural Impacts*. Academic Press, pp. 13–37. https://doi.org/10.1016/B978-0-12-812624-0.00002-8.

Zinyengere, N., Crespo, O. and Hachigonta, S. (2013) Crop response to climate change in Southern Africa: A comprehensive review. Glob. *Planet. Change*. 111, 118–126. https://doi.org/10.1016/j.gloplacha.2013.08.010.

Zomer, R.J., Neufeldt, H., Xu, J., Ahrends, A., Bossio, D., Trabucco, A., van Noordwijk, M. and Wang, M. (2016) Global tree cover and biomass carbon on agricultural land: The contribution of agroforestry to global and national carbon budgets. *Scientific Reports*. 6, 29987.

11 Crop improvement for agricultural transformation in Southern Africa

Hussein Shimelis, E. T. Gwata and Mark D. Laing

Introduction

Crop genetic improvement is one of the strategies for transforming African agriculture to meet the demand for food, feed and bioenergy. Plant breeding can deliver genetically improved and high-performing nutritionally enhanced crop cultivars, with economic benefits and environmental sustainability for human well-being, which are in alignment with the United Nations goals that aim at ending hunger, achieving food security, improving nutrition and promoting sustainable agriculture globally (Eriksson et al., 2018).

In Southern Africa, crop varieties are designed, developed or deployed by the public plant breeding programmes such as the National Agriculture Research Systems (NARS), International Research Centres, the private sector or non-government organizations (NGOs). The NARS in collaboration with international research centres breed cultivars of food security and cash crops including cereals, root tubers, legumes and oil seed crops (Table 11.1). The public sector breeding programmes are not well developed and are often under resourced due to limited investment in plant breeding education, research and infrastructure development. Conversely, some of the private sector programmes employ state-of-the-art breeding methods and biotechnological tools to develop new cultivars for the market.

Smallholder farmers in Southern Africa have had limited access to improved modern varieties that are specifically bred for cultivation under low input production systems. Most smallholder farmers continue to grow unimproved landrace crop varieties. Landraces are inheritably low yielding but stress resilient and possess various quality traits of intrinsic value for the indigenous farmers. If their new varieties are to be adopted by smallholder farmers, plant breeders have to develop and release crop varieties that meet all the key trait requirements of these farmers and the downstream value chains and that are adapted to perform well under increasingly adverse climatic conditions caused by global climate change. The objective of this chapter is to highlight the current breeding technologies, major constraints to plant breeding programmes and to present some of the reasons why there are low levels of adoption of freshly released, modern crop varieties by smallholder farmers in Southern Africa. The chapter

Table 11.1 National agricultural research institutes with crop improvement programmes and their major food security crops in the Southern African development community

Country	Research institute	Major food security crop(s)
Angola	Agricultural Research Institute of Angola	Maize, cassava
Botswana	Department of Agricultural Research	Maize, sorghum, millets, cowpea
Democratic Republic of the Congo	National Agricultural Research Institute	Maize
Lesotho	Department of Agriculture Research	Maize
Madagascar	Horticultural Technical Center of Antananarivo, Biotechnology and Plant Breeding Unit	Rice
Malawi	Department of Agricultural Research Services	Maize, cassava, pigeonpea, dry bean
Mozambique	Agricultural Research Institute of Mozambique	Maize, sorghum, cowpea, groundnut
Namibia	Ministry of Agriculture, Water and Forestry	Sorghum, pearl millet, cowpea, Bambara groundnut
Seychelles	Crop Research and Development Division	Cassava
South Africa	Agricultural Research Council	Maize, sorghum, pearl millet, wheat, barley, sweet potato, potato, fruits, dry bean, cowpea, ground nut
Swaziland	Agricultural Research Division	Maize, cowpea
Tanzania	Tanzania Agricultural Research Institute	Maize, rice, sorghum, pearl millet, soya bean, cowpea, potato, sweet potato, cassava
Zambia	Zambia Agriculture Research Institute	Maize, wheat, sorghum, pearl millet, cowpea
Zimbabwe	Department of Research and Specialists Services, Crop Breeding Institute	Maize, wheat, dry bean, ground nut, cowpea, sorghum, pearl millet, potato, sweet potato

Source: FAO, 2019

also discusses the support needs of plant breeders to guide policymakers to create enabling environments and to make investment decisions to support plant breeding as a core component of agricultural transformation.

Breeding methodologies and technologies

Various plant breeding methodologies and technologies are available, each with its advantages and limitations (Mwadzingeni et al., 2016). The public sector plant breeding programmes commonly use conventional plant breeding methods, including selections from local and exotic genetic resources, population improvement, pedigree breeding, hybrid breeding and backcrossing. These procedures typically require ten to 15 breeding generations to release an improved cultivar, unless complemented by the doubled haploid (DH) technology and other scientific innovations. Mutation breeding is rarely used in the region despite its potential to enhance genetic variation for biotic and abiotic stress tolerance and quality traits (Horn et al., 2016; Gwata et al., 2016). Tissue culture methods are useful in DH breeding and rapid mass production and multiplication for large-scale production.

In the region, genomic and proteomic tools are rarely used. Genomic tools have great potential in enhancing plant breeding in the region by complementing the conventional breeding methods, enhancing selection response, improving the accuracy of selection schemes and ensuring the efficient use of plant genetic resources. Gene editing is a relatively recent addition to genomics that is yet to be explored in the public sector plant breeding programmes for accelerated breeding and genetic gain. Initially, the use of genetic engineering to transform crops was seen as a technology of great promise. However, genetically modified organisms have been rejected by most countries in southern Africa, and there is a lack of enabling legislation in these countries. South Africa is the only country in southern Africa that has enabling legislation for the release and production of genetically modified crops such as maize, soybean, canola and cotton.

Conventional breeding programmes in the region have achieved notable successes in the release of various field crops (Walker et al., 2014). However, public breeding programmes need to develop high yielding and stress resilient crop varieties to serve the diverse needs of millions of smallholder farmers, value chains and local and regional markets. For instance, production of cereal crops in the region and SSA faces a serious threat caused by the recent arrival of the fall armyworm (FAW), a polyphagous insect pest that has more than 180 host plant species. Plant breeders urgently need to enhance host plant-resistance to FAW, which will provide an affordable, sustainable and environmentally friendly approach to minimizing its ongoing impact.

Major constraints to regional breeding programmes and farmer access to varieties

Public plant breeding programmes in the region are inadequately resourced and lack a critical mass of active plant breeders and breeding technicians. Often a high turnover of the relatively few plant breeding personnel negatively affects the continuity and impact of crop breeding projects and programmes. There is no harmonization of plant breeding programmes in the region. This has led to disjointed breeding programmes, which often results in multiple parallel projects. The cultivar development and release systems could be regionally consolidated to serve market needs (AGRA, 2015). Further, funding should be made available for research including into neglected crops, such as cowpea, Bambara groundnut and sorghum.

The adoption rate of improved varieties in SSA (excluding South Africa) is below 35%, compared with above 60% in Asia and 80% in South America (Walker et al., 2014). The low uptake of modern crop varieties in Africa is partly due to a lack of suitability of many new plant varieties to adequately meet the needs and preferences of the farmers and other actors such as processors, retailers and consumers in the value chain. The new varieties may also fail to meet the current and changing market demands. Therefore, the next generation of plant breeders should be trained to undertake demand-led breeding, focusing on the needs and preferences of the value chains, the marketplace and the stakeholders in the region. For instance, during the variety design phase, plant breeders should incorporate product profiles relevant to farmers and consumers to ensure high levels of adoption of the new varieties (Shimelis, 2017). This requires understanding the needs and preferences of smallholder farmers, processors, traders, retailers, consumers and other actors along the value chain of each crop.

Plant breeder requirements from policymakers

Many studies have shown that concerted and sustained plant breeding brings substantial returns on investment (ROI) with varied economic, social and environmental benefits. In the United Kingdom, plant breeding has reportedly attained an ROI as high as 40:1 (Webb, 2010). However, the sector requires adequate attention from policymakers who generally perceive plant breeding as a cost rather than an investment that gives substantial returns. Therefore, there is a need to educate policymakers in the critical need for increased investment in plant breeding capacity development and long-term investments, as summarized in Table 11.2. It is unlikely that any other area of investment will give sustained returns to match plant breeding, especially given the multiplier effect of agriculture on downstream value chains such as farmers, retailers, processors and consumers.

Table 11.2 Core requirements of plant breeders from policymakers in Southern Africa

Requirement	Reason/potential impact
Plant breeding education: Increased investment in plant breeding education	• Training more plant breeders and breeding technicians to breed the diverse African food security crops and to serving distinct agro-ecologies and for continuity of existing breeding projects. • Enhancing plant breeding programmes, including curricula, to train the next generation of academics and demand-led plant breeders, in Africa, with expertise in African crops.

Requirement	Reason/potential impact
Plant breeding research: Increased investment in plant breeding research	• Allocating research funds for plant breeding projects to develop farmer- and market-preferred and high performing cultivars for food security, enhanced livelihoods and return on investment. • Establishing plant breeding infrastructure (e.g., breeding nurseries, greenhouses, tissue culture and seed testing laboratories, germplasm repositories, genomics and proteomics tools, phenotyping resources, automated trait measurement resources). • Adopting demand-led plant breeding research and cultivar development based on the needs and preferences of clients and value chain and using market research, market trends and derivers, public-private sector partnership and multidisciplinary approaches. • Promoting community-based seed systems, through seed production, processing, packaging and marketing.
Policy and regulations: Introduce or reinforce enablers for plant breeders	• Enforcing African Union and regional legislation that provide for the harmonization of regulations on variety release, registration and marketing. • Establishing and supporting a regional plant breeding society and networks to exchange ideas and experiences on scientific progress, technological applications and the business of plant breeding, to contribute to the training of future plant breeders, to create a forum for communication for all stakeholders in plant breeding and to promote cooperation and closer link and involvement with agriculture. • Recognizing and rewarding plant breeders through royalty fees and award systems. • Harmonizing regional plant breeding programmes and seed policies to minimize duplication of efforts and to save resources, to accelerate the release of new varieties or new traits at reduced cost. • Enhancing cross-border germplasm exchange, variety release and seed systems within the same agro-ecological zones, across political borders. • Promoting public-private partnerships to develop new traits, new inbred lines and to breed and distribute seed of new crop varieties. • Promoting and financing small seed companies and agro-dealers, to expand the delivery system of new seed varieties and crop inputs to smallholder farmers across the entire region. • Financing smallholder farmers to buy new and improved seeds, irrigation systems, fertilizers and crop protection resources and postharvest storage facilities. • Financing infrastructure development to ensure that smallholder farmers have access to regional markets.

Breeding for value chains and marketplace

Farmers are the starting point of every crop value chain. Hence, the market potential of a new crop variety is heavily dependent on the number of farmers who are interested in growing the variety. In turn, this is dependent on demand for the product in the market by consumers and processors.

Adoption rate and commercialization of modern crop cultivars in Africa can be enhanced by integrating the breeding objectives set by NARS breeders along with the needs and preferences of the clients and market in the entire value chain. This requires well-detailed and up-to-date analyses of the value chain, market and market trends of each crop. In the past, crop breeders in SSA prioritized traits based mainly on "a priority or historical assumption", that farmers need such traits without consulting them. Furthermore, plant breeders unilaterally use quantitative and qualitative selection indices and product profiles without involving clients or the needs of the market. This form of trait prioritization and product profiling rendered low adoption rates of modern crop varieties including high yielding ones. Therefore, trait prioritization and product profile should be guided by both the market demand (proportion of growers needing the variety, or the total area grown by this variety) and price differentiation (the price premium or market share of the varieties or their traits) rather than selection indices.

Demand-led variety design should follow best practices from public and private sectors in Africa and internationally for successful variety design, product profiling and market. Partnerships between the public and private plant breeding programmes is key to learn best practices and to provide customized services needed by small-scale farmers. Partnerships between the two sectors enable access to genetic resources and modern plant breeding training services and infrastructure support for the NARS breeders. It also ensures that the private partners have access to new varieties bred in the public sector, with new traits that meet the changing needs of farmers and downstream value chain. This combined with excellent breeding science and technology, vigourous awareness campaigns with farmers and customers can lead to significant gains in adoption rates and market share of new varieties developed by public sector and small seed company breeding programmes in Africa.

Conclusions and recommendations

Plant breeding can produce improved crop cultivars with economic benefits that help in achieving food and nutrition security and sustainable agriculture globally. However, in southern Africa, the public sector breeding programmes are not well developed, and they are often poorly resourced due to limited investment in plant breeding education, research and infrastructure development. The existing public breeding programmes have developed and released many crop varieties with significant yield and quality gains. However, small-holder farmers in the region have not adopted these varieties, primarily due to a lack of access or rejection of the new varieties because they fail to carry critical quality traits. Also, the major constraints to regional breeding programmes can be attributed partly to a lack of harmonization of plant breeding programmes, restricted movement of plant germplasm resources across national borders and insufficient active plant breeding personnel in the region. The formal seed systems in the region have not engaged with smallholder farmers in seed production, distribution and marketing. In order to increase the uptake of modern crop varieties in the region, it is critical for plant breeders to incorporate quality traits that satisfy the needs and preferences of farmers and their value chains, markets and stakeholders. Success in agricultural development through crop improvement in the region is dependent on increased investment in plant breeding education, long-term research programmes and research infrastructure development. In addition, efforts should be exerted towards the development of infrastructure and markets for farmers in the region and enhancing partnerships between the public and private plant breeding programmes.

References

Alliance for a Green Revolution in Africa (AGRA). (2015) Transforming Africa's agriculture for sustainable inclusive growth, improved livelihoods and shared prosperity: A background note. *Third International Conference on Financing for Development*, 13–16 July, Addis Ababa, Ethiopia, pp. 1–10.

Eriksson, D., Akoroda, M., Azmach, G., Labuschagne, M., Mahungu, N. and Ortiz, R. (2018) Measuring the impact of plant breeding on sub-Saharan African staple crops. *Outlook on Agriculture* 47(3): 163–180.

FAO (Food and Agriculture Organization of the United Nations). (2019) *Global Partnership Initiative for Plant Breeding Capacity Building*. www.fao.org/in-action/plant-breeding/our-partners/africa.

Gwata, E.T., Shimelis, H. and Matove, P. (2016) Potential of improving agronomic attributes in tropical legumes using two mutation breeding techniques in southern Africa. In: Konvalina, P. (ed). *Alternative Crops and Cropping Systems*. IntechOpen Limited, London, pp. 71–85. ISBN 978-953-51-2279-1.

Horn, L., Ghebrehiwot, H.M. and Shimelis, H.A. (2016) Selection of novel cowpea genotypes derived through gamma irradiation. *Frontiers in Plant Science* 7: 1–13.

Mwadzingeni, L., Shimelis, H., Dube, E., Laing, M.D. and Tsilo, T.J. (2016) Breeding wheat for drought tolerance: Progress and technologies. *Journal of Integrative Agriculture* 15: 935–943.

Shimelis, H. (2017) New variety design and product profiling. In: Persley, G.A. and Anthony, V.M. (eds). *The Business of Plant Breeding: Market-led Approaches to New Variety Design in Africa*. CABI International, Wallington, UK, pp. 85–114.

Walker, T., Alene, A., Ndjeunga, J., Labarta, R., Yigezu, Y., Diagne, A., Andrade, R., Muthoni, R., De Groote, H., Mausch, K., Yirga, C., Simtowe, F., Katungi, E., Jogo, W., Jaleta, M. and Pandey, S. (2014) Measuring the Effectiveness of Crop Improvement Research in Sub-Saharan Africa from the Perspectives of Varietal Output, Adoption, and Change: 20 Crops, 30 Countries, and 1150 Cultivars in Farmers' Fields. *Report of the Standing Panel on Impact Assessment (SPIA)*, CGIAR Independent Science and Partnership Council (ISPC) Secretariat, Rome, Italy.

Webb, D. (2010) Economic Impact of Plant Breeding in the UK. *The British Society of Plant Breeders*, Ely, UK. 24 pages. www.bspb.co.uk/contact-bspb.php.

12 Integrated pest management in Southern Africa

Approaches and enabling policy issues

Mark D. Laing and Hussein Shimelis

Introduction

Crop losses and environmental damage in Africa due to pests

The United Nations Food and Agriculture Organization has defined "pest" as any organism that damages crops or the environment (FAO Glossary of Terms 2010). It includes invasive weeds, parasitic weeds, insects, nematodes, pathogens and feral animals. Across Africa pests such as the fall armyworm devastates cereal crops, rampant aquatic weeds clog water systems and thorny mesquite trees have overrun millions of hectares of agricultural lands in arid and semi-arid regions. Preharvest crop losses caused by pests are estimated at 45%, whereas postharvest losses of grain crops in storage are typically 40–70%. Possibly the most sustainable way to enhance food security in Africa would be to control the primary pre- and postharvest pests of food crops. Pest control in Africa has largely depended upon cultural control measures and agrochemicals. Cultural control alone does not cope with a scaling of the agricultural systems or the intensification of production that is required to feed the burgeoning human population in Africa. Climate change has also caused rapid shifts in pest populations in response to changing natural ecosystems in Africa. Therefore, there is a need for an integrated pest management (IPM) approach that is effective, adaptable and scalable. The objective of this chapter is to highlight the potential of IPM to enhance food security and crop productivity. Policy aspects are identified that could enhance the effective implementation of IPM in Africa.

Integrated pest management (IPM)

IPM is the integration of various pest control methods such as host-plant resistance, cultural methods, agrochemicals and biocontrol agents for the control of pests affecting crops or the environment.

Host-plant resistance

The use of pest resistant crop varieties is potentially the cheapest and most environmentally friendly approach for pest control. However, it requires the long-term investment of resources (e.g., research funding, genetic resources, capacity development and research capacity) in public and private breeding programmes that are demand-driven and generate varieties that meet farmers' preferences, as well as improved pest resistance. It also requires a downstream value chain to deliver the new, pest-resistant varieties to the farmers that need them.

Cultural methods

These include crop rotation, cover cropping, trap-cropping, soil fertilization, hand weeding and the push–pull approach (the use of trap crops to attract pests away from the main crop). These approaches may be effective but require intensive training by extension services, the commitment of substantial resources of land, labour and funds and collective action by communities of farmers (Hearne, 2009). Cultural control measures usually provide variable results. These are some of the factors that have resulted in a relatively poor adoption of most cultural control methods on their own (Parsa et al., 2014).

Agrochemicals

Agrochemicals have been crucial for crop protection for the last 70 years, making a significant contribution to world food production. Across Africa they are still the basis for crop protection on most commercial farms and many small-scale farms. However, they face a number of limitations to their role in the control of pests globally, especially in the future. In some cases, there is no effective agrochemical to control a particular pest currently, e.g., there are no fungicides currently available to control the *Fusarium* species. In other cases, target pests have evolved resistance to effective agrochemicals; e.g., grey mould (*Botrytis cinerea*) to the fungicide benomyl (Hahn, 2014). There are also strong laws and control mechanisms to prevent residues in foodstuffs that has resulted in the restriction of use of many agrochemicals. For example, the use of guazatine for the control of sour rot of fruit and vegetables was effectively banned in the EU in 2016 (EU Regulation, 2015/1910). Evidence that some agrochemicals were damaging the environment led to their global withdrawal (MeBr), while others have been withdrawn because of their extreme toxicity, and others for their carcinogenicity (Shukla and Arora, 2001; Lin et al., 2014) or their anti-androgenic activity (Uzumcu et al., 2004).

A major consideration for pest management in the long term is the extreme costs of developing new agrochemicals. Only 1 in 140,000 molecules are successfully developed into agrochemicals. The full costs were estimated to exceed US$286 million per new agrochemical in the period 2010–2014 (McDougall, 2016). Registration of agrochemicals in Africa is challenging because the costs of applying for permits for the pesticide are high relative to the potential market available in each country, except for export crops. Hence, the agrochemicals available to small-scale farmers in Africa are often older products that are cheap. However, relatively few small-scale farmers can afford to buy agrochemicals or their application equipment. Safety is also an issue because the hot and humid conditions occurring frequently in Africa make it hazardous to wear effective safety equipment (Mokhele, 2011). Extension services in most of Southern Africa are limited, making it challenging to inform the millions of small-scale farmers of the technical aspects of agrochemicals: targets, dose, frequency, timing, mixing and safety aspects. Few small-scale farmers are technically literate, which inhibits the use of information pamphlets. Information transfer then requires face-to-face extension support services or the use of mobile application technologies, using local languages.

Biocontrol

The development of biocontrol agents starts with the isolation and screening of insects, fungi, bacteria, viruses or nematodes for the control of a target pest. The chosen strain of the biocontrol agent is then manufactured, formulated and sold to farmers to treat their crops. Commercial biocontrol agents have to be applied repeatedly to the crop of concern; e.g., application of selected strains of *Trichoderma* species to seeds of many crops to control damping off of seedlings by *Rhizoctonia solani* (Yobo et al., 2004). Their relatively low costs and safety make them an attractive option for Africa. However, most of them have a limited shelf life (typically six to 24 months at 20°C). Cold storage in a deep freezer can provide for indefinite storage, but this equipment is usually absent at agro-dealers and small-scale farmers. The development and commercialization of a biocontrol product follows a value chain with defined steps and key participants (Figure 12.1).

The development of commercial biocontrol agents is relatively fast and cheap, compared to agrochemicals, an estimated US$5 million vs. US$250 million per product. However, the relatively few commercial biocontrol agents that have been developed in Africa, largely come from South Africa and Kenya (Barratt et al., 2018). Use of a biocontrol agent requires a thorough understanding of the product and the target pest. This requires competent and widely distributed extension services.

IPM – integrating the components

IPM involves the use of two or more crop protection methods, integrated for superior pest management. Ideally, IPM involves several control measures, integrated to complement each other. A model of IPM involving the integrated control of *Striga* species affecting cereals and cowpea is shown in Figure 12.2 The parasitic weeds *Striga asiatica* and *S. hermonthica* in cereal crops and *Striga gesneroides* in cowpea, cause losses of up to 80% in key food crops, especially in dry regions of Africa with low-fertility soils. Effective management of *Striga* species usually requires the integration of a spectrum of IPM measures (Hearne, 2009; Mangani et al., 2011; Shayanowako et al., 2018).

Integrated application of the four components of IPM requires positive and enabling policies. African governments need to commit to developing the value chain of IPM that will allow for the development, registration and deployment

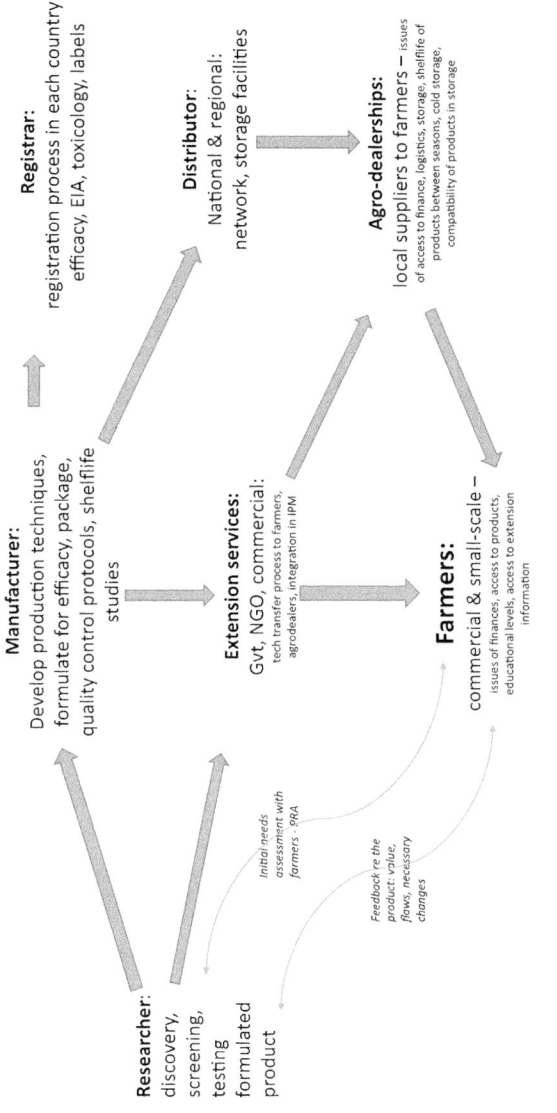

Figure 12.1 The value chain of a biocontrol product

Researcher: discovery, screening, testing formulated product

Manufacturer: Develop production techniques, formulate for efficacy, package, quality control protocols, shelflife studies

Registrar: registration process in each country: efficacy, EIA, toxicology, labels

Distributor: National & regional: network, storage facilities

Extension services: Gvt, NGO, commercial: tech transfer process to farmers, agrodealers, integration in IPM

Agro-dealerships: local suppliers to farmers – issues of access to finance, logistics, storage, shelflife of products between seasons, cold storage, compatibility of products in storage

Farmers: commercial & small-scale – issues of finances, access to products, educational levels, access to extension information

Initial needs assessment with farmers - PRA

Feedback re the product: value, flaws, necessary changes

Striga species Parasitize Major Food Crops in Africa:
maize, sorghum, pearl millet, finger millet, rice, fonio; cowpea

Breeding:
Multiple resistance genes + compatibility with biocontrol agent; mutations for resistance to SU and IR herbicides

Constraints: lack of breeders, lack of budget to breed each crop, lack of value chain to deliver seed to farmers, lack of extension training and finances

Cultural Control:
Intercrop with trap crops or catch crops; rotation with legumes; no-till + mulch; fertilizer/manure; hand-weeding

Constraints:
Lack of training be extension services, costs of processes, labour requirements, limited land, models not adapted to all agro-ecologies

Biocontrol:
Seed treatment for *Fusarium oxysporum* fsp *strigae*

Constraints: no registered products, no agrodealers to deliver products to farmers, no money, no information or training by extension services

Chemical Control:
Seed treatment with IR or SU herbicide to resistant cultivars; Apply selective herbicides to germinated *Striga*, e.g., 2,4-D

Constraints:
no agrodealers to deliver products and equipment to farmers, no money, no information or training by extension services

Overall Constraints:
access to the technologies, to information and to money, and the training and motivation to integrate these four components

Figure 12.2 IPM applied to *Striga* management

of key technologies, together with the critical technology transfer process by extension services, working with farmers. This will require the deployment of innovative communication technologies, such as the provision of video-based training shared with farmers using mobile technology. Farmers also need access to credit to be able to purchase the components of IPM: seeds, fertilizers, agrochemicals, biocontrol agents and application equipment. The Grameen Bank in Bangladesh pioneered a model for microfinancing of small-scale farmers (McWha and Willows, 2016).

Policy issues affecting access to improved seed, agrochemicals and biocontrol agents

Lack of regional harmonization

Harmonization issues affect the previous components of IPM. Given the time and cost taken to develop new crop varieties, agrochemicals or biocontrol agents, especially where the key pests affect entire regions, there is a critical need for regional harmonization in the registration of new crop varieties, agrochemicals and biocontrol agents. The African Union has published draft policies on the harmonization of pesticides (and biocontrol agents), but it has not passed the draft as a functional act (AU-IBAR, 2016). Until the Act is passed and enforced regionally, the registration processes will remain on a country-by-country basis. This is a particularly serious problem for poorer countries that usually cannot afford to employ specialist scientists to manage these processes. Similarly, the costs of development and registration, and the delays in commercializing new varieties, agrochemical or biocontrol agents, mean that these products are usually not registered in poor countries unless effective regional harmonization policies are in place. In some cases, such as the East African Community, harmonization policies have existed since 1999, but are not implemented at a country level.

Lack of resources for multiplication of insects, fungi and viruses

Across Africa, there are few research establishments that can consistently isolate, identify, store and maintain biocontrol agents, be they insects or microbes. This is a major barrier to the widespread development of biocontrol agents. There is an even greater shortage of companies that can manufacture biologicals on a large scale, register and market them so that biocontrol agents become available to all the farmers of Africa. Much of the current academic research on biocontrol is unlikely to reach the farmer because few of the research teams are working with industry partners that are committed to commercializing a wide range of novel biocontrol agents. This includes researchers at the National Agricultural Research Systems, the Consultative Group for International Agricultural Research centres and at universities across Africa, who may discover many potentially useful biocontrol agents, but unless they can be commercialized, these biocontrol studies are merely an academic exercise. A powerful policy change would be if funders and national assessors reviewing project proposals in the field of biocontrol required the researchers to include a full value chain that could deliver products into the hands of farmers (Figure 12.1).

Costs and time taken to register products

Most local seed companies, agrochemical distributors and biocontrol companies that have developed in Africa are relatively small, start-up companies. As start-up companies, they usually have a limited access to capital and their survival depends upon securing a stable cashflow before their start-up capital is spent. However, registration requirements are becoming increasingly stringent, take longer and cost more. As a result, most of the indigenous start-up companies are unable to survive this barrier. If African farmers are to have access to a wide range of modern agrochemicals and biocontrol agents, then there is a need for African governments to support indigenous crop protection companies with tax breaks and access to capital, specifically so they can complete the registration processes to international standards.

Efficacy trials

For both agrochemicals and biocontrol agents, there is a registration requirement to show efficacy in most countries. However, there is also a demand for separate registration for the same pest on every crop that the pest affects, even with a polyphagous pest or disease, e.g., twin spotted red spider mite or *Sclerotium rolfsii*, with more than 100 hosts each. However, the interaction is between the control agent and the pest, and the crop plays little part. This creates a massive barrier to registration of products for minor crops: in most cases, it is not financially viable to register a product on the minor crop because the potential income is less than the costs of registration.

Toxicology testing

While clearly these are essential, the toxicological testing requirements create a huge financial barrier for small companies. One barrier is that there are no International Organization for Standardization accredited toxicological laboratories in Africa, with the result that all toxicological testing has to be undertaken in Northern Hemisphere countries, at a cost that small start-up companies in Africa cannot afford. This requirement alone raises the barrier to development of locally registered products in Africa to an unreachable level. This means that only major agrochemical companies, or a few established biocontrol companies, can afford to develop a full spectrum of new agrochemicals and novel biocontrol agents for IPM programmes in selected countries in Africa, usually for the protection of cash crops only.

Bioprospecting policies

The Convention on Biological Diversity (CBD) was passed by the United Nations in 1993, and was followed by the Nagoya Protocol of 2010 (United Nations, 2010). These regulations directly affect the development of novel bio-control agents. The assumption is that indigenous organisms were all part of indigenous knowledge systems (IKS) and the genetic resources of one country, and therefore the company that develops a biocontrol agent must pay a royalty to the local community to whom the IKS and generic resources "belonged" (Access and Benefit Sharing – ABS). However, there is no evidence that either microbes or insects were ever used in Africa to control pests. Second, many of the microbes used as biocontrol agents, such as *Trichoderma* sp., are universal. Therefore, it is not clear what country or region should benefit from a newly discovered strain of *Trichoderma*, for example. Third, it is not clear who the "community" is who should benefit from the royalties within a country, nor are there sound administrative processes in place in much of Africa to manage these payments (Chennels, 2010; Cock et al., 2010; Millum, 2010).

Summary

Farmers in Africa have previously depended on cultural control and agro-chemicals for crop protection. However, the future of agriculture in Africa will depend on controlling pests with an integration of these measures, combined with demand-led plant breeding and biocontrol agents. The primary policy issues affecting the development and deployment of IPM include the need to:

- Develop appropriate and affordable regulatory measures for novel seeds, agrochemicals and biocontrol agents
- Ensure the actively enforced harmonization of agricultural registration requirements on a regional basis in Africa
- Create enabling environments to promote the development and distribution of locally bred crop varieties and commercial biocontrol agents
- Resolve issues associated with the Nagoya Protocol and the CBD to avoid stopping the development of novel biocontrol agents
- Support agronomic approaches that promote ecosystem services to limit pest damage as outlined in the chapter in this book.

A range of enabling policies and long-term financial commitments by African governments are needed to accelerate the development and uptake of new crop cultivars and biocontrol agents and access to novel agrochemicals. Access to these technologies, combined with appropriate training and access to credit facilities, are essential. Furthermore, there is a critical need for long term invest-ments by African governments in capacity development of scientists working on IPM, in demand-driven research, combined with the mobilization of exten-sion services and stakeholder engagement in IPM.

References

AU-IBAR. (2016) Draft Policy for the Harmonization of Pesticides Registrations in Africa https://www.google.com/search?q=harmonization+policies+African+union+pesticides& rlz=1C1ZKTG_enZA698ZA701&oq=harmonization+policies+African+union+pesti cides&aqs=chrome..69i57.15655j1j4&sourceid=chrome&ie=UTF-8. Access 2019 03 01.

Barratt, B.I.P., Moran, V.C., Bigler, F. and van Lenteren, J.C. (2018) The status of biological control and recommendations for improving uptake for the future. *BioControl* 63: 155–167.

Chennels, R. (2010) Towards global justice through benefit-sharing. *Hastings Centre Report* 40: 3.

Cock, M.J.W., van Lenteren, J.C., Brodeur, J., Barratt, B.I.P., Bigler, F., Bolckmans, K., Consoli, F.L., Haas, F., Mason, P.G. and Parra, J.R.P. (2010) Do new access and benefit sharing procedures under the convention on biological diversity threaten the future of biological control? *BioControl* 55: 199–218.

FAO Glossary of Terms (2010). In: Guide to Implementation of Phytosanitary Standards in Forestry http://www.fao.org/3/i2080e/i2080e08.pdf.

Hahn, M. (2014) The rising threat of fungicide resistance in plant pathogenic fungi: *Botrytis* as a case study. *Journal of Chemical Biology* 7: 133–141.

Hearne, S.J. (2009) Control – the *Striga* conundrum. *Pest Management Science* 65: 603–614.

Lin, C.H., Chou, P.H. and Chen, P.J. (2014) Two azole fungicides (carcinogenic triadimefon and non-carcinogenic myclobutanil) exhibit different hepatic cytochrome P450 activities in medaka fish. *Journal of Hazardous Materials* 277: 150–158.

Mangani, E.I., Ibrahim, A. and Ahom, R.I. (2011) Integrated management of parasitic plant *Striga hermonthica* in maize using *Fusarium oxysporum* (mycoherbicide) and post-emergence herbicide in the Nigerian savanna. *Tropical and Subtropical Agroecosystems* 14: 731–738.

McDougall, P. (2016) The costs of new agrochemical product discovery, development and registration in 1995, 2000, 2005-8 and 2010-2014. R&D expenditure in 2014 and expectations for 2019. *A Consultancy Study for CropLife International, CropLife America and the European Crop Protection Association* https://croplife.org/wp-content/uploads/2016/04/Cost-of-CP-report-FINAL.pdf. Accessed 2019 03 01.

McWha, J. and Willows, G. (2016) MAF05: The fight against poverty: Review of the applicability of the Grameen bank model in South Africa. In: 2016 *Southern African Accounting Association National Teaching and Learning and Regional Conference Proceedings*, pp. 357–369. ISBN number: 978-0-620-74761-5.

Millum, J. (2010) How should the benefits of bioprospecting be shared? *Hastings Centre Report* 40: 24–33.

Mokhele, T.A. (2011) Potential health effects of pesticide use on farmworkers in Lesotho. *South African Journal of Science* 107(7–8) Art. #509, 7 pages.

Parsa, S., Morse, S., Bonifacio, A., Chancellor, T.C.B., Condori, B., Crespo-Perez, V., Hobbs, S.L.A., Kroschel, J., Ba, M.N., Rebaudo, F., Sherwood, S.G., Vanek, S.J., Faye, E., Herrera, M.A. and Dangles, O. (2014) Obstacles to integrated pest management adoption in developing countries. *Proceedings of the National Academy of Science, USA* 111: 3889–3894.

Shayanowako, A.T., Laing, M., Shimelis, H. and Mwazingezi, L. (2018) Resistance breeding and biocontrol of *Striga asiatica* (L.) Kuntze in maize: A review. *Acta Scandinavica, Section B – Soil and Plant Science* 68: 110–120.

Shukla, Y. and Arora, A. (2001) Transplacental carcinogenic potential of the carbamate fungicide mancozeb. *Journal of Environmental Pathology, Toxicology and Oncology* 20: 127–131.

United Nations. (1993) Convention on biological diversity (1993) with annexes, concluded at Rio de Janeiro on 5 June 1992, Vol 1760 (30619). *United Nations Treaty Series.*

United Nation. (2010) The Nagoya protocol on access to genetic resources and the fair and equitable sharing of benefits arising from their utilization (ABS) to convention on biological diversity https://www.cbd.int/abs/about. Accessed 2019 03 01.

Uzumcu, M., Suzuki, H. and Skinner, M.K. (2004) Effect of anti-androgenic endocrine disruptor vinclozolin on embryonic testis cord formation and postnatal testis development and function. *Reproductive Toxicology* 18: 765–774.

Yobo, K.S., Laing, M.D., Hunter, C.H. and Morris, M.J. (2004) Biological control of *Rhizoctonia solani* by two *Trichoderma* species isolated from South African composted soil. *South African Journal of Plant and Soil* 21: 139–144.

13 Options for improving stored product protection in Southern Africa

Cornel Adler and Edson Ncube

Introduction

Due to the favourable climate conditions in Southern Africa, durable staple food products such as cereal grains and pulses can be attacked all year round both in the field and during storage. Damage may be caused by insects, rodents, birds and – at higher moisture content – by microorganisms. High moisture content in storage can occur locally due to incomplete product drying, pest respiration, diurnal changes in relative humidity, moisture migration and condensation, moisture seeping from the ground or coming down through a leaky roof. Once fungi are established, they also increase moisture and temperature due to respiration and often also produce mycotoxins. These are a major threat to human and animal health because fungal attack and contamination with mycotoxins is not always easy to notice. Reducing postharvest losses can become crucially important in agricultural communities with insecure harvests due to climate change. The commercial farming sector in Southern Africa has access to modern storage facilities such as privately managed storage silos where grain is delivered immediately after harvest. It is in the resource-poor smallholder farming systems where stored product protection is a major constraint.

If a harvest is left in the field for too long, seed grains may be shattered and lost, and they may be attacked by insects, birds and rodents. And if the grain is too moist, germination capacity may be lost, or microbial development may contaminate the grains before they are stored. Thus, grain should be harvested as soon as it is mature.

Losses

In 2011, the World Bank estimated that postharvest loss in cereal grains are around US$4 billion each year in sub-Saharan Africa (Sheahan and Barrett, 2017). This is a staggering amount of food loss in a continent, where there is a high prevalence of hunger due to various factors such as drought and climate change. Storage facilities in most Southern African Development Community (SADC) countries, particularly the smallholder producers, consist of traditional storage structures, which are inadequate to control insect infestations and fungal

Table 13.1 Maize postharvest losses (%) in the value chain from harvesting to farm storage in SADC countries in 2011

Country	Value chain					Total
	Harvesting/ field drying	*Platform drying*	*Threshing*	*Transport*	*Farm storage*	
Angola	6.4	4.0	1.3	2.4	4.6	18.7
Botswana	7.2	4.0	1.3	2.4	1.1	16.0
DR Congo	6.4	4.0	1.3	2.4	0.5	14.6
Lesotho	6.4	4.0	1.3	2.4	3.4	17.5
Malawi	6.3	4.0	1.4	2.4	8.2	22.3
South Africa	3.9	3.5	2.2	1.9	0.3	11.8
eSwatini (Swaziland)	6.4	4.0	1.3	2.4	4.5	18.6
Tanzania	6.4	4.0	1.3	2.4	5.0	19.1
Zimbabwe	6.4	4.0	1.3	2.4	4.8	18.9

Source: www.aphlis.net

growth during storage. Most grains are lost during postharvest handling and storage at the farm level in Africa, though the magnitude of losses varies (Sheahan and Barrett, 2017).

The scale and extent of postharvest loss varies due to different farming practices, climatic conditions, country and regional economics. The African Postharvest Losses Information System (APHLIS) provides estimates of postharvest losses in the SADC region (Table 13.1). Despite the introduction of APHLIS, direct measurements of postharvest losses are difficult and expensive and are therefore rare, with the most recent statistics dating back to almost a decade.

APHLIS estimates postharvest losses using postharvest loss profiles derived from peer-reviewed literature and seasonal factors obtained from local experts. The postharvest loss profiles (PHL profiles) quantify the expected loss – as a percentage at each stage of the postharvest chain – from harvesting to storage and market (Table 13.1). This loss data is obtained from scientific literature and broken down by crop, type of farm and climate type. Biotic and abiotic constraints as well as agricultural practices are the seasonal factors that are considered. These data are collected by APHLIS from official sources such as ministries of agriculture and statistics offices or by interviewing farmers or extension workers. The data from PHL profiles and seasonal factors in percentage losses are converted into absolute losses (in tonnes) through the use of an APHLIS algorithm (www.aphlis.net/en/page/4/how–aphlis–estimates–loss#/).

Pests

Some of the typical insect pests found in storage are the maize weevil, *Sitophilus zeamais*, as well as the lesser grain weevil or rice weevil, *S. oryzae*, the lesser grain borer, *Rhyzopertha dominica*, and in many locations also the larger grain borer, *Prostephanus truncatus*. Pulses such as beans or cowpeas cannot be attacked by these grain pests due to their protective proteins. Typical pests in stored pulses are the cowpea weevil, *Callosobruchus maculatus*, and the bean weevil, *Acanthoscelides obtectus*. A list of 30 African stored product pest species can be found in Nukenine (2010). Insects have a great capacity for multiplication and if undisturbed can destroy a complete harvest within a few months. Storage of seed grains is necessary for periods usually between three and ten months. Given that infestation with stored product pests can already occur in the field prior to harvest, a good storage should not only prevent infestation but should also control any pre-existing minor infestation. This means that in most cases pest control is a need during storage.

Product drying, threshing and transportation

Depending on region and season, drying may take place in the field. However, in higher altitudes or fluctuating weather conditions, additional drying after harvest may be necessary. In many regions, grains are just dried on the ground protected from soil moisture by a plastic tarpaulin or a concrete surface. During sunny weather, drying maize to below 14% moisture content takes about five days. But, if the harvest is not protected from occasional rain, product drying may be insufficient and there is a risk of microbial development. Drying at ground level carries the risk of contamination with soil and other foreign particles as well as the risk of attack by rodents, birds, insects or theft. Thus, simple solar grain dryers that may be supported by furnaces in a rainy period could render grain drying more independent from weather conditions. Threshing and transportation from field to storage also pose a risk of contamination and postharvest loss. In some situations, the harvested ears/cobs are left to dry on the ground while smallholder farmers are awaiting transport to ferry the grain to the homestead for storage.

Investment in handheld moisture meters that can be used by smallholder farmers to monitor grain moisture levels at harvest, during drying and storage can mitigate postharvest losses arising from poorly dried grain. However, initial purchasing costs may be high for many farmers. Another option is to weigh samples of 100g and to dry them over a heater for about two hours before weighing again. If the new weight of the dried sample is 86g the moisture content of the grain would be some 14%. In fact, drying to only 13% moisture content may be more advisable because insects develop much slower, and there is even better protection from microbials.

Pest control

For decades after World War II, chemical pest control was the preferred option in most countries of the world. However, over the last 30 years hardly any new pesticide became available for stored product protection. Many old products lost authorization in industrialized countries due to potential health risks for users and bystanders, while others caused the development of insect resistance, posed a threat to the environment or left residues that are no longer tolerated.

In Zambia and Malawi, farmers surveyed by Kamanula et al. (2011) reported serious pest damage, with *C. maculatus* and *Sitophilus* spp. being the major pests of beans and maize, respectively. It was found that 50% of farmers surveyed in eastern Zambia and all farmers surveyed in northern Malawi used synthetic pesticides. While there may be still a rather widespread use of pesticides, the mid- and long-term availability of synthetic pesticides and a prominent future role of these agents in Southern African stored product protection is not very probable.

In many rural African communities, plants and botanicals are traditionally used for both medicinal purposes and pest control. This led to a large interest in studying the efficacy of various botanicals in different formulations. In most cases where a dose dependent efficacy could be proven, plant extracts and pure compounds were identified. Of course, their cheap availability for smallholder farmers and their biodegradability are among the advantages of botanicals. A risk in the use of plant extracts is that they may change organoleptic properties such as the smell and taste of protected stored products and may also affect users and consumers. Thus, before making recommendations, the potential health risks associated with using one or another plant extract should be determined. However, in combination with improved product drying and storage, botanicals may still have a role to play in future stored product protection.

A certain efficacy against stored product pests is also attributed to plant ashes and desiccating dusts such as kaolin or diatomaceous earths (DE). Plant ashes may contain a large number of chemical compounds and may thus be a potential health hazard to users. Kaolin consists of crystalline particles and health risks associated with the inhalation of these particles must be analyzed. Thus, within this group of compounds the DE may be most promising. DE are fossil remnants of aquatic algae that occur in thick layers at sites formerly covered by fresh or saltwater. Ground DE is used to filter fruit juices or beer as well as an abrasive agent in toothpaste. Fine DE dust can remove fatty acids from the insect cuticle and thus increase the permeability of membranes leading to desiccation of stored product insects. Various DE products were authorized in Europe and North America and may be even more effective at higher temperatures and as a nontoxic insecticide on surfaces around storage structures. However, just lipophilic DE products ground to fine dusts below some 30 μm were found to be effective.

Traditional storage structures

Traditional storages that can be large clay pots with or without a lid or made solely of wood or of sticks and mud or bricks, do provide various levels of ventilation for passive drying and often some precaution against rodents but mostly little protection against insects. Improved traditional storage structures have been developed in Southern Africa such as the Tanzanian *kihenge*. The *kihenge* is constructed from bamboo or wooden sticks and sealed by plastering with a mixture of mud, cow dung and ash. It is kept under a shelter to protect it from adverse weather conditions and is raised on legs with rodent guards to protect it from rodent infestation. Another example is the improved Zimbabwe brick granary that is constructed from brick, concrete slabs, wire mesh and thatch. The *gorongosa* mud silo is another improved mud silo that is used in Mozambique. It is kept sheltered from adverse weather conditions, and contact with ground moisture is prevented by placing it on rocks or wooden pallets. Another example is the *ngokwe* traditional silo for storage in Malawi that is made of interwoven split bamboo and covered with a conical shaped grass roof, and it is also raised off the ground.

Storage in woven polypropylene bags

In many countries, woven polypropylene bags are used for storage and trade. These bags are stored in various types of buildings. Also, jute bags are available in certain regions.

For example, Kamanula et al. (2011) indicated that bags were the major storage method for 54 and 97% of maize and beans, respectively in northern Malawi, whereas in eastern Zambia, 81% of the maize grain was stored in granaries while 82% of beans was stored in bags. Studies by Kankolongo et al. (2009) indicated that most household surveyed in the southern, central, Lusaka, Copperbelt and northwestern provinces of Zambia used bags for maize storage, often without the application of chemical pesticides. Due to these storage conditions, all farmers surveyed in northern Malawi reported pest damage on maize with 93% reporting damage on beans. In eastern Zambia, almost all farmers reported pest damage to maize and beans. As much as 80% pest damage was reported in five provinces of Zambia.

Basically, woven bags are suitable for transportation but offer no protection against insect attacks. Volatiles emitted from stored products attract insects into the storage bag. Thus, if feasible, it may be advantageous to store sufficiently dried products in an insect-proof enclosure or, better, even under hermetic conditions.

Hermetic storage structures

Hermetic storage means that the harvest is stored under gas-tight conditions. A gas-tight enclosure does not emit any attractive odours that could lure pests into the storage. While rodents and birds can learn quite fast how to chew or peck open gas-tight sacks, insects are kept away from a hermetic storage if there are no attractive volatiles emitted. In case insects were sealed in with infested grain, they will reduce the oxygen available in the interstitial atmosphere until they suffocate. Young adult beetles of species like lesser or larger grain borer usually leave the grain and search for light. These adults can penetrate plastic layers after hatching from inside to outside. This would be a risk during the first weeks until anoxia leads to control of insect pests. Hermetic metal silo bins, however, could withstand rodents, birds and insects. Anoxic conditions are able to control all animal pests in grain and would lead to fungistatic conditions in a dry environment because all microorganisms tolerant to low moisture contents are fungi depending on respiration. Without oxygen even the formation of mycotoxins is stopped (Riudavets et al., 2018).

Hermetic metal silo bins (indoors/shaded)

Hermetic metal silo bins could be produced by local craftsmen in different sizes. Under hermetic conditions, respiration in the storage bins depletes the oxygen present and leads to anoxic conditions without the application of any pesticide. This effect was already known to the Romans who left written reports of underground hermetic storage in the first century AD. The typically bottle-neck shaped pits for underground hermetic storage can be traced in Northern Africa back to early Iron Ages (approx. 700 BC, Hill et al., 1983).

To benefit from the advances of a truly hermetic storage, a silo should be filled to the top in order to minimize the free headspace that would otherwise be a reservoir of oxygen. Thus, several smaller silo bins could be better than few large ones in order to balance out yearly fluctuations in harvest. Usually, the metal silo bin is built with feet that keep it standing upright and stable at some distance from the ground with a conus at the bottom from which the grain can be retrieved and a lid at the top for filling. The silo bin should be placed indoors or at least in a shaded space to avoid strong diurnal pressure changes that could damage the gas-tight seal. Strong temperature fluctuations could also

Figure 13.1 Metal silo bins from local production in Tanzania; hermetic seal feasible

Photo: C. Adler

lead to condensation within the silo bin. All openings need to be thoroughly checked and sealed. A pressure test should be used to test the gas-tightness in a silo bin prior to loading it with grain. White paint on the outside of the silo bin could reduce warming by light and the risk of condensation inside as well as improve gas-tightness around openings. After grain loading, the full silo bin should remain closed for at least two months in order to allow hermetic conditions to control insects that might have been brought in with the harvest.

Metal drums

Metal drums such as former water or oil drums could be used for grain storage, provided they are clean and can be sealed to a high degree of gas-tightness. Metal drums may be cheaper to purchase but would need to be stored safely to protect from damage or theft.

Hermetic triple layer bag

Hermetic triple layer bags such as the Purdue Improved Crop Storage (PICS) bag have been promoted in sub-Saharan Africa for a number of years now. Repeated testing showed that compared to the use of insecticides and storage in woven polyethylene bags, damage and losses by insect pests were considerably lower.

Young adult beetles like the lesser grain borer, *Rhizoperta dominica*, and the larger grain borer, *Prostephanus truncatus*, tend to leave the grain in search for mating partners and new oviposition sites. These insects can easily penetrate plastic liners from inside to outside in their search for light. If young adult beetles penetrate a hermetic bag, this would stop the depletion of oxygen inside the bag, leaving openings for emission of attractive volatiles and future infestation by stored product pests. One or a few small openings in a stored bag are not easy to detect by the grain owner; thus, faulty sacks would not be distinguished from gas-tight ones. Over time, this may lead to damage and losses.

Hermetic vacuum bags

First experiments on vacuum grain storage were reported in the 1930s (Blanc, 1938; Hyde, 1973). Tests with multilayer vacuum bags showed that vacuum can maintain a pressure of the compressed bag material on the grain for many months (Adler et al., 2016). The reduction of oxygen by application of a vacuum leads to faster control of pests infesting the grain and could thus minimize the risk of insects puncturing the bags from inside. In experiments at 20°C, all stages of *Sitophilus granarius* were controlled at a low infestation rate with 0.5 bar vacuum within five weeks, while hermetic storage alone led to control within eight weeks (Adler, unpublished data). At higher temperatures, these times would be shorter. On the other hand, punctured bags would be visible by their softness and the respective grain could be used up first or repacked in a new vacuum bag before any damage occurs.

Figure 13.2 Small hermetic vacuum bags for laboratory testing; hermetic big bags will be tested in Tanzania soon

Photo: C. Adler

Advantages and disadvantages of the various hermetic structures are given in Table 13.2.

Agriculture policies

Agricultural policies targeting smallholder farmers access to markets could reduce postharvest losses caused by biotic and abiotic constraints during storage. For instance, the Eastern Cape provincial government Rural Enterprise Development Hubs (Red Hubs) in South Africa connects production, processing and marketing to improve the competitiveness of rural economies and communities (www.ecrda.co.za). The Red Hub concept creates a communal and external market resulting in increased farm incomes that can be reinvested into the smallholder farming operations such as crop production, product drying, threshing, transportation and storage. This Red Hub concept is important given the increased maize production by smallholder farmers, which has not been matched by investments in storage infrastructure.

Table 13.2 Advantages and disadvantages of hermetic structures

Hermetic structure	Advantages	Disadvantages
Metal silo bin	Solid, long-lasting structures Multiple use Good protection vs. pests	Higher initial investment Risk of condensation
Hermetic metal drum	Smaller, flexible and cheaper structures	Needs protection against theft
Hermetic triple-layer bag	Flexible and cheaper structures Bags can be used for transportation and sale Already available to farmers	Needs regular investment Plastic waste Punctures not easily detected Less protection against birds, rodents and insects (from inside)
Hermetic vacuum bag	Vacuum leads to faster control Flexible and cheaper structures Pressure of material shows protection from pests and decay Grain from damaged bags can be utilized first Big bags available	Vacuum pump needed Plastic waste Less protection against birds and rodents Not yet available for practical use

The Malawi Farm Input Subsidy Programme is the largest known subsidy scheme of chemical protectants which subsidizes maize storage chemicals, inorganic fertilizer and improved seed varieties. However, on average, storage was found to be more profitable with hermetic triple bags compared to chemical storage protectants (Jones et al., 2014), thus indicating the need for partnerships among farmers, researchers and policymakers in order to introduce targeted and most effective interventions that reduce postharvest losses.

Lack of clear knowledge on the extent of postharvest losses and information on losses in agricultural commodities other than cereal grains is a major challenge in Southern Africa. Therefore, agricultural policies should also promote public-private partnerships and cooperation with international development partners that undertake such assessments. For example, the Swiss Agency for Development and Cooperation partnered with Tanzania in a Grain Postharvest Loss Prevention Project for smallholder farmers, while the German Corporation for International Cooperation GmbH (GIZ) has conducted a rapid assessment of food losses along the maize value chains in South Africa, Malawi and Zambia.

Indigenous knowledge systems should be taken into consideration through participatory action research in order for interventions to be more readily adopted in reducing postharvest losses in Southern Africa. Understanding cultural practices is also important, for example, in some countries or regions farmers may prefer to display their harvest openly while in others farmers prefer to keep their harvest indoors. Participatory action research also ensures responsiveness to the needs of both women and men farmers in reducing postharvest losses across agricultural commodities and helps in monitoring the effectiveness of the interventions. From today's perspective, hermetic storage and hermetic vacuum storage appear the most promising technologies in order to reduce postharvest losses, particularly in staple foods produced by smallholder farmers in Southern Africa.

References

Adler, C.S., Ndomo-Moualeu, A.F., Begemann, J. and Münzing, K. (2016) Effect of vacuum storage of wheat (*Triticum aestivum*) grain on the granary weevil, *Sitophilus granarius* and wheat quality. In: S. Navarro, D.S. Jayas, and K. Alagusundaram (Eds.) *Proceedings of the 10th International Conference on Controlled Atmosphere and Fumigation in Stored Products (CAF2016)*. Winnipeg, Canada: CAF Permanent Committee Secretariat, pp. 287–290.

Blanc, A. (1938) Assais de conservation de blé en atmosphère confinée. *Comptes Rendus de l'Academie d'Agriculture de France*, vol 24, pp. 625–630, 71p, supplément.

Hill, R.A., Lacey, J. and Reynolds, P.J. (1983) Storage of barley grain in iron age type underground pits. *Journal of Stored Products Research*, vol 19, pp. 163–171.

Hyde, M.B. (1973) Storage of grain in airtight silos or under vacuum. *Annales de Technologie agricole*, vol 22, no 4, pp. 707–718.

Jones, M., Alexander, C. and Lowenberg-DeBoer, J. (2014) A simple methodology for measuring profitability of on-farm storage pest management in developing countries. *Journal of Stored Products Research*, vol 58, pp 67–76.

Kamanula, J., Sileshi, G.W., Belmain, S.R., Sola, P., Mvumi, B.M., Nyirenda, G.K.C., Nyirenda, S.P. and Stevenson, P.C. (2011) Farmers' insect pest management practices and pesticidal plant use in the protection of stored maize and beans in Southern Africa. *International Journal of Pest Management*, vol 57, no 1, pp. 41–49.

Kankolongo, M.A., Hell, K. and Nawa, I.R. (2009) Assessment for fungal, mycotoxin and insect spoilage in maize stored for human consumption in Zambia. *The Journal of the Science of Food and Agriculture*, vol 89, no 8, pp. 1366–1375.

Nukenine, E.N. (2010) Stored product protection in Africa: Past, present and future. In: M.O. Carvalho, P.G. Fields, C.S. Adler, F.H. Arthur, C.G. Athanassiou, J.F. Campbell, F. Fleurat-Lessard, P.W. Flinn, R.J. Hodges, A.A. Isikber, S. Navarro, R.T. Noyes, J. Riudavets, K.K. Sinha, G.R. Thorpe, B.H. Timlick, P. Trematerra, and N.D.G. White (Eds.) *Proceedings of the 10th International Working Conference on Stored Product Protection.* Estoril, Portugal: Julius Kühn-Archive 425, pp. 26–41.

Riudavets, J., Pons, M.J., Messeguer, J. and Gabarra, R. (2018) Effect of CO2 modified atmosphere packaging on aflatoxin production in maize infested with *Sitophilus zeamais*. *Journal Stored Product Research*, vol 77, pp. 89–91.

Sheahan, M. and Barrett, C.B. (2017) Review: Food loss and waste in Sub-Saharan Africa. *Food Policy*, vol 70, pp. 1–12.

Further readings

http://solarmaize.com/
http://cleanleap.com/new-open-sourced-grain-drying-technology-east-africa
https://postharvest.nri.org

14 Technology as a transformative tool in livestock production

Annelin Molotsi, Tinyiko Edward Halimani,
Phetogo I. Monau and Kennedy Dzama

Introduction

In Southern Africa, livestock production is lagging behind the demand for live-stock products and services due to the increasing human population, urbaniza-tion, competition between food and feed and alternative and competing access to natural resources (land, water, etc.). In addition, there are several production constraints including climate change and the need to operate in a carbon con-strained economy, animal welfare concerns and changing socioeconomic values (Thornton, 2010), lack of resources and skills, endemic and emerging diseases and parasites, lack of record keeping and lack of an enabling policy environ-ment. The Southern African Development Community (SADC) has identified several challenges, including:

> low productivity, lack of efficient and effective animal disease control, lack of marketing infrastructure, poor market access of livestock products, together with lack of availability of information, and other associated factors.
>
> (SADC, 2014)

Southern Africa has an estimated human population of 277 million people. Sixty percent of the land area is non-arable and is mainly used for livestock pro-duction. There are an estimated 64 million cattle, 39 million sheep, 38 million goats, 7 million pigs, 1 million horses and 380 million poultry, 75% of which are produced in extensive smallholder systems (SADC, 2014). Livestock produc-tion in the region ranges from extensive resource-poor smallholder systems to highly organized business-oriented intensive systems. Women and children are the major caretakers of livestock resources in smallholder systems.

Because of these differences in the size and scale of these systems, it will be impossible to prescribe a one-size-fits-all type of solution. There are many technologies that are currently employed in livestock globally and in some parts of the region. However, there is no one technology that can be a silver bullet in transforming the Southern African livestock industry. As a matter of fact, we would argue that simple changes in the way we manage livestock (e.g., changing the herd composition to favour more breeding females in ruminant production) coupled with a few technological and policy interventions are likely to have a bigger impact than simply introducing one or two technologies haphazardly. Many technologies have been introduced into this sector in that manner and have failed spectacularly (Muzari et al., 2012).

Progress in technology and innovation can help in overcoming or mitigating production constraints, especially in developing countries in the global south. Such technology include key multiuse enablers, assisted reproductive technologies, molecular biotechnologies (Jutzi et al., 1999) and social innovation. Social innovation refers to new ideas (products, services and models) that meet social needs more effectively than alternatives and create new social relationships or collaborations. In this case, there is need to create new synergies within the livestock production value chain including sustainable intensification, changes in production objectives and new market linkages.

Enabling technologies

Most critical enablers of livestock systems will be those that enhance information acquisition, analysis, dissemination and use within the value chains. Such enablers include mobile phone technology (information capturing, sharing and management tools), data science, artificial intelligence and machine learning. Though seemingly unrelated to livestock systems, mobile phones are not just about communication but have become a means of enhancing business and socioeconomic connections. This, coupled with an increase in mobile phone penetration in Sub-Saharan Africa (GSMA, 2018), has the potential of improving the livestock systems through reducing market information asymmetry, improving access to production information and enhancing overall participation of hitherto neglected smallholder farmers. Tools can now be developed to enable smallholder farmers, who own most of the livestock in Southern Africa, to participate in livestock recording and traceability schemes, access veterinary and animal health information, access farm management and weather data and, overall, better organize themselves. Such inclusivity will open the window for data science, artificial intelligence and machine learning to be applied to livestock systems in a way that has not been possible before. Mobile phone technology can also be used to spread key information and practices that can transform livestock systems. One of the largest impediments to increasing livestock production in communal areas where large numbers of livestock are kept is the poor offtake associated with these herds. Restructuring the herd composition such that more breeding stock are kept and inclusion of smallholders in the formal markets are important targets here.

Creative use of phone applications (apps) can help in real-time mapping of livestock related issues (diseases and parasites, epidemiology, weather and climate, etc.) that enable quicker interventions. Near real-time livestock health systems have been implemented successfully before, e.g., in Thailand (Yano et al., 2018), and they are evolving (Holmstrom and Beckham, 2017).

Feeds and feeding

Livestock production in Southern Africa is season sensitive due to changes in quality and quantity of feed throughout the year. This unfortunately makes it vulnerable to droughts and climate change. There are many feeding technologies that have been trialled across several countries, but this is an area where there is very little adoption due to competing issues at farm level. There are conflicts between food production on the one hand and feed and forage production on the other. The promotion of proven technologies is usually fragmentary and does not recognize the impact of the technology on the production system as a whole. This is especially true for those aspects of the system that can be negatively impacted by the new technology (Kebebe, 2018). Livestock production cannot improve if the issues related to feeds and feeding are not addressed. Technologies that can be pursued include growing and conservation of forages, use of crop residues, use of biotechnology, grassland and pasture management, new and nonconventional feeds and matching the breeds to available feed resources, including the use of locally adapted and smaller livestock breeds. These are mainly indigenous farm animal genetic resources (AnGR).

Closely related but distinct from the issue of feeds is the issue of water management. Deutsch et al. (2010) and FAO (2018) reported that livestock use approximately 11,900 km^3 of global freshwater annually, which is equivalent of 10% of global water flow. The Southern African region is projected to become increasingly water stressed (UN-ECA n.d.). Climate change is a reality in this part of the world as it is in many other parts of the world (see Chapter 2.3). However, the effects of climate change are likely to be felt much more in this region than in many other parts of the world, primarily because of poor mitigation and adaptation strategies. There is competition for water and increasing human and animal populations will, concomitantly, increase demand and competition (FAO, 2018). In livestock production, water use depends on the size, age, ration consumed, activity, productivity and temperature. Some of these can be managed while others are beyond farmer control. The size of the animal takes the argument back to use of locally adapted AnGR.

Assisted reproductive technologies (ART)

Velazquez (2008) defines assisted reproductive technologies as, "techniques that manipulate reproductive-related events and/or structures to achieve pregnancy with the final goal of producing healthy offspring in . . . females". The technologies have been around for a long time and are being refined and improved. From the definition, ART encompasses a whole suite of unrelated techniques such as manipulation of reproductive physiology (especially female), artificial insemination, multiple ovulation and embryo transfer, in vitro fertilization, cloning, preimplantation diagnostics and sperm sexing. Use of various aspects of ART is common, but not widely adopted, in Southern Africa, particularly oestrus synchronization and artificial insemination (Mugwabana et al., 2018).

However, like other technologies there are social, technological and institutional barriers to its use despite its obvious benefits to herd improvement and breeding.

Molecular biotechnology

Veterinary and nutritional applications

There is a suite of techniques and procedures within molecular biotechnology that lend themselves easily to enhance livestock production and health. These include preventative techniques (vaccines, medicines and metabolic modifiers), nutrition related (in-feed enzymes, probiotics, prebiotics and related products), advanced diagnostics, immuno-castration, etc. (Borroto, 2008). Use of these is confined to the intensive production systems and almost completely absent in the more widespread smallholder systems. There are opportunities for manipulation of the rumen environment given the contribution of ruminants to greenhouse gas emissions.

Genetic selection technologies

For many years, selection for traits of economic importance in livestock has been responsible for the increase in yields that we are observing currently. A classical animal example to mention is that of the poultry industry, where a broiler increased in size at 28 days of age from weighing 0.32 kg in 1957 to 1.39 kg in 2005 (Zuidhof et al., 2014). This was achieved mainly by traditional animal breeding selection. For genetic improvement to be feasible a few things need to be in place. First, there should be genetic diversity within the population. The genetic diversity of smallholder livestock populations in Southern Africa is generally unknown but is directly related to their potential for genetic improvement. Second, good record keeping should be in place. Record keeping allows farmers or breeders to identify the phenotypic variation within the population and also to keep track of changes occurring within the population as a result of the selection applied. Record keeping is important to keep track of the pedigree information. Pedigree allows tracing of gene flow from one generation to the next and to measure its influence on the specific traits of interest that are expressed. With the previously mentioned background in mind we will discuss advanced technologies available in the field of breeding and genetics, the "omics" and the artificial intelligence (AI) revolution as reported earlier.

Genomics

Presently, genomics, proteomics and nutrigenomics are very popular among scientists worldwide. Genomics includes the use of genetic markers whole genome sequencing, RNA sequencing and genome-wide association studies. Genetic markers can be used for various analysis which includes population genetic studies, genetic diversity studies, marker assisted selection identification of causal variants, genome wide association studies and genomic selection (Kijas et al., 2012; Demars et al., 2013; Qwabe et al., 2013). Genomic selection is a tool that can be used to improve genetic gain in animal or plant populations. The inclusion of genotypic data based on genetic markers and specifically single nucleotide polymorphism (SNPs) allow the estimation of genomically enhanced estimated breeding values (GEBVs). Genomic selection involves the use of phenotypic and genotype records to derive GEBVs for traits of economic importance. The use of genomics in livestock will be of benefit to the traits that are lowly heritable and expressed later during an animal's life. This will enable the ability to predict a sire/dam performance for a specific trait at an early age and so increase genetic gain (Kijas et al., 2012) by decreasing the generation interval. The constraints concerning the implementation of these technologies include obtaining reliable records for traits of economic relevance of an appropriate reference population. Another challenge would be the genotyping costs involved, particularly start-up costs, even though those are falling.

The policy environment

Southern Africa is characterized by different levels of development of policies related to livestock production. There are a lot of policy deficits and conflicts that hinder the development of the sector. Some of the key policy gaps can be summarized, as in Table 14.1.

Further, there are no policies in place in any Southern African country which protect the valuable indigenous AnGR intellectual property rights from illegal exploitation and looting.

To illustrate the issues surrounding policy gaps, we look at two contrasting case studies:

> In Malawi there is a new livestock policy (as of 2019) being crafted which is almost complete but not yet available for publication. In discussions with the national focal point and key experts and opinion makers in AnGR it was very clear that the new policy deals with issues around promotion and utilisation of indigenous AnGR. The country has no formal selection and breeding program and the hope was that one would be set up once funding was made available. The major benefit from setting up such a scheme would be to further highlight the benefits of local AnGRs.

In South Africa, the Livestock Development strategy developed a decade ago by the Department of Agriculture is in place. It recognizes the value of genetic evaluation programmes for improvement of livestock. It also recognises adapted local AnGRs but fell short of putting together a strategic plan for utilization of these resources. Selection and breeding programmes are further reinforced by more important legislation like the Animal Improvement Act 62 of 1998

Table 14.1 An analysis of the policy environment and key strategies to mitigate deficits

Policy gap	Strategy
Lack of livestock policies (including improvement and conservation policies) Lack of regulations/regulatory bodies to monitor and evaluate the impact of breeding programmes and other interventions Non-functional policies due to poor implementation environment	Better implementation of recommendations by the SADC Food, Agriculture and Natural Resources (FANR) Directorate's recommendations to "develop, promote, coordinate and facilitate the harmonisation of policies and programmes in order to increase agricultural and natural resources production and promote trade, and ensure food security and economic development in the region on a sustainable basis" (SADC, 2011)
Lack of harmonization of government legislation and policies (within and between ministries and departments)	Complete review of policies within and between countries to ensure policy consistency and coherence
Lack of strategic plans and action plans and budget allocation for livestock research, animal genetic resource conservation and management	Develop strategic and action plans to implement various livestock related policies, in particular, those that are related to sustainable use and conservation of animal genetic resources
Lack of ring-fenced budget allocation for animal genetic resources by national governments	Invest more in agriculture as envisaged in the Maputo Declaration (AU-NEPAD, 2003)

and Genetically Modified Organisms Act 15 of 1997. A major weakness of these important pieces of legislation is that they do not specifically mention indigenous AnGRs. The Animal Improvement Act creates the Office of the Registrar of Animal Improvement, who oversees the implementation of the act. In discussions with key policy and technical informants, it was evident that even though there were important tools and policies around genetic evaluation, the implementation seemed to leave a lot to be desired in particular due to lack of capacity to execute these instruments.

Potential positive and negative impacts

Positive impacts

Increasing animal yields from smallholder farmers, can result in an increase in income and food security for Southern Africa. Increased consumption of animal-sourced foods can be highly beneficial to households. The positive impacts of suggested technologies can be felt more strongly if they are targeted at livestock species kept by resource-poor farmers including women and youths such as poultry and small ruminants. Furthermore, sufficient local production can curb imports of animal products and help to stabilize regional economies with an added benefit of saving foreign exchange.

Negative impacts

Employing these technologies are expensive and the money could be used elsewhere. As has been observed with commercial agriculture, increasing production through increased technology can have negative effects on the environment (Lichtenberg, 2000). By implementing strict selection to favour specific trait, genetic diversity can be lost; therefore, it is important to ensure that genetic diversity is maintained within genomic selection programmes. Some technologies could result in a decrease of manual labour needed on farms resulting in joblessness and poverty.

References

AU-NEPAD. (2003) *Comprehensive Africa Agriculture Development Programme (CAADP)*, Available at: www.nepad.org/file-download/download/public/15057.

Borroto, C.G. (2008) Biotechnology and its application to veterinary science. In *19th Conference of the OIE Regional Commission for the Americas (Havana, Cuba, 17–21 November)*. Havan, Cuba, pp. 231–240. Available at: www.oie.int/doc/ged/D6070.PDF [Accessed March 14, 2019].

Demars, J. et al. (2013) Genome-wide association studies identify two novel BMP15 mutations responsible for an atypical hyperprolificacy phenotype in sheep. *PLOS Genetics*, 9(4), p. e1003482. https://doi.org/10.1371/journal.pgen.1003482 [Accessed March 14, 2019].

Deutsch, L., Falkenmark, M., Gordon, L., Rockström, J. and Folke, C. (2010) Water-mediated ecological consequences of intensification and expansion of livestock production. In: Steinfeld, H., Mooney, H.A., Schneider, F. and Neville, L.E. (eds.). *Livestock in a Changing Landscape*. Washington, DC: Island Press, pp. 97–111.

FAO. (2018) *Water Use of Livestock Production Systems and Supply Chains Guidelines for Assessment (Draft for Public Review)*. Rome: Livestock Environmental Assessment and Performance (LEAP) Partnership, FAO. Available at: www.fao.org/3/I9692EN/i9692en.pdf [Accessed March 14, 2019].

GSMA. (2018) *The Mobile Economy: Sub-Saharan Africa 2018*. Available at: www.gsmaintelligence.com/research/?file=809c442550e5487f3b1d025fdc70e23b&download [Accessed March 13, 2019].

Holmstrom, L.K. and Beckham, T.R. (2017) Technologies for capturing and analysing animal health data in near real time. *Revue Scientifique Et Technique De L Office International Des Epizooties*, 36(2), pp. 525–538. Available at: https://pdfs.semanticscholar.org/b262/d9b13e81bd723a60618394c7513f0862cab2.pdf [Accessed March 14, 2019].

Jutzi, S.C., Otte, J. and Wagner, H-G. (1999) *The Potential Impact of Biotechnology on the Global Livestock Sector*. Rome. Available at: www.fao.org/ag/againfo/resources/en/publications/agapubs/pproc04.pdf.

Kebebe, E. (2018) Bridging technology adoption gaps in livestock sector in Ethiopia: A innovation system perspective. *Technology in Society*. Available at: https://doi.org/10.1016/j.techsoc.2018.12.002 [Accessed March 14, 2019].

Kijas, J.W. et al. (2012) Genome-wide analysis of the world's sheep breeds reveals high levels of historic mixture and strong recent selection. *PLoS Biology*, 10(2), p. 1001258.

Lichtenberg, E. (2000) Agriculture and the environement. Working Paper No. 00–15. Available at: https://tind-customer-agecon.s3.amazonaws.com/47823c9d-52b6-4197-bfab-612d913559a0?response-content-disposition=inline%3B%20filename%2A%3DUTF-8%27%27

wp00–15.pdf&response-content-type=application%2Fpdf&AWSAccessKeyId=AKIAXL
7W7Q3XHXDVDQYS&Expires=1559052062&Signature=WfziK23q1LZGOAJjoMZr
EteECCc%3D [Accessed May 28, 2019].

Mugwabana, T.J. et al. (2018) The effect of assisted reproductive technologies on cow productivity under communal and emerging farming systems of South Africa. *Journal of Applied Animal Research*, 46(1), pp. 1090–1096. Available at: www.tandfonline.com/action/journal Information?journalCode=taar20 [Accessed March 14, 2019].

Muzari, W., Gatsi, W. and Muvhunzi, S. (2012) The impacts of technology adoption on smallholder agricultural productivity in Sub-Saharan Africa: A review. *Journal of Sustainable Development*, 5(8). Available at: http://dx.doi.org/10.5539/jsd.v5n8p69 [Accessed May 28, 2019].

Qwabe, S.O., van Marle-Köster, E. and Visser, C. (2013) Genetic diversity and population structure of the endangered Namaqua Afrikaner sheep. *Tropical Animal Health and Production*, 45(2), pp. 511–516. Available at: http://link.springer.com/10.1007/s11250-012-0250-x [Accessed March 14, 2019].

SADC. (2011) *Regional Agricultural Policy (RAP) Country Summary Agricultural Policy Review Reports*. Available at: www.sadc.int/files/7113/5293/3509/Regional_Agricultural_Policy_Review_Reports_2011.pdf [Accessed March 25, 2019].

SADC. (2014) *Southern African Development Community: Livestock Production*. Available at: www.sadc.int/themes/agriculture-food-security/livestock-production/ [Accessed March 14, 2019].

Thornton, P.K. (2010) Livestock production: Recent trends, future prospects. *Philosophical Transactions of the Royal Society B: Biological Sciences*, 365(1554), pp. 2853–2867.

UN-ECA. *Climate Change and the Rural Economy in Southern Africa: Issues, Challenges and Opportunities*. Available at: www.uneca.org/sites/default/files/PublicationFiles/climate-change-and-the-rural-economy-in-southern-africa.pdf [Accessed March 14, 2019].

Velazquez, M.A. (2008) Assisted reproductive technologies in cattle: Applications in livestock production, biomedical research and conservation biology. *Annual Review of Biomedical Sciences*, 10(February), pp. 36–62.

Yano, T. et al. (2018) A participatory system for preventing pandemics of animal origins: Pilot study of the Participatory One Health Disease Detection (PODD) system. *JMIR Public Health and Surveillance*, 4(1), p. e25. Available at: www.ncbi.nlm.nih.gov/pubmed/29563079 [Accessed March 14, 2019].

Zuidhof, M.J. et al. (2014) Growth, efficiency, and yield of commercial broilers from 1957, 1978, and 2005. *Poultry Science*, 93, pp. 1–13. Available at: http://dx.doi.org/ 10.3382/ps.2014-04291.

15 Technologies for agricultural transformation

Animal health

Delia Grace

Introduction

Animals in Southern Africa are essential for nutrition, income, livelihoods and ecosystem services. However, animal diseases are a threat both to the performance of the livestock sector and to the ability of countries to benefit from wildlife resources. Animal diseases can also have direct and indirect impacts on human health. In this chapter we first summarize the different categories of animal disease and their relevance to Southern Africa. Next, we review the technologies that are presently available, and which are being taken up, or could be taken up, by livestock keepers. We then discuss the approaches that can help close the gap between technologies and widespread adoption. Finally, we make recommendations for research and policy to overcome these barriers.

Priority animal diseases in Southern Africa

As is the case in many low- and middle-income countries (LMIC), information on the burden of animal disease in Southern Africa is lacking. Here we draw on a survey of state veterinary services (Grace et al., 2015a), the information officially reported to the World Animal Health Organisation (www.whahid.org) and the literature (Grace et al., 2012; World Bank, 2012) to identify the priority diseases under different categories.

- Epidemics are defined as occurrence of a certain disease in a population at levels higher than expected. The most important livestock epidemic diseases are caused by rapidly transmitting pathogens that produce acute and serious disease in large numbers of hosts. According to state veterinary services, the priority epidemic disease in Southern Africa is foot and mouth disease (FMD) followed by contagious bovine pleuropneumonia (CBPP). Other priority epidemics are *peste des petits ruminants* (PPR), Newcastle disease and lumpy skin disease (LSD).
- Endemic diseases are constantly present in a population. Although livestock endemic diseases are less dramatic than epidemics, some believe that the overall impact is greater. Even though a disease is endemic in an area,

seasonal or sporadic outbreaks may occur. Endemic diseases important in Southern Africa are clostridial diseases, ticks and tick-borne diseases (TTBD), helminth infections and African animal trypanosomosis (AAT).

• Zoonoses are animal diseases that are transmissible to people. Over 60% of human pathogens are zoonotic (Taylor et al., 2001), but a smaller number of zoonoses are responsible for most illness. The priority zoonoses in Southern Africa are rabies, followed by brucellosis and anthrax. Other zoonoses are emerging. For these diseases, human infection is currently rare, but as these pathogens evolve they may become better adapted to humans: priority emerging diseases are highly pathogenic avian influenza (HPAI) and Rift Valley fever (RVF).

• More than half of the priority foodborne diseases are zoonotic (Havelaar et al., 2015), and animal source foods are an important source of both zoonotic and foodborne diseases. The human health impact of foodborne disease is comparable to that of HIV/AIDs, malaria or tuberculosis (Havelaar et al., 2015). The economic costs for LMICs are at least US$115 billion a year (Jaffee et al., 2018). Foodborne disease is likely to worsen in Southern Africa over the next decades (Grace, 2015).

• The priority wildlife diseases are FMD and anthrax, which are also diseases of livestock and the priority aquatic diseases is epizootic ulcerative syndrome.

The distribution of infectious diseases (human, animal and plant) and the timing and intensity of disease outbreaks is often closely linked to climate and weather. Associations are strongest for diseases that are vector-borne, soil associated, water or flood associated, rodent associated or air temperature/humidity associated, and most of the priority animal diseases in Southern Africa are considered climate-sensitive (Grace et al., 2015b).

Technologies for better managing animal disease

Technological advances have revolutionized our ability to detect, diagnose, cure and prevent animal diseases.

Diagnostics are used to understand infection and epidemiology, in monitoring disease, in discovering pathogens, in developing and evaluating control strategies and in treating individual animals. Advances in diagnostics include the use of recombinant technology, the development of lateral flow tests and real-time polymerase chain reaction (PCR) on field lab platforms (Howson et al., 2017). However, many of these tests are not yet routinely used or commercially available. Their introduction will depend upon investment in the technology, leading to performance and cost advantages over the existing approaches used to control disease outbreaks which in turn depends upon developing a commercial market.

In tropical developing countries, there is also a need for field-friendly diagnostic tests. A good example of this is the FAMACHA test for anaemia in sheep to detect animals who need treatment for haemonchosis. Animals are restrained, and the eyes are examined and scored against a standardized set of five colours ranging from red-pink (normal) to white (terminal anaemia). Developed for use in sheep in South Africa, the method has been extended to other animal systems and used in other countries (O'Brien et al., 2018).

Vaccines are one of the most effective means of controlling disease, and there are more than 300 veterinary vaccines registered around the world (Barrett, 2016). Although vaccines exist for many priority diseases, technological advances can improve uptake and usability. Thermostable vaccines exist for Newcastle disease and are under development for other diseases. "DIVA" (differentiation of vaccinated from naturally infected animals) vaccines allow vaccinated and infected animals to be distinguished so the latter can be culled. Molecular epidemiology allows the development of vaccines that are safer and cheaper and give long-lasting immunity. Insertion of protective antigens into a live but apathogenic vector organism (vector-based vaccines) has been used successfully against viral diseases but are still only emerging for bacterial diseases. Multivalent vaccines can protect against several diseases and are attractive to farmers.

There is also a rapidly growing concern about increasing antimicrobial resistance in human pathogens. One landmark study predicted that by 2050, 10 million deaths worldwide will be attributable to antimicrobial resistance (O'Neill, 2016). The use of antimicrobials in agriculture is considered to contribute to this, and there is much interest in innovations that would allow reduction of antimicrobials in livestock. As well as vaccines, research in Africa is investigating the potential of prebiotics, probiotics, phages, heavy metals, phytochemicals, organic acids, engineered peptides, nanoantibiotics, highly effective chicken and plant immunoglobulins and genetically resistant animals (Marquardt and Li, 2018).

Information and communication technologies (ICT) and eAgriculture has been one of the fastest growing areas in recent years and has many applications to animal health (also see Chapter 4.4 in this volume). Several projects in Africa have used mobile telephones to send information to producers and to support disease reporting. Use of electronic tags and readers can transform paper-based livestock traceability systems into an ICT-compliant system that is more secure and transparent. Model systems have been used in South Africa and Namibia and are considered to have wide applicability (Gitonga, 2017). Blockchain also has potential to revolutionize livestock value chains.

Accurate information on presence, level and impacts and the costs for controlling disease is needed to plan disease control. Disease surveillance is an information-based activity that involves collection and analysis of information on disease occurrence. Well-functioning surveillance systems and timely responses may reduce the cost of outbreaks by 95% (Grace, 2014). Most developing countries currently lack capacity to detect diseases. Promising surveillance and reporting opportunities for poor countries include:

- Risk based (targeted) surveillance: traditional surveillance assumes that the probability of disease is constant across all individuals, but this is rarely the case. By concentrating surveillance on the diseases, sectors, sub-populations or areas most likely to be affected, costs can be reduced and efficiency increased.

- mSurveillance: mobile phones have reached widespread cover in developing countries. Pilot programmes involving veterinarians, community animal health workers and farmers have been successful in several countries.
- Participatory disease surveillance (PDS) was originally developed in Africa to harness the skills of local communities in detection and reporting of rinderpest. It has subsequently been used for several diseases including avian influenza. It often reaches further and costs less than traditional surveillance. However, reports typically require confirmation by other means.
- Satellite data are increasingly being used to aid disease prediction, especially for those diseases that occur in epidemics such as Rift Valley fever. There is huge potential to calibrate these data, based on the local Meteorology Station data, so they can be used in short-term disease prediction and longer-term forecasting. These can be combined with mathematical models to better understand options for disease control.

Translating technologies to better animal health

Technologies by themselves will not improve animal health. Mechanisms are needed by which they can be deployed at scale. Recent years have seen the development of a series of approaches that bring together processes, technologies with enabling policies and incentives to bridge the gap between innovation and adoption. Three key approaches are sustainable intensification, progressive disease control and risk-based approaches for food safety.

Sustainable intensification implies increasing livestock productivity but not at the expense of the environment, or economic or social well-being. Countries in Southern Africa, like many LMIC, are forecast to experience significant growth in demand for livestock and fish products over the next decades. At the same time, there is increasing concern over the environmental externalities of livestock especially their contribution to greenhouse gas, pollution and environmental degradation. In Southern Africa, much of the livestock is kept by smallholders or by farmers who keep large numbers of animals but operate low-input, low-output systems. For these farmers, intensifying farming – as opposed to industrializing it – can be supported by adoption of a package of technologies. These include enhancing feed, better matching genetics with environment and improving health (ILRI, 2019). Producing more from less can also reduce the per kilogram carbon footprint of livestock products.

Progressive disease control, with the ultimate aim of eradication, has become prioritized as a result of the successful eradication of rinderpest or cattle plague. This catastrophic disease of ruminants was the second disease to be eradicated from the planet (after smallpox).

Eradication led to nearly a billion dollars in annual economic benefits in Africa alone, bringing immense benefits to livestock keepers. Global eradication may not always be feasible, but many diseases can be controlled by a combination of treatment, vaccination, culling and reduction of transmission. Control is usually staged with initial measures used to reduce prevalence progressing to more rigorous and expensive methods to eliminate infection. These staged approaches bring together stakeholders to develop a road map for control. They have been developed for foot and mouth disease (OIE and FAO, 2012), trypanosomosis (Diall et al., 2017), cysticercosis and other priority diseases present in Southern Africa. Control activities are most advanced for peste des petits ruminants (PPR) and rabies (Jarvis, 2016). PPR eradication is expected to cost US$2.26 billion over 15 years, which will create US$76.5 billion in benefits to farm communities, nearly 34 times the original investment and equal to 25% of the annual agricultural output of sub-Saharan Africa (Jones et al., 2016).

South Africa experienced the world's largest ever recorded outbreak of listeria (Listeria monocytogenes), with 209 human deaths between January 1, 2017, and June 5, 2018. Domestically processed ready-to-eat meat was identified as the probable source (Hunter-Adams et al., 2018). Managing food safety is best done through use of risk analysis: this combines risk assessment (what is the risk to human health?), risk management (what best to do about it?) and risk communication (the two way and iterative engagement among stakeholders). Although the gold standard for managing food safety, risk analysis has not been widely adopted in LMICs. In the last decade, participatory methods have been developed to make risk analysis easier to apply and have been successfully used in several countries in Southern Africa including Tanzania, Mozambique and South Africa (Roesel and Grace, 2014).

Policy and processes to improve animal health in Southern Africa

The increasing importance of the human health externalities of agriculture, including emerging diseases, zoonoses and antimicrobial resistance, means animal health has to go beyond impacts on livestock and fish. The best practice for managing these is an approach known as "One Health" or Ecohealth. This assumes that the health of humans, animals and the planet are interdependent and problems at the intersection of human and health require solutions based on cross-disciplinary collaborations. Community animal health programmes have been successfully implemented in many countries but require an enabling national animal health policy, which is not always present. Governments can establish and support cross-ministerial One Health units, apply One Health methods to the control of zoonotic diseases and AMR and support community-based animal health services (Munyua et al., 2016).

Societies around the world increasingly recognize the obligation to treat animals humanely. Animal welfare fits naturally into health discussions: poor animal health causes great animal suffering, and reduction in animal disease also reduces disease in humans. In addition, animal welfare is related more broadly to livestock production. Providing adequate nutrition, husbandry and housing for livestock is critical for their welfare as well as for their productivity. Adequate livestock transport and competent slaughter processes reduce both animal suffering and losses from damaged carcasses. In developing and emerging economies, improvements in livestock welfare often simultaneously improve livestock productivity, presenting a win–win opportunity. Governments need to ensure animal welfare legislation is present and raise awareness on the need for, and benefits of, improving animal welfare,

Veterinary Services (VS) comprise all actors, public and private, who collaborate in the domain of animal health under the overall control and direction of the Chief Veterinary Officer. Veterinary services are a global public good and are essential to safeguarding and improving the health of animals and animal-related health and nutrition of people. They are essential to global trade in livestock and livestock products. There is considerable evidence that these services have been underinvested in (OIE, 2019) and that adequate funding of VS has considerable benefits for animal and human health (Jaffee et al., 2018). The Performance of Veterinary Service Pathways supported by the OIE offers an appropriate and sustainable way for Southern African countries to strengthen VS (see www.oie.int/solidarity/pvs-evaluations/).

Conclusions

We are currently in an era of unprecedented interest and advances in animal disease research. This livestock sector is growing rapidly in response to demand, and the consequent intensification brings about needs for new and adapted technologies. Advances in epidemiology, molecular epidemiology, genomics, diagnostics, vaccines and ICT have great potential for controlling disease and improving productivity in livestock. At the same time, the growing concern about human health externalities of livestock production (especially emerging zoonotic disease, foodborne disease and antimicrobial resistance), substandard animal welfare and the environmental impact of livestock is stimulating new investments in research to tackle these problems. Much of the growth in demand for livestock products and generation of negative externalities occurs in LMIC, and these will be at the forefront of future research. A One Health perspective that understands the importance of livestock in the context of animal, human and environmental health can help ensure a sustainable transformation of the livestock sector.

References

Barrett, A. (2016) 'Vaccinology in the twenty-first century', *NPJ Vaccines*, p. 1.

Diall, O., Cecchi, G., Wanda, G. et al. (2017) 'Developing a progressive control pathway for African animal trypanosomosis', *Trends in Parasitology*, vol 33, no 7, pp. 499–509.

Gitonga, P. (2017) 'Livestock management, identification and traceability solution', unpublished, https://doi.org/10.13140/rg.2.2.28790.19523.

Grace, D. (2014) 'The business case for one health', *The Onderstepoort Journal of Veterinary Research*, vol 81, p. 2.

Grace, D. (2015) 'Food safety in low and middle income countries', *International Journal of Environmental Research and Public Health*, vol 12, no 9, pp. 10490–10507.

Grace, D., Bett, B., Lindahl, J. et al. (2015b) 'Climate and livestock disease: Assessing the vulnerability of agricultural systems to livestock pests under climate change scenarios', CCAFS Working Paper no. 116., *CGIAR Research Program on Climate Change,* Agriculture and Food Security, Copenhagen.

Grace, D., Mutua, F., Ochungo, P. et al. (2012) *Mapping of poverty and likely zoonoses hotspots,* International Livestock Research Institute, Nairobi.

Grace, D., Songe, M. and Knight-Jones, T. (2015a) 'Impact of neglected diseases on animal productivity and public health in Africa', paper written for 21st conference of the *World Organisation for Animal Health (OIE) Regional Commission for Africa, Rabat, Morocco, 16–20 February,* International Livestock Research Institute, Nairobi.

Havelaar, A.H., Kirk, M.D., Torgerson, P.R. et al. (2015) 'World health organization global estimates and regional comparisons of the burden of foodborne disease in 2010', edited by von Seidlein, L., *PLOS Medicine, Public Library of Science (PLoS)*, vol 12, no 12, p. e1001923.

Howson, E., Soldan, A., Webster, K. et al. (2017) 'Technological advances in veterinary diagnostics: Opportunities to deploy rapid decentralised tests to detect pathogens affecting livestock', *Revue Scientifique et Technique de l'OIE*, vol 36, no 2, pp. 479–498.

Hunter-Adams, J., Battersby, J. and Oni, T. (2018) 'Fault lines in food system governance exposed: Reflections from the listeria outbreak in South Africa', *Cities & Health*, vol 2, pp. 17–21.

ILRI. (2019) *Options for the livestock sector in developing and emerging economies to 2030 and beyond,* Meat: The Future – Economic Forum, Geneva.

Jaffee, S., Henson, S., Unnevehr, L. et al. (2018) *The safe food imperative: Accelerating progress in low- and middle-income countries,* Agriculture and Food Series, World Bank, Washington.

Jarvis, S. (2016) 'Aiming for elimination of dog-mediated human rabies cases by 2030', *Veterinary Record, BMJ*, vol 178, no 4, pp. 86–87.

Jones, B.A., Rich, K.M., Mariner, J.C. et al. (2016) 'The economic impact of eradicating peste des petits ruminants: A benefit-cost analysis', edited by Chakravortty, D., *PLoS One, Public Library of Science (PLoS)*, vol 11, no 2, p. e0149982.

Marquardt, R.R. and Li, S. (2018) 'Antimicrobial resistance in livestock: Advances and alternatives to antibiotics', *Animal Frontiers*, Oxford University Press (OUP), vol 8, no 2, pp. 30–37.

Munyua, P., Bitek, A., Osoro, E., Pieracci, E.G., Muema, J., Mwatondo, A. et al. (2016) Prioritization of zoonotic diseases in Kenya, 2015. *PLoS One*, vol 11, no 8, p. e0161576.

O'Brien, D., Schoenian, S., Semler, J., Gordon, D. and Bennett, M. (2018) 'PSVI-19 Consistency of FAMACHA© scores to fecal egg counts and gain in meat goat kids', *Journal of Animal Science*, Oxford University Press (OUP), vol 96, suppl_3, pp. 460–461.

OIE. (2019) https://www.oie.int/fileadmin/Home/eng/Media_Center/docs/pdf/SG2018/PVS_BUSINESS_CASE_FINAL.pdf.

OIE and FAO. (2012) *The global foot and mouth disease control strategy,* Food and Agriculture Organisation of the United Nations, Italy and World Organisation for Animal Health, Paris.

O'Neil, J. (2016) Tackling drug-resistant infections globally: Final report and recommendations. *The Review on Antimicrobial Resistance.* https://amr-review.org/publications (accessed 2016), 84p.

Roesel, K. and Grace, D. (2014) *Food safety and informal markets: Animal products in Sub-Saharan Africa*, Routledge, Oxford.

Taylor, L.H., Latham, S.M. and Woolhouse, M.E.J. (2001) 'Risk factors for human disease emergence', edited by Woolhouse, M.E.J. and Dye, C., *Philosophical Transactions of the Royal Society of London. Series B: Biological Sciences*, The Royal Society, vol. 356, no 1411, pp. 983–989.

World Bank. (2012) *People, pathogens and our planet*, World Bank, Washington.

Part IV

Emerging technologies

Emerging technologies that are relatively new and that could be used in transforming agriculture in Southern Africa in the near future are outlined in this section. In some cases, these emerging technologies are not yet registered for use by Southern African national government agencies; in other cases, they will require industrial investment for development.

16 Harnessing ecosystem services in transforming agriculture in Southern Africa

Barbara Gemmill-Herren, Florence Mtambanengwe, Paul Mapfumo, Gisèle L. Herren, Tlou S. Masehela, Philip C. Stevenson and Jeremy K. Herren

Introduction

Southern Africa is a region that is extraordinarily rich in natural resources and biodiversity and is equally an area of striking food insecurity. Agriculture in Southern Africa has evolved over tens of thousands of years, and for most of this time it has been small-scale, labour intensive and low-tech. Over the last half-century or more, however, new forms of agriculture have emerged which make extensive use of inputs – improved seeds, fertilizers and pesticides – to increase production to meet the food needs of a growing global population. But there is growing evidence that these agricultural techniques – both in rich and poor countries – are helping to undermine the natural resource base of the communities and economies that depend upon them, including extensive degradation of soils (IPBES, 2018). In many regions of Southern Africa, conventional high-input agriculture has not taken hold. In many such regions, resource-poor farmers contend with issues of marginal high-risk environments and experience poor yields just where food security is most vulnerable.

In contending with these issues, other pathways than conventionally intensified agriculture are not just possible but increasingly becoming an imperative in ensuring sustainability. In transforming agriculture in Southern Africa, numerous approaches exist to work with nature, rather than against it, in harnessing key biological processes that sustain and enhance production while also generating other multiple benefits.

These sets of processes are collectively known as "ecosystem services" (ES), defined as "the conditions and processes through which natural ecosystems, and the species that make them up, sustain and fulfil human life" (Daily, 1997). In this chapter we review three which directly underpin agricultural production: soil, pollination and natural pest regulation services. We explore the current state of knowledge of these services, in Southern Africa where possible and beyond where relevant. We ask how farmer access to knowledge of these services can be better promoted and disseminated, given that all of them are knowledge-intensive, rather than input-intensive routes to sustain productivity. And finally, for each ES, we review existing research needs.

Approaches to harnessing soil ecosystem services in Southern Africa

For a smallholder farmer in Southern Africa, soil remains a key natural resource-base for the provision of food production. Soil is fundamental to a wide range of ES – from provisioning services including food, fibre and fuel production, through regulating services (carbon sequestration), cultural services (e.g., building materials, pottery making) and supporting services (source and sink of essential plant growth nutrients) (Millennium Ecosystem Assessment, 2005). Farmers can enhance these services for direct and indirect benefits to their daily livelihoods through agricultural activities.

Agricultural management practices known to positively impact soils and their provisioning of ES include reducing soil disturbance, maintaining ground cover, using organic amendments, optimizing timing and rate of fertilizer application, water management and improved grazing management (reviewed in Smith et al., 2015). The actual mechanism by which practices impact or enhance soil ecosystem services and overall ecosystem resilience is mediated by the response of soil organism to these practices. Targeted management of soil community composition through the concept of soil ecological engineering is a promising approach to enhance agricultural sustainability (Bender et al., 2016). Soil management practices such as cover crops and no–till can impact microbial communities in ways that enhance stress tolerance and resilience (Schmidt et al., 2018).

In Southern Africa, farmers enhance soil productivity by application of organic nutrient resources such as cattle manure, woodland litter, compost and crop residues (Mapfumo and Giller, 2001), practices that in turn boost the soil nutrient reserves and enhance soil quality. To use compost efficiently, attention must be given to the choice of feedstocks and the management of the composting process (Bernal et al., 2017).

Two soil-related ES that have wider societal benefits across a landscape are carbon sequestration and watershed functions. Carbon fixation in agroecosystems is primarily a function of crop type, density and/or mixtures, a common practice in the region. Farmers' practices of soil conservation through organic mulching and/or cover crops (Ngwira et al., 2012) may have positive downstream effects of provision of clean water through erosion control.

Promoting farmer access to knowledge on harnessing soil ecosystem services through policy in Southern Africa: Communities in Southern Africa have long recognized the value of soil ecosystems as reflected in traditional soil management and agricultural production systems and practices. But rising concerns about sustainability of contemporary technologically driven agricultural production systems in the face of land degradation and a decline in the natural resource base upon livelihoods for many rural and peri-urban populations in the world has triggered a strategic research focus on ecosystems services over the last two decades (e.g., Barrios, 2007; Adhikari and Hartemink, 2016). There is a need for global attention to the value of ecosystem services. Three main agricultural approaches in Southern Africa that build on ecosystem services are:

1   Conservation agriculture (CA), seeking to minimize soil tillage, maximize cover and promote agricultural diversification
2   Integrated soil fertility management (ISFM), seeking to increase soil productivity and achieve more efficient nutrient cycling at field and farm scales through better targeting and allocation of production resources – including combinations of organic and inorganic fertilizers, selection of appropriate crop types and varieties and prudent agronomic practices
3   Agro-ecology, seeking to deepen the application of ecological principles in agricultural production systems to harness natural systems and processes otherwise underpinning resilience and sustainability natural ecosystems for the benefit of diversified agriculture

While farmer-centred research and extension approaches hold promise in promoting these knowledge-intensive and complex technical approaches, the response by policymakers has not been consistent and is often still entrenched in top-down agricultural and environmental management policies of "Green Revolution" type production packages, despite their failures in turning the region into a breadbasket case overall. Development of transformative agricultural and environmental policies informed by emerging evidence on the value of soil ecosystem service is a critical work in progress for Southern Africa.

Approaches to managing pollinator forage for pollination services in Southern Africa

Animal-mediated pollination is a regulating ecosystem service, both for natural ecosystems and for agriculture. The service is provided by pollinators, primarily wild and managed wild bees and other insects. Recent global assessments have stressed the importance of this service, with nearly 90% of all the flowering plants of the world depending, in part if not entirely, on animal pollination (IPBES, 2016). Declines of pollinators have been noted in many regions, although thorough documentation tends to be limited to Europe and North America. Simultaneously, global agriculture has become increasingly pollinator-dependent (Aizen and Harder, 2009).

The Southern African region encompasses a diversity of environments that support a wide variety of pollinators across various taxa. The abundance, diversity and richness of these taxa depends on their interaction with natural ecosystems whereby requirements tend to differ among species depending on their biology, behaviour and even migration patterns (Chapman et al., 2011).

For forage, both nectar and pollen providing plants need to be both diverse and of high quality throughout the pollinator's life cycle. It is not always possible to secure the full suite of resource needs for pollinators within a farm; for example, some farms have little to no natural vegetation around them to support the pollinators. The forage and nesting requirements for these pollinators will have to be met elsewhere within the surrounding landscapes or adjacent environments. Thus is not always easy to ensure due to competing land uses, such as the demand for land to cater for urban development, housing and even expanding agricultural activities that may provide little resources for pollinators (Sayer et al., 2013). Thus, managed pollinators such as honey bees become crucial in providing pollination services for agriculture (Masehela, 2017). Even if their forage needs may be challenged due to environmental pressures, managed honey bees can be moved around in hives over long distances to compensate for lack of forage in certain areas.

Forage for pollinators can be preserved or even enhanced by 1) not burning, cutting or applying herbicides to bee-friendly plants (weeds) unnecessarily and 2) encouraging the planting of multiple crops and trees with good nectar and pollen rewards on farms, in home gardens and in open park spaces. The ultimate goal is for a holistic approach in planning and management that supports and promotes pollinator friendly land use practices at all times.

Promoting farmer access to knowledge on harnessing pollination services through policy in Southern Africa: Agriculture and environmental policies must recognize the importance of pollinators for food production and environmental sustainability. To communicate these values to stakeholders in Southern African, regional examples that resonate with farmers and their daily practices are needed. Key policy messaging should address overall health, habitat requirements and management of pollinators in different landscapes across farm levels. The knowledge and information that farmers need to manage pollinator forage resources both within their farms but also across landscapes needs to be practical and flexible to respond to the many different interfaces of farmers with natural areas. Conveying the rather complex ecological information of managing forage resources across landscapes, in a way that is both timely and scientifically valid, is a challenge. Extension officers, environmental educators, farmer associations and nongovernmental organizations, as key communicators of such messages, need to take into account different literacy levels, language barriers across regions, various forms of media distribution and the relevance and applicability of such knowledge to farmers.

Approaches to managing natural pest control services in Southern Africa

Biodiversity underpins agricultural ecosystem services and food security, livelihoods and economic development by provisioning natural enemies of crops pests (Gurr et al., 2012). Field margins and non-crop agricultural habitat, if well managed, can provide food (alternative prey and nectar) and refuge for predators and parasitoids, increasing diversity (Bianchi et al., 2006) and abundance (Chaplin-Kramer et al., 2011) of natural enemies and consequently enhances natural pest regulation. Yet, agricultural intensification in Southern Africa has led to biodiversity losses with consequences for the ecosystem services of natural pest control.

Nonetheless, there is substantial evidence that natural pest control services can be enhanced through management practices in Southern Africa. For example, natural pest regulation (NPR) of stem borer pests contributes substantial economic benefit to maize production in Southern and East Africa (Midingoyi et al., 2016), while the field release of the ichneumonid wasp *Diadegma semiclausum* reduced diamondback moth (*Plutella xylostella* L.) damage on cabbage by 50% in farms where chemical insecticides were not applied (Kennedy et al., 2016). The role of non-crop habitats in supporting beneficials is less well resolved. One example reported that Tephritid fruit fly abundance increased and rates of parasitism decreased in Mango orchards with distance from non-crop habitats (Henri et al., 2015), although the role of non-crop habitats in supporting longer-term control was not determined.

Landscape heterogeneity can enhance NPR, although the specific drivers still need to be established in order to develop appropriate interventions. For example, in the wine growing areas of the Cape floristic region, old fields provide high plant and prey diversity and subsequently natural enemies (Gaigher et al., 2016a), although this doesn't guarantee natural spill over into adjacent crops (Gaigher et al., 2016b). While natural enemy abundance and NPR are limited by dispersal this could be compensated for by landscape management; it is the quality and complexity of field margins along with the spatial arrangement that facilitate natural enemy dispersal to agricultural land (Griffiths et al., 2008). Synchronization between predator and prey population are also important factors for effective biological control (Macfadyen et al., 2015). Simply enhancing non-crop habitats may not be enough to support NPR. Landscape composition has been shown to explain significant variation in assessed natural enemy and pest abundance, predation rates and crop damage. But while each varied (both increased and decreased) in landscapes with more non-crop habitat, there was no consistent trend (Karp et al., 2018). A better understanding of the impacts of specific interventions and the benefits of specific insects will likely improve the impact of NPR in Southern Africa.

In addition to the dynamics between crops, pests and natural enemies, there are ecological processes occurring within organisms which could be the target of novel forms of biological control. Understanding the microbiome of insects can lead to the development of new sustainable control strategies for pests and strategies to augment the health of beneficial insects. Recent research has indicated, for instance, that symbiont infection can be the difference between invasive and noninvasive insect species; using this knowledge in predicting risks or learning to manipulate such infections in pest populations offers a potential new approach (Himler et al., 2011).

Many crop diseases rely on insect vectors for transmission. Instead of relying on insecticides, it may be possible to block the transmission of diseases by insect vectors by increasing the presence of symbiotic microbes which may be more economical since they are self-sustaining. For example, studies have shown that retention of cassava mosaic begomoviruses (CMBs) was significantly reduced in lines carrying inherited bacterial endosymbionts such as *Rickettsia* and *Arsenophonus* compared to controls, and there was an associated reduction in transmission of CMBs – possibly as a consequence of observed elevated immune gene expression in the presence of these symbionts (Ghosh et al., 2018). Endosymbiont-based strategies have also been implemented to suppress vector populations. They may as well have a role in supporting populations of beneficial insects. Insect gut microbiota has been shown to have a vital role in host metabolism, protection from parasites and pathogens and modulation of immune responses (Engel and Moran, 2013).

Promoting farmer access to knowledge on managing natural pest control services through policy in Southern Africa: Natural pest regulation (NPR) is a knowledge-intensive pathway to improved pest management and agricultural production that is sustainable and economically viable. Successful implementation is highly dependent on knowledge transfer to farmers and their primary points of contact – often government technical advisers. Therefore, investments in NPR will engage farmers to ensure effective change and resilient agriculture that enhances livelihoods and buffers production against future threats and risks. One major hurdle in promoting ecological intensification that harnesses ecosystem services such as NPR is fundamental knowledge about the natural enemies. It is often challenging for farmers to understand the underlying concept of beneficial insects and to distinguish between good and bad insects let alone understand how land management might influence this service detrimentally (such as field margin clearance). The training of farmers with examples and practical tools can improve understanding about beneficial insects and change how farmers manage field margins (Mkenda et al., in review; Elisante et al., 2019).

Interactions between ecosystem services

Recent studies also point to interactions between these different ecosystem services, thus suggesting multiple win-wins; for example, there is evidence that soil-born microbes can influence aboveground plant-herbivore interactions, suggesting that soil microbial management could be adapted for pest management strategies (Pineda et al., 2017). Compost application used to remediate degraded soils can enhance the ability of soils to suppress plant pathogens (Pane and Zaccharelli, 2014). Another example of pest management are field margin plants that support natural enemies and pollinators and providing complementary ESs. Another example would be botanical pesticides that have been shown to be effective in promoting NPR in certain pest crop interactions when compared to synthetic pesticides that work against NPR (Mkenda et al., 2015).

Research requirements

Research needs for the relatively new area of inquiry on ecosystem services underpinning agricultural production systems in Southern Africa are summarized in Table 16.1. The list is indicative, not comprehensive, noting that only 6% of studies worldwide on ecosystem services have been carried out in Africa to date (Adhikari and Hartemink, 2016).

Table 16.1 Recommendations for improved impact of ecosystem services in Southern Africa

Ecosystem service	Research
Soil ecosystem services	Promotion of soil management recommendations
	Support use of diverse and efficient crops and cover crops
	Integrate crop breeding and with rhizosphere microbiome engineering, as a technology on the horizon
	Improved support for biofertilizers and biocontrol products
Pollination services	Support the study of pests and diseases of pollinators
	Improve regulation of pesticide use in agricultural landscapes
Natural pest regulation	Support field margin and non-crop habitats improvement
	Introduce behaviour modification of natural enemies in the field
	Research insect microbiome dynamics to control pests and enhance health of beneficial insects

Conclusion

Biodiversity and ecosystem services lie at the centre of many solutions for sustainable increases in agricultural productivity that reduce the negative externalities of production and at the same time increase positive externalities – such as watershed protection and creation of biodiversity-friendly habitats on-farm. Throughout these profiles of three key ecosystem services underpinning agricultural production, certain commonalities come to the fore. The functioning of the services in each case is quite complex, and it requires a high level of knowledge on the part of researchers but also of farmers and advisors. There are intriguing areas of interactions and synergies between the services. However, the allocation of funding to such ecologically based solutions, and to the training and dissemination of relevant management practices, has been quite minor in comparison to more technologically based approaches, yet their potential to provide multiple benefits and less environmental costs is substantial.

References

Adhikari, K. and Hartemink, E. (2016) Linking soils to ecosystem services – A global review. *Geoderma*, 262, 101–111. https://doi.org/10.1016/j.geoderma.2015.08.009.

Aizen, M.A. and Harder, L.D. (2009) The global stock of domesticated honey bees is growing slower than agricultural demand for pollination. *Current Biology*, 19(11), 915–918.

Barrios, E. (2007) Soil biota, ecosystem services and land productivity. *Ecological Economics*, 64(2), 269–285.

Bender, S.F., Wagg, C. and van der Heijden, M.G. (2016) An underground revolution: Biodiversity and soil ecological engineering for agricultural sustainability. *Trends in Ecology & Evolution*, 31(6), 440–452.

Bernal, M.P., Sommer, S.G., Chadwick, D., Qing, C., Guoxue, L. and Michel Jr, F.C. (2017) Current approaches and future trends in compost quality criteria for agronomic, environmental, and human health benefits. In *Advances in Agronomy* (Vol. 144, pp. 143–233). Academic Press.

Bianchi, F.J.J.A., Booij, C.J.H. and Tscharntke, T. (2006) Sustainable pest regulation in agricultural landscapes: A review on landscape composition, biodiversity and natural pest control. *Proc Biol Sci*, 273, 1715–1727.

Chaplin-Kramer, R., O'Rourke, M.E., Blitzer, E.J. and Kremen, C. (2011) A meta-analysis of crop pest and natural enemy response to landscape complexity. *Ecol Lett*, 14, 922–932.

Chapman, B.B., Brönmark, C., Nilsson, J.Å. and Hansson, L.A. (2011) The ecology and evolution of partial migration. *Oikos*, 120(12), 1764–1775.

Daily, G. (1997) Introduction: What are ecosystem services? In Daily, G., ed. *Nature's Services: Societal Dependence on Natural Ecosystems*. Washington, DC: Island Press.

Elisante, F. Ndakidemi, P.A., Arnold, S.E.J., Belmain, S.R., Gurr, G.M., Darbyshire, I., Xie, G., Tumbo, J. and Stevenson, P.C. (2019) Knowledge gaps on the role of pollinators and value of field margins among smallholders in bean agri-systems. *Journal of Rural Studies*, 70, 75–86.

Engel, P. and Moran, N.A. (2013, September) The gut microbiota of insects-diversity in structure and form. *FEMS Microbiol Rev*. 37(5), 699–735. doi:10.1111/1574-6976.12025.

Gaigher, R., Pryke, J.S. and Samways, M.J. (2016a) High parasitoid diversity in remnant natural vegetation, but limited spillover into the agricultural matrix in South African vineyard agroecosystems. *Biological Conservation*, 186, 69–74.

Gaigher, R.J.S. and Samways, M.J. (2016b) Old fields increase habitat heterogeneity for arthropod natural enemies in an agricultural mosaic. *Agriculture Ecosystems and Environment*, 230, 242–250.

Ghosh, S., Bouvaine, S., Richardson, S.C.W., Ghanim, M. and Maruthi, M.N. (2018) Fitness costs associated with infections of secondary endosymbionts in the cassava whitefly species. *J. Pest Sci*, 91(1), 17–28.

Griffiths, G.J.K., Holland, J.M., Bailey, A. and Thomas, M.B., (2008) Efficacy and economics of shelter habitats for conservation biological control. *Biological Control*, 45, 200–209.

Gurr, G.M., Wratten, S.D. and Snyder, W.E. (eds) (2012) *Biodiversity and Insect Pest Management: Key Issues for Sustainable Management*. West Sussex: Wiley.

Henri, D.C., Jones, O., Tsiattalos, A., Thebault, E., Seymour, C.L. and van Veen, F.J.F. (2015) Natural vegetation benefits synergistic control of the three main insect and pathogen pests of a fruit crop in southern Africa. *Journal of Applied Ecology*, 52, 1092–1101.

Himler, A.G., Adachi-Hagimor, T., Bergen, J.E., Kozuch, A., Kelly, S.E., Tabashnik, B.E., Chief, E., Duckworth, V.E., Dennely, T.J., Zchori-Fein, E. and Hunter, M.S. (2011) Rapid spread of a bacterial symbiont in an invasive whitefly is driven by fitness benefits and female bias. *Science*, 332(6026), 254–256.

IPBES. (2016) *Assessment Report on Pollinators, Pollination and Food Production*. Bonn, Germany: UN Intergovernmental Platform on Biodiversity and Ecosystem Services.

IPBES. (2018) *Summary report, Land Degradation*. Bonn, Germany: UN Intergovernmental Platform on Biodiversity and Ecosystem Services.

Karp, D.S. et al. (2018) Crop pests and predators exhibit inconsistent responses to surrounding landscape composition. *PNAS*, 115, E7863–E7870.

Kennedy, C.M. et al. (2016) A global quantitative synthesis of local and landscape effects on wild bee pollinators in agroecosystems. *Ecol. Lett.*, 16, 584–599.

Macfadyen, S., Hopkinson, J., Parry, H., Neave, M.J., Bianchi, F.J.J.A., Zalucki, M.P. and Schellhorn, N.A. (2015) Early-season movement dynamics of phytophagous pest and natural enemies across a native vegetation-crop ecotone. *Agriculture, Ecosystems and Environment*, 200, 110–118.

Mapfumo, P. and Giller, K.E. (2001) *Soil Fertility Management Strategies and Practices by Smallholder Farmers in Semi-Arid Areas of Zimbabwe*. International Crops Research Institute for the Semi-Arid Tropics (ICRISAT)/ Food and Agriculture Organization of the United Nations (FAO), 60p.

Masehela, T.S. (2017) An assessment of different beekeeping practices in South Africa based on their needs (bee forage use), services (pollination services) and threats (hive theft and vandalism) (Doctoral dissertation, Stellenbosch: Stellenbosch University).

Midingoyi, S., Affognon, H.D., Macharia, I., Ong'amo, G., Abonyo, E., Ogola, G., De Groote, H. and LeRu, B. (2016) Assessing the long-term welfare effects of the biological control of cereal stemborer pests in East and Southern Africa: Evidence from Kenya, Mozambique and Zambia. *Agriculture, Ecosystems and Environment* 230, 10–23.

Millennium Ecosystem Assessment. (2005) Ecosystems and *Human Well-Being: Synthesis Report*. Washington, DC: Island Press.

Mkenda, P., Mwanauta, R., Stevenson, P.C., Ndakidemi, P., Mtei, K. and Belmain, S.R. (2015) Field margin weeds provide economically viable and environmentally benign pest control compared to synthetic pesticides. *PLoS One*, 10, e0143530.

Ngwira, R.A., Thierfelder, C. and Lambert, D.M. (2012) Conservation agriculture systems for Malawian smallholder farmers: Long-term effects on crop productivity and soil quality. *Renewable Food Systems*. doi:10.1017/S1742170512000257.

Pane, C. and Zaccardelli, M. (2014) Principles of compost-based plant diseases control and innovative new developments. In *Composting for Sustainable Agriculture* (pp. 151–171). Cham: Springer.

Pineda, A., Kaplan, I. and Bezemer, T.M. (2017) Steering soil microbiomes to suppress aboveground insect pests. *Trends in Plant Science*, 22(9), 770–778.

Sayer, J., Sunderland, T., Ghazoul, J., Pfund, J.L., Sheil, D., Meijaard, E., Venter, M., Boedhi-hartono, A.K., Day, M., Garcia, C. and van Oosten, C. (2013) Ten principles for a land-scape approach to reconciling agriculture, conservation, and other competing land uses. *Proc. Nat. Acad. Sci.*, 110(21), 8349–8356.

Schmidt, R., Gravuer, K., Bossange, A.V., Mitchell, J. and Scow, K. (2018) Long-term use of cover crops and no-till shift soil microbial community life strategies in agricultural soil. *PloS One*, 13(2), e0192953.

Smith, P., Cotrufo, M.F., Rumpel, C., Paustian, K., Kuikman, P.J., Elliott, J.A., McDowell, R., Griffiths, R.I., Asakawa, S., Bustamante, M., House, J.I., Sobocká, J., Harper, R., Pan, G., West, P.C., Gerber, J.S., Clark, J.M., Adhya, T., Scholes, R.J. and Scholes, M.C. (2015) Biogeochemical cycles and biodiversity as key drivers of ecosystem services provided by soils. *SOIL Discussions*, 2(1), 537–586.

17 The role of mechanization in transformation of smallholder agriculture in Southern Africa

Experience from Zimbabwe

Frédéric Baudron, Raymond Nazare and Dorcas Matangi

This chapter will explore "the why", "the what" and "the how" of mechanization for smallholder farmers in Southern Africa, using Zimbabwe as an example, and is based on the experience of the ACIAR-funded project "Farm Mechanization and Conservation Agriculture for Sustainable Intensification" (FACASI, www.facasi.act-africa.org) and the UNDP-funded project "Program for Resilience and Growth" (PROGRESS).

Why mechanization? Evidence of labour as a major limiting factor to the productivity of farming systems in Southern Africa

Per capita food production in Southern Africa has declined dramatically in the last half-century (Pretty et al., 2011), highlighting the need for intensification in the region. In addition, the need to foster a new form of intensification – often coined "sustainable intensification" (SI) – one that increases agricultural production and productivity while minimizing detrimental economic, social and environmental outcomes – is widely recognized (Vanlauwe et al., 2014). By definition, intensification (whether conventional or sustainable), is a process of increasing agricultural output. This increase is generally accompanied by an increase in farm power demand, to handle greater volumes during harvest, transport and processing (Clarke and Bishop, 2002). In addition, the implementation of SI technologies tends to increase management and precision requirements, thus resulting in increased labour demand in a context of low mechanization. For example, precise application of fertilizer manually (as in the case of micro-dosing) increases labour demand compared with fertilizer banding (Babiker et al., 2017). Similarly, timely weeding is often conditioned by labour availability (Orr et al., 2002). Management practices intended to improve the quantity and quality of manure also tend to be highly labour intensive (Harris, 2002). In manual conservation agriculture (CA) production systems, labour demand for land preparation and weeding is much higher than for conventional production systems (Rusinamhodzi, 2015). Finally, the adoption of agroforestry technologies – such as alley cropping – has been found to be limited by high labour demand for, e.g., pruning (Hoekstra, 1987). The positive impact on productivity of precise fertilizer application, timely weeding, CA, agroforestry and many other SI technologies is well known, but their impact on labour demand, which may limit their adoption in the context of Southern Africa where mechanization levels are low, is rarely acknowledged. This lack of consideration for labour issues emanates from the perception that labour in smallholder systems of Southern Africa is abundant and thus nonlimiting. This view is also fueled by macroeconomic analyses (of e.g., land:labour ratio; Headey and Jayne, 2014), which are based on national data that may be too aggregated to reveal farm-level dynamics (Baudron et al., 2019).

Several lines of evidence point to the fact that labour and farm power are increasingly becoming major limiting factors to the productivity of smallholder systems in Southern Africa (Baudron et al., 2015) and most likely a significant constraint to the adoption of SI technologies (which are labour intensive, as demonstrated earlier). To illustrate this point, we use data collected under the FACASI project in two contrasting sites of Zimbabwe: Domboshawa and Makonde.

The attainable land productivity (maximum land productivity a farm can achieve considering all grains, roots and tubers produced on-farm) appears correlated to total labour used on-farm in these two sites (Figure 17.1), which clearly demonstrates that labour/farm power is a major limiting factor to the productivity of the production systems considered. This, associated with the growing scarcity (and cost) of rural labour – in particular because of rural-urban migration (Collier, 2017) – points to the need for mechanization to increase

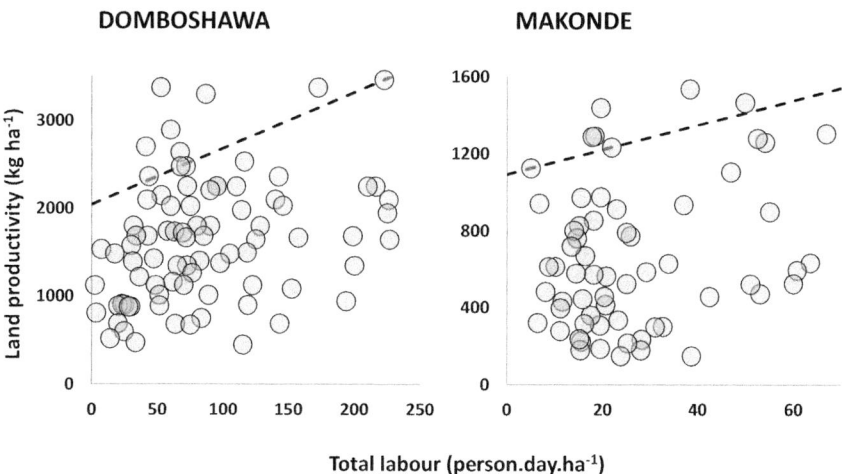

Figure 17.1 Land productivity (considering all grains, roots and tubers produced on-farm) as a function of total labour (per unit of farm area) for Domboshawa and Makonde; dashed lines represent linear regressions fitted through the 90th percentile (i.e., "boundary line")

the productivity of smallholder agriculture in Southern Africa. Mechanization is also expected to reduce the postharvest losses currently experienced by smallholders in the region (Tefera, 2012) and reduce drudgery, which is disproportionately placed on women (Baudron et al., 2019).

The form mechanization should take in the smallholder farming systems of Southern Africa is the subject of much debate. Smallholder farms in the region tend to be small and fragmented (0.79 ± 0.46 ha in Domboshawa and 4.63 ± 3.10 ha in Makonde for example). The use of (relatively) large (two axle) tractors would thus require land consolidation. Some authors have argued that consolidation is a prerequisite to mechanization, to use large tractors efficiently (e.g., Asiama et al., 2017). In contrast, others have argued for a concept of "appropriate mechanization", whereby machines are adapted to farm size and not the opposite (e.g., Baudron et al. 2015). This is because of the negative social (e.g., labour displacement; Binswanger et al., 1995) and environmental (e.g., loss of landscape heterogeneity; Benton et al., 2003) consequences of land consolidation and because of negative farm size productivity relationships often reported in smallholder farming systems in Africa (Ali and Deininger, 2015; Baudron et al., 2019).

The use of animal traction is part of appropriate mechanization. However, draught animals are uncommon in large parts of Southern Africa, with most oxen concentrated in the central plateau of Zimbabwe, Southern Zambia and the highlands of Malawi. Elsewhere, diseases such as trypanosomiasis restrict the presence of oxen. Even in regions where draught animals are commonly used, their numbers tend to decline because of the combined effect of epidemics (in particular tick-borne diseases such as the East Coast fever), recurring droughts and feed shortages (Mapiye et al., 2009; Moyo et al., 2017).

Thus, it could be argued that the need for smallholder mechanization in Southern Africa combined with the presence of small and fragmented fields and the diminishing availability of animal traction calls for small (less than 25 horsepower) motorized solutions. Such mechanization pathways successfully took place in countries like Bangladesh. Despite very small and fragmented fields, Bangladesh agriculture is highly mechanized, but power is delivered by hundreds of thousands of small (single axle) tractors and other small engines, not large (two-axle) tractors (Biggs et al., 2011). Such pattern of mechanization guided the projects FACASI and PROGRESS in exploring the potential impact of appropriate mechanization based on two-wheel tractors for smallholder in Zimbabwe (and other areas of Eastern and Southern Africa). The next section will explore what tasks should be mechanized in priority and what commercially available two-wheel tractor ancillary equipment is available for that.

What task(s) to mechanize? With what commercially available machines?

Land preparation is the most power-intensive farming operation in rainfed agriculture (Lal, 2004). It is also one of the most critical operation in Southern Africa, as delayed land preparation and delayed planting often result in severe yield penalties in the region (Nyagumbo et al., 2017). In addition, during focus group discussions organized under the FACASI project, it was stated consistently in all sites – including Domboshawa and Makonde in Zimbabwe – and by both men and women that mechanizing land preparation and crop establishment is a priority (Baudron et al., 2019). Indeed, although land preparation is a men's task, the quality and timeliness of this operation was said to also affect weeding intensity, one of the main tasks carried out by women. Un-mechanized land preparation was also said to take several days, affecting the labour burden of women who have to prepare food and transport it to men in the field during that period.

Two-wheel tractors can be used to plough light (e.g., sandy) soils (Kebede and Getnet, 2016) but do not produce enough traction to plough heavier soils in rainfed conditions (Holtkamp and Lorenz, 1990; Singh, 2006). However, two-wheel tractors could be used to establish a crop in these soils providing energy requirements for tillage are reduced. This can be achieved by simplifying land preparation i.e., using reduced or no tillage, which cuts energy requirements by about half compared to conventional (i.e., mouldboard or disc ploughing) land preparation (Lal, 2004). Therefore, it could be argued that reduced or no tillage could make the use of two-wheel tractors for crop establishment viable in most of Southern Africa.

Several direct seeders (i.e., placing seed and fertilizer without prior tillage) for two-wheel tractors are now commercially available, from countries such as China and Brazil and can be used to seed most large grain (e.g., maize, cotton) and small grain crops (e.g., wheat, rice). Local manufacturers – including Zimplow LTD (Figure 17.2a) have also started manufacturing direct seeders for two-wheel tractors.

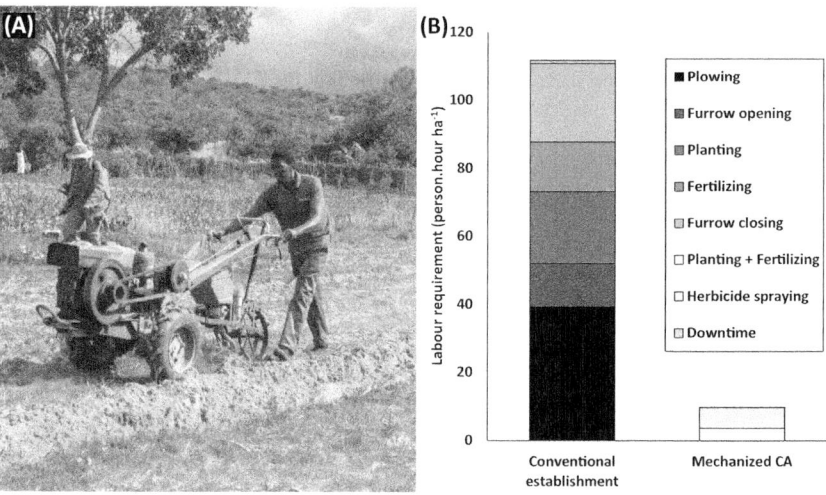

Figure 17.2 a) direct seeding with a Zimplow single row seeder powered by a 12 HP two-wheel tractor and b) labour per task required to establish a maize crop conventionally vs. using mechanized CA (Zimplow single row seeder)

During the 2017/18 season in Eastern Zimbabwe, (Figure 17.2b) the use of a Zimplow single row seeder was found to reduce labour requirements to establish a crop of maize from 111.8 person.hour ha^{-1} to 9.7 person.hour ha^{-1} compared to conventional crop establishment (ox ploughing and lining followed by manual placement of seeds and fertilizers). This corresponds to a reduction by a factor 11.5. The mean fuel consumption was low, at 3.4 ± 1.2 L ha^{-1}.

Direct seeding using a two-wheel tractor was not found to impact maize yield significantly, except in few cases. This may be due to the fact that only low quantities of crop residues tend to be used as mulch. The improvement of soil conditions under CA also tend to take several years (Thierfelder et al., 2015). Nevertheless, by saving time, labour and cost, direct seeding using a two-wheel tractor appears to be profitable for users. Shelling and transport were other mechanized operation considered by FACASI and PROGRESS, for small engines to be in productive use for a greater part of the year and to increase the profitability of mechanization (and reduce the unit cost of custom work in the case of service provision).

From the results presented in this section, small mechanization clearly has a place in the SI basket of technologies for smallholders in Southern Africa. The next section will explore modalities to deliver small mechanization to smallholders in the region for the greatest number to benefit.

How to deliver mechanization in different contexts?

An ex ante analysis conducted in Makonde and Domboshawa revealed that the use of two-wheel tractors and their accessories was not economically viable for farmers as individual owners, operating solely on their own farms, or on their own farms and neighbouring farms (Table 17.1). However, the same analysis revealed that small mechanization could be an attractive investment for individual farmers if they provide services to neighbouring farmers at commercial rates on a full-time basis and ensure demand. Such a model – based on service

Table 17.1 Net present values (NPV), benefit cost ratios (B/C) and internal rates of return (IRR) calculated from ex ante analysis for three business models, providing for a combination of planting, shelling and transporting operations in Zimbabwe

Business models	Indicators	Makonde	Domboshawa
Farmer operator solely working on his/her farm	NPV (US$)	−9207	−8649
	B/C ratio	0.2	0.11
	IRR (%)	–	–
Farmer operator working on neighbouring farms	NPV (US$)	−4154	−4862
	B/C ratio	0.89	1.8
	IRR (%)	2	9
Full time service provider	NPV (US$)	18446	4307
	B/C ratio	1.3	1.2
	IRR (%)	51	27

hiring – appears viable in Zimbabwe as the large majority of farmers in the region currently hire labour, and many of them also hire animal traction services or tractor services. For example, 94 and 97% of farming household hire labour in Domboshawa and Makonde, respectively.

After five years of implementation of the FACASI project in Makonde, an ex–post analysis was also conducted to evaluate the performance of various hired service business models combining different operations (Table 17.2). Results of the study show that, when averaged for two years, all the business models are profitable as demonstrated by positive net present values and benefit-cost ratios greater than 1. Profitability is particularly high during years with maximum business (Table 17.2). These high performances are achieved through aggressive marketing of services and an established clientele base. In contrast, years with low business typically have low profit margins and even losses for some business models. This is particularly so for planting services, which is timebound and may end-up being unprofitable in very dry or very wet years that only allow planting for a few days

Additionally, the results also indicate that service providers offering more than one service i.e., more implements, tend to be profitable in all scenarios considered. This suggests that businesses can complement each other, as demonstrated by the combination of planting and shelling (Table 17.2). The findings of this study are consistent with Kahan et al. (2017) who observed that profitability was higher for a service provider offering a range of services compared to one offering a single service. Bundling of services increases capacity utilization of the tractor, as it is the major source of power for these services. Offering more than one service is, however, dependent on the capacity of the SP to invest in additional implements, affordability and access to financial resources. Thus, business model performance is dependent on the actors and is context specific.

Table 17.2 Average, minimum and maximum profitability – assessed by net present values (NPV), benefit cost ratios (B/C) and internal rates of return (IRR) for business models combining different operations in Makonde

Indicators	2WT-power sheller	2WT Planter + sheller	2WT Planter
Average			
NPV (US$)	17679.2	21778.9	1763.2
B/C ratio	4	2.6	1.17
IRR (%)	172%	106%	30%
Minimum			
NPV (US$)	18033.2	16740.9	−3628.8
B/C ratio	4.1	2.4	0.58
IRR (%)	175%	88%	-6%
Maximum			
NPV (US$)	17325.2	26816.9	7155.2
B/C ratio	4	2.8	1.62
IRR (%)	169%	106%	58%

Table 17.3 Cost of planting and shelling using draught animals and labour (conventional practices) or using two-wheel tractor-based mechanization

Operations	Conventional practices	Mechanized practices	% Reduction
Planting (US$)	120	56	53%
Shelling (US$)	129.6	97.5	25%

The ex-post analysis also evaluated the benefits that accrue to farmers in terms of cost saving through receiving mechanized services (Table 17.3). The results demonstrate that mechanized planting is more attractive to farmers (clients) than postharvest operations. The fact that the opposite is true for service providers (Table 17.2) raises questions as to the interventions that are necessary to scale small mechanization in Zimbabwe. Considering the high cost of planters, government may need to intervene with some incentives for service providers to acquire them in order for the largest number of farmers to benefit from this service (Benin et al., 2013).

Conclusions: lessons learned

Transforming smallholder agriculture in Southern Africa will require an improvement in access to farm power. Evidences presented in this chapter demonstrate that mechanization in the region can be delivered by service providers using small engines. This approach should be prioritized in area characterized by: 1) relatively commerce-oriented agriculture (e.g., presence of cash crops); 2) agriculture constrained by labour shortages, at least seasonally; 3) high cost of maintaining draught animals (e.g., feed shortage); 4) field accessibility (e.g., feeder roads); 5) existence of hire services (e.g., ox ploughing); 6) relatively deep and stone-free soils; and 7) small and fragmented fields. For small mechanization interventions to be successful, it is crucial to involve private sector stakeholders – dealers, manufacturers, etc. – in every step. Incentive schemes (matching grants, soft loans, guarantee funds, etc.) are necessary to set up service providers in business. It should also be noted that an approach centred on the private sector alone may not work when targeting marginal areas (e.g., rain fed systems dominated by staples), marginal groups (e.g., resource-constrained smallholders), technologies providing public goods (e.g., conservation agriculture) or complex technologies (not a "product"). In such circumstances, the public sector has a crucial role to play in commercialization, in particular through the creation of a conducive business environment to attract private sector actors.

References

Ali, Daniel Ayalew, and Klaus Deininger. 2015. "Is There a Farm Size–Productivity Relationship in African Agriculture? Evidence from Rwanda." *Land Economics* 91 (2): 317–343. https://doi.org/10.3368/le.91.2.317.

Asiama, Kwabena Obeng, Rohan Bennett, and Jaap Zevenbergen. 2017. "Land Consolidation for Sub-Saharan Africa's Customary Lands – The Need for Responsible Approaches." *American Journal of Rural Development* 5 (2): 39–45. https://doi.org/10.12691/AJRD-5-2-2.

Babiker, Monirah, Mohamed Arbab, Yassin Mohmad, and Ibrahim Dagash. 2017. "Micro-dosing Technology of Fertilizer for Sorghum Production At." *Cell Biology & Development* 1 (1): 18–22. https://doi.org/10.13057/cellbioldev/t010104.

Baudron, Frédéric, Brian Sims, Scott E Justice, David G Kahan, Richard Rose, Saidi Mkomwa, Pascal Kaumbutho, et al. 2015. "Re-Examining Appropriate Mechanization in Eastern and Southern Africa: Two-Wheel Tractors, Conservation Agriculture, and Private Sector Involvement." *Food Security* 7 (4): 889–904. https://doi.org/10.1007/s12571-015-0476-3.

Baudron, Frédéric, Michael Misiko, Bisrat Getnet, Raymond Nazare, John Sariah, and Pascal Kaumbutho. 2019. "A Farm-Level Assessment of Labour and Mechanization in Eastern and Southern Africa." *Agronomy for Sustainable Development* 5. https://doi.org/10.1007/s13593-019-0563-5.

Benin, Samuel, Michael Johnson, Emmanuel Abokyi, Gerald Ahorbo, Kipo Jimah, Gamel Nasser, Victor Owusu, Joe Taabazuing, and Tenga Albert. 2013. "Revisiting Agricultural Input and Farm Support Subsidies in Africa: The Case of Ghana's Mechanization, Fertilizer, Block Farms, and Marketing Programs." *IFPRI Discussion Paper no. 01300*, November: 1–121. https://doi.org/10.13140/RG.2.2.23891.17447.

Benton, Tim G, Juliet A Vickery, and Jeremy D Wilson. 2003. "Farmland Biodiversity: Is Habitat Heterogeneity the Key?" *Trends in Ecology & Evolution* 18 (4): 182–188. https://doi.org/10.1016/S0169-5347(03)00011-9.

Biggs, Stephen, Scott E Justice, and David Lewis. 2011. "Patterns of Rural Mechanisation, Energy and Employment in South Asia: Reopening the Debate." *Economic & Political Weekly* xlvi (9): 78–82.

Binswanger, H., K. Deininger, and G. Feder. 1995. "Power, Distortion, Revolts, and Reform in Agricultural Land Relations." *Handbook of Development Economics* 3 (3): 2659–2772.

Clarke, Lawrence, and Clare Bishop. 2002. "Farm Power – Present and Future Availability in Developing Countries." In *Special Session on Agricultural Engineering and International Development in the Third Millennium*. ASAE Annual International Meeting/CIGR World Congress, July 30, 2002. Chicago, USA. https://doi.org/10.1007/s11023-007-9060-8.

Collier, Paul. 2017. "African Urbanization: An Analytic Policy Guide." *Oxford Review of Economic Policy* 33 (3): 405–437. https://doi.org/10.1093/oxrep/grx031.

Harris, F. 2002. "Management of Manure in Farming Systems in Semi-Arid West Africa." *Experimental Agriculture* 38 (02): 131–148. https://doi.org/10.1017/S0014479702000212.

Headey, Derek D., and T. S. Jayne. 2014. "Adaptation to Land Constraints: Is Africa Different?" *Food Policy* 48: 18–33. https://doi.org/10.1016/j.foodpol.2014.05.005.

Hoekstra, Dirk A. 1987. "Economics of Agroforestry." *Agroforestry Systems* 5 (3): 293–300. https://doi.org/10.1007/BF00119127.

Holtkamp, R., and J. Lorenz. 1990. *Small Four-Wheel Tractors for the Tropics and Subtropics: Their Role in Agriculture and Industrial Development*. Technical. Bonn, Federal Republic of Germany.

Kahan, David, Roger Bymolt, and Fred Zaal. 2017. "Thinking Outside the Plot: Insights on Small-Scale Mechanisation from Case Studies in East Africa." *Journal of Development Studies* 00 (00): 1–16. https://doi.org/10.1080/00220388.2017.1329525.

Kebede, Laike, and Bisrat Getnet. 2016. "Performance of Single Axle Tractors in the Semi-Arid Central Part of Ethiopia." *Ethiopian Journal of Agricultural Sciences* 27 (1).

Lal, Rattan. 2004. "Carbon Emission from Farm Operations." *Environment International*. https://doi.org/10.1016/j.envint.2004.03.005.

Mapiye, C., M. Chimonyo, and K. Dzama. 2009. "Seasonal Dynamics, Production Potential and Efficiency of Cattle in the Sweet and Sour Communal Rangelands in South
Africa." *Journal of Arid Environments* 73 (4–5): 529–536. https://doi.org/10.1016/j.
jaridenv.2009.01.003.

Moyo, I.A., T.N. Mudimba, D.N. Ndhlovu, S. Dhliwayo, S.M. Chikerema, and G. Matope.
2017. "Temporal and Spatial Patterns of Theileriosis in Zimbabwe: 2000–2014." *Bulletin of
Animal Health and Production in Africa* 65 (3): 569–575. https://www.ajol.info/index.php/
bahpa/article/view/167662.

Nyagumbo, Isaiah, Siyabusa Mkuhlani, Walter Mupangwa, and Daniel Rodriguez. 2017.
"Planting Date and Yield Benefits from Conservation Agriculture Practices across Southern Africa." *Agricultural Systems* 150: 21–33. https://doi.org/10.1016/j.agsy.2016.09.016.

Orr, A., B. Mwale, and D. Saiti. 2002. "Modelling Agricultural 'Performance': Smallholder
Weed Management in Southern Malawi." *International Journal of Pest Management* 48 (4):
265–78. https://doi.org/10.1080/09670870210149808.

Pretty, Jules, Camilla Toulmin, and Stella Williams. 2011. "Sustainable Intensification in African Agriculture." *International Journal of Agricultural Sustainability* 9 (1): 5–24. https://doi.
org/10.3763/ijas.2010.0583.

Rusinamhodzi, Leonard. 2015. "Tinkering on the Periphery: Labour Burden Not Crop
Productivity Increased under No-till Planting Basins on Smallholder Farms in Murehwa
District, Zimbabwe." *Field Crops Research* 170: 66–75.

Singh, Gyanendra. 2006. "Agricultural Machinery Industry in India (Manufacturing, Marketing and Mechanization Promotion)." *Status of Farm Mechanization in India*, 154–174.
http://scholar.google.com/scholar?hl=en&btnG=Search&q=intitle:Agricultural+Machi
nery+Industry+in+India+(+Manufacturing+,+marketing+and+mechanization+promo
tion+)#0.

Tefera, Tadele. 2012. "Post-Harvest Losses in African Maize in the Face of Increasing Food
Shortage." *Food Security* 4 (2): 267–277. https://doi.org/10.1007/s12571-012-0182-3.

Thierfelder, Christian, Rumbidzai Matemba-Mutasa, and Leonard Rusinamhodzi. 2015.
"Yield Response of Maize (Zea Mays L.) to Conservation Agriculture Cropping System in Southern Africa." *Soil and Tillage Research* 146 (PB): 230–242. https://doi.org/
10.1016/j.still.2014.10.015.

Vanlauwe, B., D. Coyne, J. Gockowski, S. Hauser, J. Huising, C. Masso, G. Nziguheba,
M. Schut, and P. Van Asten. 2014. "Sustainable Intensification and the African Smallholder Farmer." *Current Opinion in Environmental Sustainability* 8: 15–22. https://doi.org/
10.1016/j.cosust.2014.06.001.

Further reading

www.fao.org/waicent/FAOINFO/AGRICULT/againfo/programmes/documents/livatl2/
draftoxenmap.htm.

18 Advanced genetic technologies for improving plant production

Jennifer A. Thomson, Sylvester O. Oikeh,
Idah Sithole-Niang and Leena Tripathi

Introduction

Genetically modified (GM) crops have been commercialized since 1996. The rapid adoption of these crops by both large and smallholder farmers in industrial and developing countries reflects their substantial multiple benefits (ISAAA, 2017). Unfortunately, these benefits are not being shared by most African farmers and consumers as only South Africa, Sudan and, very recently, Nigeria, have commercialized GM crops. The main crops are insect tolerant and/or herbicide resistant maize, soybean and cotton. However, African crops suffer from many biotic and abiotic stresses not found elsewhere. Some examples include maize streak virus (MSV), banana Xanthomonas wilt (BXW), cassava mosaic and brown streak viruses and cowpea pod borer insects. In addition, drought is a common phenomenon in many parts of Africa, and it is worsening due to climate change. Standard breeding techniques, including modern improvements such as marker assisted breeding (MAB) have been unable to produce the desired resistant varieties, but many local scientists have succeeded using genetic modification techniques. Some of these will be discussed in this chapter.

State of the art and constraints

The only countries in Africa which have commercialized GM crops are South Africa, Sudan and Nigeria, the latter in 2018 for cotton and in January 2019 for cowpea. In South Africa, 70 events have been approved since 1998 for food, feed and planting, including 42 maize, 12 soybean, ten cotton, five canola and one rice (ISAAA, 2017). In 2017, the estimated GM maize area was 85%, of which 66% was stacked insect resistant (IR) and herbicide tolerance (HT), and the rest contained single IR or HT traits in equal proportions. In 2017, 95% of soybeans planted were HT. All the cotton planted was GM with 95% stacked and 5% used for refugia to prevent insects becoming resistant to the IR trait. The IR trait that has been commercialized is due to the expression in the plant of the toxin gene derived from the soil bacterium, *Bacillus thuringiensis*. Hence, the IR trait is often referred to as Bt.

Sudan is in its sixth year of commercialization of GM crops, with an estimated 90,000 farmers growing insect resistant cotton in 2017 on farms with an average size of 2.1 hectares (ISAAA, 2017). In 2012, only one variety was planted, but continuous research over the last six years has resulted in the approval of two new hybrids, gradually increasing the acreage from an initial modest launch of 20,000 hectares in 2012, to 192,000 hectares in 2017. The use of GM cotton hybrids has raised yields by two to three times higher than those of conventional varieties.

Burkina Faso commercialized insect resistant cotton for a short period from 2008 to 2016. The reason for the discontinuation of these crops was not a failure of the GM technology but rather that the varieties used had shorter fibre lengths than conventional varieties (ISAAA, 2016). This shows the importance of ensuring that the right varieties of a GM crop are introduced. It is more important to wait for introgression into the best varieties than to forge ahead only to discover later that the wrong varieties were chosen.

While a number of other African countries, such as Kenya, Mozambique, Ethiopia, Ghana, Malawi and Nigeria, are conducting multilocation field trials on a variety of GM crops, none of them has approved commercialization. One of the main reasons that Africa has not accepted GM crops is due to the adverse influence of Europe to this technology. Africa receives most of its aid from Europe, and many Africans study in Europe; therefore, the opinions of Europeans matter to African decision makers. When organizations such as Green Peace, Food Rights Alliance and many more put out statements such as a radio advertisement from the UK-based ActionAid in Uganda, which stated, "Did you know that GMOs can cause cancer and infertility?" (Lynas, 2018), who can blame people from reacting negatively? Even though the ActionAid head office in London later disavowed the advertisement, the damage remained, and local activists are still active in their opposition.

African crops suffer from many biotic and abiotic stresses not found elsewhere. Standard breeding techniques, including modern improvements such as marker assisted breeding (MAB) have been unable to produce resistant varieties, but many local scientists have succeeded using genetic modification techniques. For instance, maize resistant to MSV (Shepherd et al., 2007), bananas resistant to BXW and nematodes (Tripathi et al., 2017), virus resistant cassava (Beyene et al., 2017), pod borer resistant cowpeas (Popelka et al., 2006; Bett et al., 2017; Bosibori et al., 2017) and drought-tolerant and insect-protected maize through the Project Water Efficient Maize for Africa (WEMA; Oikeh et al., 2014; Edge et al., 2018) have all been developed.

However, even though some of these have been field tested in a number of countries, only South Africa has commercialized WEMA products trademarked TELA® maize in 2016 (ISAAA, 2017). As Mark Lynas writes: "[It is] somewhat peculiar that some non-governmental organizations (NGOs), which are ostensibly concerned with poverty reduction, should doggedly oppose such a basic poverty-reducing measure as better seeds for farmers" (Lynas, 2018). These NGOs have led to the development of local equivalents which, together with some organic farmers' organizations, are, in our experience, similarly opposed to advances in agricultural biotechnology. In our opinion, at the very least farmers should be allowed to test these crops and make their own decisions.

To give some examples of what African farmers are missing out on, herewith are some of the relevant data from field trials.

Drought-tolerant and insect-protected (TELA®) maize through the WEMA Project (now known as the TELA Maize Project)

Through the Water Efficient Maize for Africa (WEMA) Project (now known as the TELA [TELA is derived from the Latin word *tutela* which means "protection"] Maize Project), three transgenic traits encoded by the drought-tolerant transgene (DroughtGard®, MON87460), the cold-shock protein gene (*CspB*) and three *Bt* genes (MON810 (*Cry1Ab*) and MON89034 (*Cry1A.105* and *Cry2Ab2*)) were accessed royalty-free for smallholder farmers in Africa by the African Agricultural Technology Foundation (AATF). The *Bt* genes have been stacked with the DroughtGard® gene to provide several variety options for the farmers who operate in drought-prone environments.

Confined field trials (CFTs) carried out in Kenya and Uganda for five seasons to test the efficacy of the *Bt* MON810 gene in controlling the spotted stemborer (*Chilo partellus*) and the African stemborer (*Busseola fusca*) under conditions of artificial infestation indicated that maize hybrids containing the *Bt* gene yielded, on average, 52% more than the isogenic hybrids without the gene (Kyetere et al. in press). Similarly, the CFTs carried out with the stacked drought-tolerant and insect-protection (DroughtGard® + *Bt* MON810) traits under natural infestation of the fall armyworm (*Spodoptera frugiperda*; FAW) and stemborers in Ethiopia, Mozambique and Uganda, and unnatural infestation of FAW and artificial infestation with stemborer larvae in Kenya and Tanzania showed that the *Bt* MON810 trait gave partial but significant control against the FAW, with full control of stemborers. For example, preliminary results of the trials carried out in Mozambique under natural infestation of both FAW and stemborers, showed that all the transgenic maize hybrids except one realized 21–98% yield advantage over the non-transgenic isogenic hybrids based on the level of infestation (Figure 18.1).

The FAW is a new insect pest, native to the Americas but recently reported in Africa, where it is ravaging staple crops, particularly maize, causing serious crop destruction with estimated maize yield losses of 8.3–20.6 million tonnes, worth US$2.48–6.19 billion in 12 African countries (CABI, 2017). It was first reported in Nigeria in 2016, but it has since spread to over 40 countries in Africa, thus posing a major threat to food and nutrition security in Africa.

In South Africa which has a long history of commercializing biotech crops, five TELA® hybrids have been commercialized to smallholder farmers since 2016. Farmers are currently growing these with good protection against stemborer and FAW.

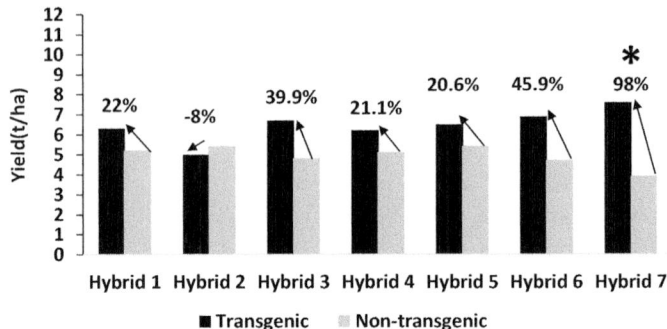

Figure 18.1 Performance of stacked drought-tolerant and insect-resistant traited hybrids (*Bt*; red bars) and *isogenic* hybrids (non-*Bt*; green bars) under natural infestation of stem borer and FAW in Mozambique, 2017

Source: Own presentation

Bt cowpea

Cowpea (*Vigna unguiculata* Walp.) is the most important legume food crop grown in sub-Saharan Africa. Cowpea serves as a major source of dietary protein, being consumed as a fresh leafy vegetable, soft pods as well as grain. In West Africa it is also the main forage crop. Cowpea is drought tolerant and enriches the soil by fixing nitrogen. The major production constraints include a wide range of biotic pests and diseases.

The Network for the Genetic Improvement for Africa, together with the AATF, has been working on cowpea resistant to the pod borer (*Maruca vitrata*) for a number of years. As mentioned earlier, the AATF obtained the *Bt Cry1Ab* gene from Monsanto, and the transformation of cowpea, variety IT86D-1010, was carried out by Commonwealth Scientific and Industrial Research Organization (CSIRO) in Australia. Event 709A was identified as the lead event. It has a single copy of the Cry1Ab gene, which is inherited as a dominant gene giving a segregation ration of 3:1. The efficacy of this event was tested in a confined environment under severe artificial infestation of Maruca in Nigeria, Burkina Faso, Ghana and Malawi. It gave near complete control and increased the number of pods per plant by 1.6- to 13-fold and grain yield by several fold.

Event 709A has been backcrossed into farmers' preferred varieties, and, depending on the pressure of Maruca, these out-yielded conventional cowpeas by 20% to more than 100%. In order to prevent resistance build-up to a single gene, a second *Bt, Cry2Ab*, has been used to transform cowpea, and the best six events are undergoing efficacy tests in West Africa.

At the end of January 2019, the Nigerian National Biosafety Management Agency announced the approval for commercial production of *Bt* pod borer resistant cowpea (IITA 2019). A real breakthrough for Africa!

Banana resistant to banana Xanthomonas wilt

Banana Xanthomonas wilt (BXW), caused by *Xanthomonas campestris* pv. *musacearum* (*Xcm*), is one of the major diseases of banana that is prevalent in the Great Lakes region of East and Central Africa, where banana is a staple food crop and a source of income for smallholder farmers. There is no known source of resistance within the *Musa* germplasm except for diploid wild type banana *Musa balbisiana*. Currently, the control of BXW relies upon improved phytosanitary practices and cultural practices like de-budding, use of clean farming tools and use of pathogen free planting material (Tripathi et al., 2009). These practices can limit the spread of BXW, but the adoption of such practices has been inconsistent, as they are labour intensive.

Use of disease-resistant varieties has been an effective and economically viable strategy for management of plant diseases. In the absence of natural host plant resistance among banana cultivars, researchers have developed transgenic banana expressing the Hypersensitive Response Assisting Protein (*Hrap*) or Plant Ferredoxin Like Protein (*Pflp*) gene originated from sweet pepper (*Capsicum annuum*). These transgenic banana plants have exhibited strong resistance to BXW in the screen house evaluation (Tripathi et al., 2010; Namukwaya et al., 2012). The best 65 resistant lines were further tested in a confined field trial at the National Agricultural Research Laboratory (NARL), Kawanda, Uganda (Tripathi et al., 2014). Twelve transgenic events have been shown to be completely resistant to BXW under confined field trials and also showed flowering and yield (bunch weight and fruit size) characteristics comparable to non-transgenic varieties. The transgenic plants did not exhibit any difference from its non-transformed controls, suggesting that constitutive expression of these genes does not seem to alter plant physiology. To minimize the potential for resistance to a single gene trait, transgenic plants with stacked genes (*Hrap-Pflp*) are being developed for durable high resistance to BXW disease.

Crops developed using CRISPR-type technologies

All the genetically modified crops mentioned earlier have been produced using *Agrobacterium tumefaciens*, a naturally occurring soil bacterium, to insert genes into plants. A new technique, however, is now available, called Clustered Regularly Interspaced Short Palindromic Repeats (CRISPR). This is essentially a short piece of RNA which can locate a specific site in a plant's genetic material and, together with an enzyme called Cas9, make a double-stranded cut in the DNA. The plant's own repair mechanism then either repairs it to be the same as before or, for plant improvement use, introduces a few random nucleotides, resulting in a mutation. This is called genome editing and can be used to mutate genes.

A recent article shows how this technique can benefit cassava and banana, staple crops for many Africans (Gomez et al., 2018; Tripathi et al., 2019). Cassava brown streak disease (CBSD) is a major constraint on yields in East and

Central Africa and threatens its production in West Africa. CBSD is caused by two species of virus which require the interaction of the viral genome-linked protein (VPg) with host translational initiation factor 4E, of which there are five isoforms. By mutating two of these isoforms, cassava plants displayed delayed and attenuated CBSD symptoms, as well as reduced severity and incidence of storage root necrosis. The ability to simultaneously change multiple genes in cassava using CRISPR-Cas9 and achieve these results shows the potential of this technology for Africa.

Banana streak virus (BSV) is a badnavirus of the family *Caulimoviridae*, affecting production of plantain (*Musa* spp., AAB genome). Upon infection BSV integrates in the host genome, mainly in the B genome of banana and is known as endogenous BSV (eBSV). The eBSV gets activated under stress conditions like propagation through tissue culture, hybridization or/and unfavourable environmental conditions such as temperature and water stress. Therefore, BSV is considered as one of the major challenges in plantain breeding and dissemination of hybrids having at least one B genome. Recently, it has been demonstrated that the eBSV sequences integrated in the B genome of plantain can be inactivated by creating targeted mutations in the viral sequences (Tripathi et al., 2019). The CRISPR/Cas9 system editing multiple targets in the integrated virus sequences may serve as a solution of inactivating the eBSV into the infectious virus particles.

By comparison, many crops in current use have been developed by random mutagenesis using either chemicals, such as ethyl methane sulphonate (EMS), 1-methyl-1-nitrosourea and 1-ethyl-1-nitrosourea, or irradiation, such as X-rays, Gamma rays and neutrons. Treatments such as these introduce random mutations throughout the plant's DNA. Breeders will select the trait they are looking for but have no idea what other, potentially harmful, mutations may also reside in these plants (Oladosu et al., 2016). Despite these obvious potential problems, crops developed by random mutagenesis have never required regulation. Food that is commonly consumed that have been developed by these methods include ruby red grapefruit, along with some 3,000 other crop varieties consumed by millions every day, such as mutant wheat used for bread and pasta and barley for beer and whiskey (Kastrinos, 2016).

Government policies

Laws governing GM crops are in place (or pending) in a number of African countries, but the various committees and authorities tasked with implementing the regulations are, in our experience, often subject to political pressures. Political will is, therefore, essential, but it tends to ebb and flow. The Ethiopian government, for example, was once highly skeptical about biotechnology but recently changed its position, and in 2017, they petitioned to join the WEMA Project to access biotech maize (Getnet, 2018). Various outreach efforts appear to have had a positive influence. In Tanzania, political will is trending in a positive direction, but significant hurdles remain (Abdu, 2019). In Kenya, political will to approve the cultivation of GM cotton appears to be growing with the ongoing national variety performance trials (NPTs) that will culminate in variety registration and commercialization (Meeme, 2019). Thus, the approval of *Bt* cotton for commercialization could create a positive precedent for GM maize approval in the country.

Impacts

It is estimated that the economic gains from biotech crops for South Africa for the period from 1998 to 2016 was ~US$2.3 billion, with US$330 million for 2016 alone (Brookes and Barfoot, 2018). A study focusing on gender-aggregated benefits by Gouse et al. (2016) found that female smallholder farmers and household members value GM herbicide tolerant maize higher than their male counterparts because of the labour-saving benefits the technology brings. The researchers found that females in HT maize seed adopting households were able to save ten–12 days of manual weeding per hectare, compared to their conventional and GM insect resistant maize planting and traditional manual weeding counterparts. Interestingly, females spent most of their extra time doing housework (cleaning and cooking) and working in their own or community vegetable gardens.

Conclusion

It is clear that Africa can gain great benefit from more widespread adoption of GM crops. However, it will require political will from leaders on the continent to allow their own farmers and consumers to participate in growing and using these highly advantageous commodities.

References

Abdu, F. (2019) GMO research not prohibited in Tanzania – expert. *Daily News* June 29, 2019. https://www.dailynews.co.tz/news/2019-06-295d1719a911818.aspx.

Bett, B., Gollasch, S., Moore, A., James, W., Armstrong, J., Walsh, T., Harding, R. and Higgins, T.J.V. (2017) Transgenic cowpeas (*Vigna unguiculata* L. Walp) expressing *Bacillus thuringiensis* Vip3Ba protein are protected against the *Maruca* pod borer (*Maruca vitrata*). *Plant Cell, Tissue and Organ Culture* **131**, 335–345.

Beyene, G., Chauhan, R.D., Ilyas, M., Henry Wagaba, H., Claude M. Fauquet, C.M., Douglas Miano, D., Alicai, T. and Taylor, N.J. (2017) A virus-derived stacked RNAi construct confers robust resistance to Cassava Brown Streak Disease. *Frontiers in Plant Science* **7**. doi:103389/pls.2016.02052.

Bosibori, B., Gollasch, S., Moore A., James, W., Armstrong, J., Walsh, T., Harding, R. and Higgins, T.J.V. (2017) Transgenic cowpeas (*Vigna unguiculata* L. Walp) expressing *Bacillus thuringiensis* Vip3Ba protein are protected against the *Maruca* pod borer (*Maruca vitrata*). *Plant Cell Tissue Organ Culture* **131**, 335–345. doi:10.1007/s11240-017-1287-3.

Brookes, G. and Barfoot, P. (2018) Farm income and production impacts of using GM crop technology 1996–2016. *GM Crops & Food* **9**, 59–89. doi:10.1080/21645698.2018.1464866.

Centre for Agriculture and Bioscience International – CABI (2017) www.invasive-species. org/fawevidencenote (accessed November 2017).

Edge, M., Oikeh, S.O., Kyetere, D., Mugo, S. and Mashingaidze, K. (2018) Water efficient maize for Africa: A public-private partnership in technology transfer to smallholder farmers in Sub-Saharan Africa. In: Nicholas Kalaitzandonakes, Elias G. Carayannis, Evangelos Grigoroudis and Stelios-Rozakis (eds) *From Agriscience to Agribusiness: Theories, Policies and Practices in Technology Transfer and Commercialization.* Springer Publication, New York, USA. ISSN 2197-5701 (electronic). 490p.

Getnet, T. (2018) *GMO corn test starts in Ethiopia.* https://www.capitalethiopia.com/featured/ gmo-corn-test-starts-ethiopia/ (accessed 7 October 2019).

Gomez, M.A., Lin, Z-J.D., Moll, T., Chauhan, R.D., Renninger, K., Beyene, G., Taylor, N.J., Carrington, J.C., Staskawicz, B.J. and Bart, R. (2018) Simultaneous CRISPR/Cas9-mediated editing of cassava *elF4E* isoforms *nCBP-1* and *nCBP-2* reduces cassava brown streak disease symptom severity and incidence. *Plant Biotechnology Journal* **17**, 421–434 doi:10.1111/pbi.12987.

Gouse, M., Sengupta, D., Zambrano, P. and Zepeda, J.F. (2016) Genetically modified maize: less drudgery for her, more maize for him? Evidence from smallholder maize farmers in South Africa. *World Development* **83**, 27–38.

IITA (2019) *Major breakthrough for farmers and scientists as Nigerian biotech body approves commercial release of genetically modified cowpea.* https://www.iita.org/news-item/major-breakthrough-for-farmers-and-scientists-as-nigerian-biotech-body-approves-commercial-release-of-genetically-modified-cowpea/ (accessed 7 October 2019).

ISAAA (2016) Global status of commercialized biotech/GM crops (2016) *ISAAA Brief* No. 52. ISAAA, Ithaca, NY.

ISAAA (2017) Global status of commercialized biotech/GM crops in 2017: Biotech crop adoption surges as economic benefits accumulate in 22 years. *ISAAA Brief* No. 53. ISAAA, Ithaca, NY.

Kastrinos, A. (2016) Delicious mutant foods: Mutagenesis and the genetic modification controversy. *Generic Literacy Project,* June 13, 2016.

Kyetere, D.T., Okogbenin, E., Okeno, J., Sanni, K.A., Oikeh, S., Munyaradzi, J., Nangayo, F., Kouko, E., Karuiki, A. and Issoufou, A. (In press) The role and contribution of plant breeding and plant biotechnology to sustainable agriculture in Africa. *Afrika Focus.*

Lynas, M. (2018) *Seeds of Science: Why We Got It So Wrong on GMOs.* Bloomsbury Sigma, London.

Meeme, V. (2019) Kenya reconsidering GMO crop ban to support food security. *Alliance for Science* April 30, 2019 https://allianceforscience.cornell.edu/blog/2019/04/kenya-recon sidering-gmo-crop-ban-support-food-security/ (accessed 7 October 2019).

Namukwaya, B., Tripathi, L., Tripathi, J.N., Arinaitwe, G., Mukasa, S.B. and Tushemereirwe W.K. (2012) Transgenic banana expressing *PFLP* gene confers enhanced resistance to Xanthomonas wilt disease. *Transgenic Research* **4**, 855–865. doi:10.1007/s11248-011-9574-y.

Oikeh, S.O., Nganyamo-Majee, D., Mugo, S.I.N., Mashingaidze, K., Cook, V. and Stephens, M. (2014) Chapter 13: Water efficient maize for Africa: An example of public-private partnership. In: D.D. Songstad, J.L. Hatfield and D.T. Tomes (eds) *Biotechnology in Agriculture and Forestry: Convergence of Food Security, Energy Security, and Sustainable Agriculture.* Vol. 67, Springer Publication, New York, USA. ISBN 978-3-642-55261-8. 372p.

Oladosu, Y., Rafii, M.Y., Abdullah, N., Hussin, G., Ramli, A., Rahim, H.A., Miah, G. and Usman, M. (2016) Principle and application of plant mutagenesis in crop improvement: A review. *Biotechnology and Biotechnological Equipment* **30**, 1–16. https://doi.org/10.1080/ 13102818.1087333.

Popelka, J.C., Gollasch, S., Moore, A., Molvig, L. and Higgins, T.J.V. (2006) Genetic transformation of cowpea (Vigna unguiculata L.) and stable transmission of the transgenes to progeny. *Plant Cell Reports* **25**, 304–312.

Shepherd, D.N., Mangwende, T., Martin, D.P., Bezuidenhout, M., Kloppers, F.J., Carolissen, C.H., Monjane, A.L., Rybicki, E.P. and Thomson, J.A. (2007) Maize streak virus-resistant transgenic maize: A first for Africa. *Plant Biotechnology Journal* **5**, 759–767.

Tripathi, J.N., Ntui, V.O., Ron, M., Muiruri, S.K., Britt, A. and Tripathi, L. (2019) CRISPR/Cas9 editing of endogenous *banana streak virus* in the B genome of *Musa* spp. overcomes a major challenge in banana breeding. *Communications Biology* doi:10.1038/s42003-019-0288-7.

Tripathi, L., Atkinson, H., Roderick, H., Kubiriba, J. and Tripathi, J.N. (2017) Genetically engineered bananas resistant to Xanthomonas wilt disease and nematodes. *Food and Energy Security* **6**, 37–47. doi:10.1002/fes3.101.

Tripathi, L., Mwaka, H., Tripathi, J.N. and Tushemereirwe, W.K. (2010) Expression of sweet pepper *Hrap* gene in banana enhances resistance to *Xanthomonas campestris* pv. *musacearum*. *Molecular Plant Pathology* 11, 721–731. doi:10.1111/J.1364-3703.2010.00639.X.

Tripathi, L., Mwangi, M., Abele, S., Aritua, V., Tushemereirwe, W.K. and Bandyopadhyay, R. (2009) A threat to banana production in east and central Africa. *Plant Disease* **93**, 440–451. doi:10.1094/PDIS-93-5-0440.

Tripathi, L., Tripathi, J.N., Kiggundu, A., Korie, S., Shotkoski, F. and Tushemereirwe, W.K. (2014) Field trial of Xanthomonas wilt disease-resistant bananas in East Africa. *Nature Biotechnology* **32** (9), 868–870. doi:10.1038/nbt.3007.

19 Unleashing the power of vegetables and fruits in Southern Africa

Thomas Dubois, Thibault Nordey and Umezuruike Linus Opara

Vegetables and fruits for nutrition and income generation

Vegetables and fruits are a vitally important source of micronutrients, vitamins and minerals and therefore essential components of balanced and healthy diets. Yet, production and demand are still too low to provide the population in many countries in Southern Africa with the minimum per capita consumption of 400 g/person/day required for good health (FAO/WHO, 2004). In Africa, most countries do not reach even half of this minimum requirement (Ambrose-Oji, 2009). Of particular importance are traditional African vegetables such as amaranth, African eggplant, roselle, okra and many others that have been cultivated in African gardens for decades. These "superfoods" provide much higher amounts of provitamin A, vitamin C and several important minerals than staples and globally traded vegetables (WorldVeg, 2018). They also contain antioxidants and other health-related phytochemicals that prevent chronic diseases such as cancer and cardiovascular disease. For growers, vegetables and fruits are often more profitable per unit volume than staples and command higher profit margins and farm gate prices per unit area of production, especially when access to farmland is limited and labour is surplus (Gabre-Madhin and Hagglade, 2003), like in many rural parts of Southern Africa. As such, vegetables and fruits are great income generators and profit can be achieved on relatively small land units. Nevertheless, sustainable and market-oriented production and consumption of safe vegetables and fruits is not being achieved in many parts of Africa, including Southern Africa. The predominant staple-based diets in the region, based on maize, wheat and rice need to be diversified with vegetables and fruits to increase nutrition and income.

Key drivers for horticulture in Southern Africa: constraints and opportunities

Climatic conditions

Climatic conditions (temperature and solar radiation) in Southern Africa are suitable for year-round production of vegetables and tropical fruits (such as mango, banana, citrus and avocado) and temperate fruits (such as apple, pear and grape) at higher elevations. The main limiting factors, however, are pests, diseases, weeds, low soil fertility and lack of year-round water supply. These constraints are exacerbated by the effects of climate change. Whereas temperatures and carbon dioxide will gradually increase, more worry that rainfall will become more erratic with increased incidence of floods and droughts. A major effect is likely increases in pest and disease pressures.

Urbanization and the supermarket revolution

Africa is experiencing the world's most rapid rate of urbanization at ~3.5% per year, with over 40% of its population already living in cities. In the near future, food demand in Africa will come from urban populations, and agricultural systems, including horticulture, will need to become more efficient, with lower transaction costs and increased capital per worker, facilitating a sharp increase in the use of purchased inputs and specialized capital provided by local agri-businesses (Masters et al., 2013). Production will be concentrated in or around urban areas. Because of their short value chains, scale neutrality and high profitability on small plots of land, horticulture is well positioned to become a key driver of future African agriculture, which is set to shift towards urban and peri-urban systems.

Consumption habits are expected to change with urbanization. Consumers in Southern Africa will increasingly demand uniformity, packaging and convenience offered by formal retail outlets as opposed to traditional markets, triggering a supermarket revolution as seen recently in Asia and Latin America. By 2004, South Africa had almost 2,000 large supermarkets, from very few during the Apartheid era, and Lusaka, Zambia, has more than 100 large supermarkets, from virtually none three decades earlier (Reardon et al., 2004). It is hoped that this transformation in Southern Africa will improve market institutions, break monopolies, improve farmers' terms of trade and create a competitive market for inputs and farm credit, as has happened during Asia's supermarket revolution (Reardon, 2013).

Postharvest

Postharvest losses of horticultural produce in Southern Africa are staggering, e.g., 55% for mangoes, 43% for leafy vegetables and 33% for tomatoes (Affognon et al., 2015), and as such, the lack of suitable postharvest handling and processing is likely among the key constraints for horticulture in sub-Saharan Africa. Short shelf-life is a major hurdle, especially for vegetables. High-value perishable products such as vegetables and fruits are presently almost exclusively supplied by urban and peri-urban farmers because of resulting short value chains (Chagomoka et al., 2016).

Cold chain technologies and infrastructure remain the holy grail for fresh horticultural value chains in Southern Africa. An alternative to the need of cold chains is value addition through processing, including pickling, drying and sauce and jam making. Numerous preharvest and postharvest technologies are developed for smallholder farmers, including improved varieties that have longer shelf-life or can withstand transport better; technologies related to field handling and harvesting procedures to reduce damage; sanitizing pretreatments such as the use of calcinated calcium; and modified atmospheric packaging. Although these technologies generally work well under local conditions and would be easily adoptable by smallholder horticulturalists, they are difficult to pull into the market due to several factors, including lack of finance and technical know-how. Even the simplest of postharvest technologies, such as sorting and grading, can increase market value by 20–60%, merely by segregating produce by size or applying ethylene on climacteric fruit to control ripening. Yet few smallholder farmers engage in these improved postharvest quality management activities and therefore are limited in their access to regional and international markets.

Dichotomy between local low-input and regional high-input production

In Southern Africa, smallholder cultivation of fruits and vegetables for home and local market consumption differs significantly from medium- and large-scale business-oriented production. Each system has different investment capacity and requires different production technologies, especially for vegetables where greenhouses may be needed. Only more intensive and profitable production systems can ensure compliance with regulations on maximum pesticides residue levels and target export market standards. Although most horticultural products are still sold in local markets, exports from Southern Africa have increased more than 2.5-fold between 2006 and 2016 (FAO, 2017). Whereas regional exports are much more important than exports to Europe, Asia and the U.S., exports to the latter areas will develop further under the economic partnership agreement signed between the EU and six SADC members (Bertelsmann-Scott and Markowitz, 2018).

Food safety

A major barrier to vegetable and fruit production is access to sufficient, year-round supply of clean water for quality irrigation and postharvest cleaning. Use of unclean water, such as wastewater in urban areas or water used in wet markets, poses serious health risks due to microbial contamination, especially for fresh produce such as green leafy vegetables. A microbial risk assessment in Ghana's major cities estimated a possible loss of about 12,000 disability-adjusted life years annually resulting from the consumption of wastewater-irrigated lettuce alone (Drechsel and Seidu, 2011). In addition, green leafy vegetables accumulate heavy metals present in industrially contaminated water in high concentrations (Arora et al., 2008) to the potential detriment of human and livestock health.

Consumer health is further compromised by excessive and erroneous use of pesticides. Their application is relatively cheap for high-value horticulture crops, and they are seldom applied according to label specifications (proper dose, active ingredient, timing, mixture, application method and/or safety equipment) (Dinham, 2003). Runoff causes contamination of soil and especially groundwater, further compromising the water supply. Farmers do not have proper access to training in chemical use from extension services.

Technologies that can be implemented

Moving from technological to institutional and organizational interventions

Numerous efficient technologies are available for smallholder horticulturalists (i.e., irrigation systems, fertilization, improved varieties, integrated pest management, protected cultivation). However, enticing adoption of these technologies is difficult, partly due to the high input requirements and complexity of horticultural value chains as explained previously. It is important that technological innovations are embedded in institutional (laws, regulations and standards but also soft assets such as social and cultural norms) and organizational contexts and innovations.

One example is implementing postharvest-related technologies at scale. Postharvest is a complex problem, especially for smallholder farmers, and complex problems need innovative solutions. Most researchers, when focusing on developing, testing and validating new postharvest technologies, focus on the technological aspects for smallholder farmers, yielding new products and processes. However, actors, especially in complex systems such as postharvest systems for smallholders in Africa, are not isolated and never act alone. For postharvest technology to work, the technological, institutional and organizational aspects are equally important. Many functions along the value chain, including processing, packaging, wholesaling and retailing, are becoming separated in space and time, with different facilities and infrastructure where these functions take place. In Africa, most vegetable and fruit farmers are small-scale, often women, with little access to inputs, finance, information and markets. Aggregation of these many informal or semi-formal growers is an essential organizational innovation that remains a huge challenge.

Another example is the supermarket revolution that Southern Africa is currently experiencing. These supermarkets are leading the way in overhauling the traditional procurement model of sourcing products from the traditional wholesalers and wet markets. The supermarket chains seek constantly to lower product and transaction costs and risk – with the result of selecting only the most capable farmers, and in many African countries these are mainly the medium and large farmers. Moreover, as supermarkets compete with each other and with the informal sector, they will not allow consumer prices to increase, and ultimately horticultural farmers cannot avoid paying for costs such as safe water and record keeping systems. This will be a huge challenge for small operators and consequently, retail concentration will cascade into supplier concentration (Reardon et al., 2004; Reardon, 2013).

Seed and seed systems

A major bottleneck is lack of selection and breeding research related to vegetables and fruits that are adapted for Southern Africa, and this is especially the case for traditional African vegetables (Dinssa et al., 2016). Efforts related to selection and breeding need to be coupled with effective seed supply systems, proper agronomic practices and an adequate regulatory and policy framework, supported by a growing private seed supply sector.

The horticultural seed sector in Southern Africa is not as regulated nor effective as that of staple crops. Few companies are present that have active breeding and seed dissemination programmes targeting smallholder farmers (Access to Seed Foundation, 2016). Also, seed companies do not have good geographic penetration and only serve directly the large metropolitan areas. In addition, low quality and even counterfeit seeds flood local markets.

The seed sector needs to function using a value chain approach and ensure strategic linkages of seed producers to key value chain actors such as extension (for technical advice), regulation (for seed inspection and certification), research (for provision of varieties and foundation seed), market (as seed/off-takers) and credit service providers. Only when driven by market forces will seed production and marketing be sustainable. Governments still play a crucial role in helping this transition, for example, through introduction of quality declared seed systems into national regulations.

Needed policies and their processes for implementation

Increasing consumer demand for vegetables and fruits

Many towns and cities in sub-Saharan Africa are characterized by unsustainable consumption patterns. They are undergoing a nutritional transformation away from traditional diets to increased consumption of dairy products and meat, less complex carbohydrates and a general decline in dietary diversity and nutritional value (Ambrose-Oji, 2009). Such unsustainable consumption patterns are especially worrying among children, as dietary patterns are well established by the age of 13, and dietary patterns that develop in childhood often are maintained into adulthood (Nicklas, 1995). Governments are pivotal in reversing this worrisome decline and should actively stimulate demand for locally produced vegetables and fruits. Of particular focus are children, for whom the importance of fruit and vegetable consumption should be included in school curricula.

Quality standards through voluntary labels

Food safety is of high concern in horticultural products. However, in Southern Africa, there are few standards such as those related to maximum residue limits, and if these standards are present, they are mostly not implemented or adhered to, except for export overseas. However, large outlets in sub-Saharan Africa, such as international supermarkets, are starting to pay attention to maximum residue limits and traceability. Together with the move to more global crops, this may force them to import vegetables and fruits for African markets from more advanced economies, at the detriment of local farmers. This scenario is already happening in the large cities within Kenya, Mozambique and Tanzania (Dinham, 2003). To implement good agricultural practices and remain within maximum residue limits, policy regulatory frameworks need to be in place and enforced by officials. At present, lack of human, infrastructure and financial capacity will likely not make this possible in the near future in many countries in Southern Africa.

More important, the driver for ensuring safely produced vegetables and other horticultural products needs to come from the demand side – consumers. However, unlike in developed countries, consumers in sub-Saharan Africa do not yet demand or are not willing to pay a premium for safely produced food. Through Southern Africa's nascent supermarket revolution, transformation from local, low-value and unbranded foods to branded, high-value and some-times processed goods is set to happen. It is important that smallholders reap the benefits of this revolution. A solution is to slowly and stepwise implement pri-vate safety standards for horticultural products and selling these to increasingly food safety-aware consumers using local, voluntary labels (e.g., by the Common Market for Eastern and Southern Africa Comprehensive African Agriculture Development Programme for its member countries).

Improving market access through farmer aggregation, with a focus on youth and women

Horticultural producers usually sell their products to village collectors or petty traders at low farm-gate prices. Producer groups exist but are too small to negotiate prices with traders or processors directly and only serve as collec-tion centres for larger traders or processors. Also, limited or no access to formal sources of finance creates an impediment to vegetable and fruit production for youth and women because they have insufficient or no collateral. Commercial banks, due to high transaction costs associated with credit monitoring and pro-duction risks of individual smallholders or informal groups, may be reluctant to provide credit or invest. Finally, governmental extension systems are often poorly skilled in other than staple crop systems, are geared towards men and have little or no access to new technologies and practices that are relevant for smallholder horticulture farmers.

Women and youth are especially disadvantaged in relation to market access. Women are the custodians of local horticultural value chains in Southern Africa, functioning as the gatekeepers of the nutritional benefits of vegetables and fruits for local communities, while youth are attracted to their income-earning potential. As such, women and youth have a huge potential to contribute to and benefit from horticulture, yet there are multiple reasons why this potential is not being realized. A primary reason is lack of skills and knowledge, which are in general more elevated than those of other crops and for which youth and women are at a disadvantage compared to men, often resulting in adhering to suboptimal and subsistence-based agricultural practices. Youth and women also lack access to high quality inputs such as improved seeds and fertilizers. Provided capital is available, the majority of smallholder horticultural producers buy their inputs and technologies from local agro-dealers. The quality of these inputs, particularly seeds and fertilizers, is often very poor, and accreditation through governmental bodies is in most countries absent or not enforced.

Beyond staples: a shift in agricultural research towards horticulture

Although agriculture is currently deemed an essential driver for development in sub-Saharan Africa, research on horticulture is treated as secondary to that of staple crops, despite the nutritional and income-generating potential of vegetables and fruits. For instance, large African bodies setting the agricultural research agenda, such as the Alliance for a Green Revolution in Africa, do not include horticulture in their research agendas. Granted, unlike staple crop production, horticulture in Southern and sub-Saharan Africa as a whole is blessed with the attention of a vibrant private sector. However, this private sector is focused on production of mainly global vegetables and fruits and is heavily biased towards overseas export markets. Another reason for lack of investment in horticultural research may be the absence of precise data. Vegetables are largely produced and sold informally, and production, sales and consumption figures are underreported or not documented at all. In FAOSTAT and other databases on which governments and donors rely to make decisions, there is a strong focus on staples, with horticulture largely ignored or grouped together, making it impossible to make valid judgments for any horticultural group, especially local vegetables. Horticultural research requires urgent donor attention and a proactive research agenda, based on disaggregated and reliable production and consumption data. Especially postharvest investments are urgently needed; currently, < 5% of funding for horticultural research and extension has been allocated to postharvest issues over the past 20 years (Kitinoja et al., 2011).

Focus on traditional African vegetables for regional trade instead of commoditized export crops

Because of a fast-increasing population and rapid urbanization, natural resources will face increased pressure to produce adequate and nutritious food, and this pressure is further compounded by climate change. Crops are required that can withstand these environmental shocks, increasing resilience of households that grow them. Tomato, cabbage and onion, global commodity vegetables currently grown in Southern Africa as well as French bean and pepper produced for export markets, are relatively sensitive to environmental constraints. On the other hand, traditional African vegetables, which are grown in specific localities where they are economically important, may be better suited to thrive under suboptimal and even marginal conditions faced by many vulnerable households. Many of these traditional African vegetables are more resistant to biotic and abiotic stresses (requiring less inputs like fertilizers and pesticides), are easy and fast to grow (allowing for multiple cropping cycles per year on the same plot) and are much more nutritious. Nevertheless, a clear shift exists from traditional to global vegetables in consumption patterns in sub-Saharan Africa (Weinberger and Msuya, 2004). It is important for governments and regional bodies to pay particular attention to these neglected crops.

References

Access to Seed Foundation (2016) *Access to Seeds Index Report 2016*. Access to Seed Foundation, Amsterdam.

Affognon, H., Mutungi, C., Sanginga, P. and Borgemeister, C. (2015) 'Unpacking postharvest losses in sub-Saharan Africa: A meta-analysis', *World Development*, vol 66, pp. 49–68.

Ambrose-Oji, B. (2009) 'Urban food systems and African indigenous vegetables: Defining the spaces and places for African indigenous vegetables in urban and peri-urban agriculture', in, C. M. Shackleton, M. W. Pasquini, and A. W. Drescher (eds) *African Indigenous Vegetables in Urban Agriculture*. Earthscan, London.

Arora, M., Kiran, B., Rani, S., Rani, A., Kaur, B. and Mittal N. (2008) 'Heavy metal accumulation in vegetables irrigated with water from different sources', *Food Chemistry*, vol 111, pp. 811–815.

Bertelsmann-Scott, T. and Markowitz, C. (2018) *Policy Briefing 174. The Impact of the SADC EPA on South Africa's Agriculture and Agro-Processing Sectors*. South African Institute of International Affairs, Johannesburg.

Chagomoka, T., Unger, S., Drescher, A., Glaser R., Marschner, B. and Schlesinger, J. (2016) 'Food coping strategies in northern Ghana. A socio-spatial analysis along the urban – rural continuum', *Agriculture and Food Security*, vol 5, pp. 1–18.

Dinham, B. (2003) 'Growing vegetables in developing countries for local urban populations and export markets: Problems confronting small-scale producers', *Pest Management Science*, vol 59, pp. 575–582.

Dinssa, F. F., Hanson, P., Dubois, T., Tenkouano, A., Stoilova, S., Hughes, J. D. A. and Keatinge, J. D. H. (2016) 'AVRDC – The World Vegetable Center's women-oriented improvement and development strategy for traditional African vegetables in sub-Saharan Africa', *European Journal of Horticultural Science*, vol 81, pp. 91–105.

Drechsel, P. and Seidu, R. (2011) 'Cost-effectiveness of options for reducing health risks in areas where food crops are irrigated with treated or untreated wastewater', *Water International*, vol 36, pp. 535–548.

FAO (2017) *FAOSTAT*. FAO, Rome.

FAO/WHO (2004) *Fruits and Vegetables for Health*. FAO, Rome.

Gabre-Madhin, E. Z. and Hagglade, S. (2003) *Successes in African Agriculture: Results of an Expert Survey. Markets and Structural Studies Division Discussion Paper No. 53*. IFPRI, Washington, DC.

Kitinoja, L., Susanta, S. S. and Kader, A. A. (2011) 'Postharvest technology for developing countries: challenges and opportunities in research, outreach and advocacy', *Journal of the Science of Food and Agriculture*, vol 91, pp. 597–603.

Masters, W. A., Djurfeldt, A. A., De Haan, C., Hazell, P., Jayne, T., Jirström, M. and Reardon, T. (2013) 'Urbanization and farm size in Asia and Africa: implications for food security and agricultural research', *Global Food Security*, vol 2, pp. 156–165.

Nicklas, T. A. (1995) 'Dietary studies of children: the Bogalusa heart study experience', *Journal of the Academy of Nutrition and Dietetics*, vol 95, pp. 1127–1133.

Reardon, T. (2013) *The Economics of Urbanization, Farm Technology, and Farm Size Distribution in Asia.* CGIAR, Montpellier.

Reardon, T., Timmer, P. and Julio Berdegue, J. (2004) 'The rapid rise of supermarkets in developing countries: Induced organizational, institutional, and technological change in agrifood systems', *Electronic Journal of Agricultural and Development Economics*, vol 1, pp. 168–183.

Weinberger, K. and Msuya, J. (2004) *Indigenous Vegetables in Tanzania: Significance and Prospects.* WorldVeg, Shanhua.

WorldVeg (2018) *Annual Report 2017.* WorldVeg, Shanhua.

20 Going digital

Harnessing the power of emerging
technologies for the transformation
of Southern African agriculture

Heike Baumüller and Muhammadou M.O. Kah

Introduction

The story of the impressive expansion of mobile phones across the developing
world has often been told. Southern Africa is no exception to this trend. The
region has been digitalizing fast, albeit unevenly. The average mobile penetration rate across the SADC (Southern African Development Community) region
has reached 88 per 100 people in 2018 and 30% of the population has access
to the web primarily through mobile phones.[1] However, these numbers hide
significant differences between countries. Looking at the statistics by income
quartile, we find that the share of mobile phone, internet and social media users
is considerably higher in higher-income countries than in the lowest-income
countries (Figure 20.1). Whereas mobile phones are the dominant means to
access the web across the region, users in higher income countries also use
computers for much of their web surfing, and more of them benefit from faster
internet connectivity through broadband (Figure 20.2). We are also seeing a
continued rural-urban divide in particular with regard to internet access and
network speeds (Mothobi et al., 2017).

These discrepancies may go some way towards explaining why digitalization
has not yet led to the transformation of Southern African agriculture that it
promises. Although there are numerous digital services available for smallholder
farmers, they are often disjointed, short-lived, lacking in scale or financially
unsustainable. Farmers have proven difficult to reach because of low levels of
digital literacy, technological sophistication and purchasing power. Information
and communication technologies (ICT) can undoubtedly play a role in the
toolbox of measures to assist smallholder farmers. However, the transformative
power of digital technologies will in the short- to medium-term show itself in
segments higher up in the value chain and among larger producers and industry
players, including in areas such as production planning, processing, supply chain
management, quality control and financial services.

Opportunities for advanced digital technologies
in African agriculture

We recognize four digital technology trends that could help to revolutionize
agriculture in Southern Africa. What these trends have in common is that

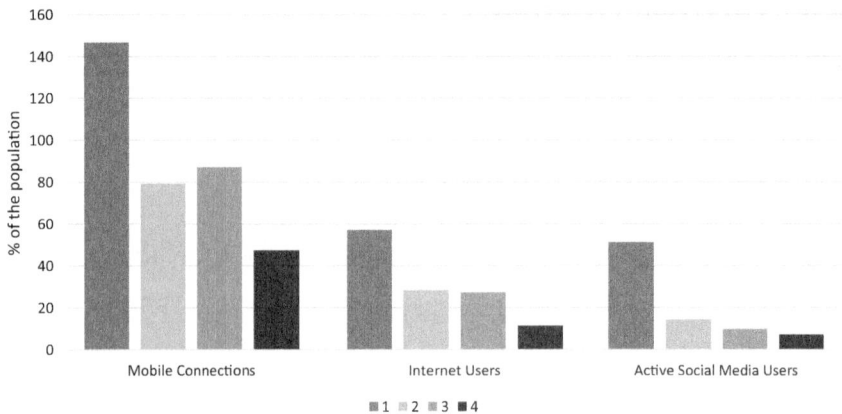

Figure 20.1 Internet use and mobile connections in Southern Africa

Source: Digital in Africa 2018, Hootsuite and We Are Social, www.slideshare.net/wearesocial (accessed 10 Feb 2019)

Note: Countries categorized by GDP/capita: 1 – Seychelles, Mauritius, Botswana, South Africa; 2 – Namibia, Swaziland, Angola, Zambia; 3 – Lesotho, Tanzania, Zimbabwe; 4 – Comoros, Madagascar, Mozambique, Malawi, Democratic Republic of the Congo

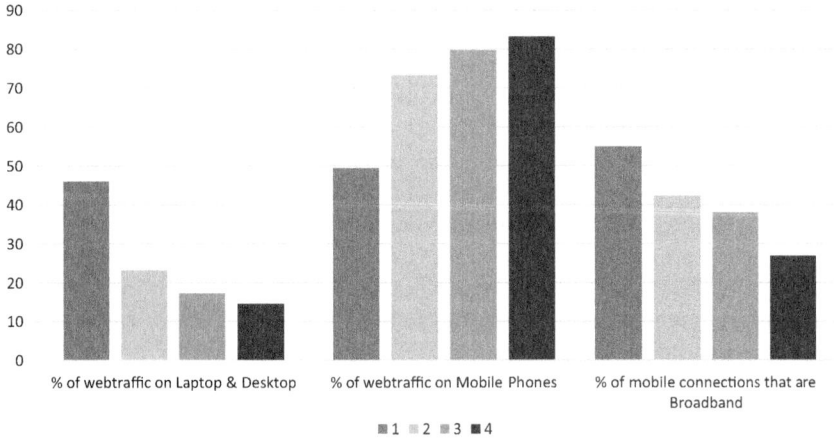

Figure 20.2 How Southern Africans are accessing the internet

Data source: Digital in Africa 2018, Hootsuite and we are social, www.slideshare.net/wearesocial (accessed 10 Feb 2019)

Note: Countries categorized by GDP/capita: 1 – Seychelles, Mauritius, Botswana, South Africa; 2 – Namibia, Swaziland, Angola, Zambia; 3 – Lesotho, Tanzania, Zimbabwe; 4 – Comoros, Madagascar, Mozambique, Malawi, Democratic Republic of the Congo

they often make use of mobile phones but also incorporate other advanced digital technologies and backend solutions. These trends should not be seen in isolation, however. Rather, their true value will come from their integration.

Data for decision making

Diverse devices that are interconnected through the so-called "Internet of Things", such as radio-frequency identification (RFID) tags, sensors, drones, satellites or meters, are making it possible to collect a wide range of data. At the backend, sophisticated analytical tools can integrate, analyze and visualize the data, including through data mining, artificial intelligence, machine learning and smart algorithms. This could be a leapfrogging technology in Africa where access to data for decision making remains a challenge. Automating a larger part of the data collection can reduce data collection costs, increase accuracy and enable the integration and joint analysis of larger data sets. Areas of application include, for instance, smart water and farmland management, pest and disease monitoring, recording and tracking of goods and animals, credit scoring or automated processing. Among the technology trends presented here, these are the areas where most progress has been made in Africa.

Examples of related applications can be found across the continent, often developed by local start-ups. Many applications focus on data collection at the farm level, using a range of tools to gather data, such as drones, satellites, sensors or mobile phones. For instance, the Nigerian company BeatDrone[2] employs drones for crop supervision and farmland mapping, while the South African company DroneScan[3] uses them for inventory management in warehouses. Also, in South Africa, the company Aerobotics[4] uses satellite and drone images for pest and disease detection on farms. Data analytics are also assisting in the provision of financial services, including credit[5] and insurance[6]

Platforms for goods, services and people

Digital technologies are offering new opportunities to set up networks of users to exchange goods, services, labour, money or knowledge. Platforms such as Facebook, WhatsApp and Uber are prominent examples of this trend, which has also been described as platform economy, gig economy, microtask platforms, sharing economy or uberization. Areas of application in food and agriculture include hiring labour, home food delivery, sharing machinery (with or without operator), trading agricultural products, sharing knowledge among farmers[7] or accessing finance through crowdfunding[8].

In the so-called gig economy, "paid work [is] allocated and delivered by way of internet platforms without an explicit or implicit contract for long-term employment" (Graham et al., 2017, p. 2). This could offer new employment opportunities for the self-employed even in remote areas as long as they are well-connected to the internet, thereby removing constraints of local labour markets (ibid). Such platforms already exist in Africa, albeit not in the agriculture sector. The taxi platform Uber is probably the most well-known example. Another model is being applied to the sharing of machinery, such as Hello Tractor in Nigeria.[9] A study of seven African countries has shown that most of the services provided through such platforms are manual or physical labour e.g., laundry, driving (Onkokame et al., 2017).

Such networks can also be used to set up virtual trading platforms for food and agriculture products. They would lend themselves in particular to the sale of goods where the quality is consistent and subject to certain standards that can be easily certified or verified, such as seeds, agro-inputs or processed food products. Trade in fresh products would require additional quality control mechanisms. ICTs can be used to ensure quality and reliability of such virtual trading, for instance GPS for tracking the goods or animals, cameras to take pictures or videos or sensors to measure moisture content. QualiTrace[10] in Ghana, for instance, allows farmers to check the authenticity of agro-inputs by entering the serial number on their mobile phone. Digital payment systems make such platforms possible and can also assist in trust building, where the payments are held by the platform providers until the quality has been verified.

Blockchain technologies for traceability, transparency and security

Blockchain technologies have attracted much attention even though commercial applications in Africa are still limited (Pollock, 2018). The first and most well-known application is the cryptocurrency Bitcoins launched in 2008. However, the utility of blockchains in the agri-food industry goes far beyond Bitcoins (Tripoli and Schmidhuber, 2018). Also referred to as Distributed Ledger Technology, blockchains promise to revolutionize record keeping, product tracing and contracting. Simply put, a blockchain allows users to move from centralized to distributed storage and verification of data on multiple computers, which makes verification cheaper and faster, transactions traceable and transparent, and records immutable (ibid). Areas of application include, for instance, smart contracts, management of registries, supply chain management and financial services.

Only a few examples of blockchain applications in food and agriculture can be found in Africa, but they already give an indication of the technology's potential. The provision of financial services beyond crypto-currencies, such as loans or insurance, is one of the promising areas of application. For instance, the Kenyan company Twiga Foods,[11] which links farmers and vendors through supply chain management, partnered with IBM to use a blockchain to manage micro-lending to small food kiosks in Kenya (Kinai, 2018). Insurance providers can also take advantage of blockchain-enabled smart contracts which are used to automate the contracting process by auto-executing certain actions when specific conditions are met, such as insurance pay-outs. Such contracts have the potential to simplify dealing with a large number of dispersed customers in rural areas. They could also reduce the need for litigation and courts, which can be useful in countries with weak judicial systems (Shadab, 2014).

Some companies are experimenting with distributed ledgers to trace products from African farmers to the end consumer. The German software company SAP is piloting its SAP Cloud Platform Blockchain[12] to manage the entire supply chain in collaboration with various food companies (Milano, 2018). Another example is AgriLedger,[13] which has piloted its blockchain-based traceability system with Kenyan wheat farmers to assist them with record-keeping on their own farm and tracking the produce along the value chain. Land registries are another promising area of application. For instance, the Rwandan government is collaborating with Medici Land Governance[14] to set up a blockchain-based land management and property rights platform (Overstock.com, 2018).

Autonomous systems for greater efficiency in production and processing

The use of autonomous machines for automation in food and agriculture also holds great promise, for instance to assist in farming operations, logistics and processing. The widespread use of such machines may still be some way off in Southern Africa, but a few examples can give a glimpse of what is possible. Drones and robots, for instance, can be used for seeding, weeding[15] or the application of fertilizer or pesticides[16]. These machines can also store information for future decision making, for instance to apply fertilizer at the specific location where the seed was planted. Automation can also assist in the operation of farm machinery. Automated tractor steering, for example, reduces the need for experienced machine operators and mitigates the risk of equipment damage.

The food industry offers many more opportunities for digital automation, also referred to as Industry 4.0. It can help to increase efficiency, reduce waste, ensure consistent quality and meet safety standards. Automation in the industrial sector is of course not new, but the arrival of digital technologies is opening up new opportunities by offering access to a wider range of data and analytical tools for machine learning that can facilitate autonomous cyber-physical systems of connected devices (Siemens, 2017). Realistically, simpler automation will be the first step for the African food industry. Even in industrialized countries advanced digital manufacturing technologies are still in their early days and the drivers that incentivize their application, such as increasing labour costs, a sizeable middle class, well-developed internet infrastructure and high-tech skills, are less pronounced in Africa than in other global regions (Dinham, 2017).

Creating an enabling environment for digitalization

Many of these ICTs may seem far out of reach for Southern Africa now, but as the technologies improve, simplify and become cheaper, they will become increasingly applicable in the region. To assist adoption of transformative digital technologies in the food and agriculture sector, it is essential that Southern African countries provide the necessary framework including policies, infrastructure, skills development and start-up support.

Policies are required to enable the provision of digital services in the first place. At the regional level, the SADC countries put ICTs high on the political agenda already in 2001 by adopting the Declaration on Information and Communication Technology, which committed member states to work towards bridging the digital divide in the region. However, as the numbers provided have shown, the divide continues to exist. Implementation of relevant regulations at the country level are needed to move the digital agenda forward, for instance in areas such as taxation, licensing, spectrum management, rules to govern network access and universal access funds.[17] Conducive competition policies are also essential. The Mobile Network (MNO) market in SADC is still concentrated and dominated by a few big players such as Airtel, MTN and Vodacom (Mothobi et al., 2017). More competition could help to reduce prices and improve service quality and diversity.

At the same time, policies to ensure access to digital technologies and services will need to be accompanied by policies that govern their use. An area that has not received enough attention in Africa is regulating the collection and protection of data. There is also a need to mitigate any downsides of digitalization. For instance, potential job losses that may result from automation require contingency plans to offer alternative employment. At the same time, policies will need to ensure that ICTs narrow rather than widen the existing digital divides (e.g., rural/urban, men/women). A survey conducted in seven African countries has shown that only 17% of respondents in rural areas had used the internet compared to 43% in urban areas (RIA et al., 2018). Usage was considerably lower among women.[18]

Policies need to be accompanied by investments in infrastructure to ensure connectivity, including sea cables, fibre optic cables, internet exchange points and last mile infrastructure. Not only the reach of mobile networks is important but also their speed and usage costs. Mobile data prices vary widely across Southern Africa ranging from US$2.3 per 1 GB in Tanzania to US$35.3 in Swaziland in 2017 (Mothobi et al., 2017). In particular, the last mile infrastructure to reach end users requires attention. Decentralized solutions could prove particularly useful, such as WLAN, WiMAX and WiBack. Equally important is the expansion of electricity access. Appropriate and sustainable alternative energy sources, especially solar energy, will be a key to the adoption of emerging technologies in "smarting" Africa's agriculture.

As emerging digital technologies are rolled out, African countries need to ensure they have adequate human capacities to operate and maintain these technologies. This, undoubtedly, will require radical shifts in policy, mind-sets and awareness, as well as continuous investment to provide the appropriate learning infrastructure for smart-farming and agri-tech entrepreneurship. It will need to include early interventions in the education ecosystem as well as intensive and comprehensive extension and advisory services. Here, there is an opportunity to attract young people to work in the food and agriculture sector and create much needed jobs for Africa's growing youth population. The establishment of agri-tech learning and entrepreneurship hubs may help to reorient young Africans towards agriculture and build the requisite "smart agri-talent pool" to drive smart agriculture in Africa.

The right policies, investments and skills will help to build a conducive innovation environment for local start-ups to thrive. Private sector engagement is essential to promote digitalization in food and agriculture since companies possess the necessary technical know-how and can keep up to date with the fast-developing ICT sector. Other supportive measures can include innovation hubs that offer a space to meet mentors and investors, incubation centres to develop ideas into full-fledged businesses, access to start-up funding through competitions or investors and, importantly, access to mid-level funding that allows start-ups to scale. Such initiatives have been shown to have driven the emergence of a local ICT sector in Kenya (Baumüller, 2016).

Frugal ICT innovation to cover the "last mile"

It is widely established in the literature that the ever-evolving nature of digital technologies makes it easy to get caught up in the excitement of new innovations. It is important that we do not overlook the value of simple, pragmatic and appropriate ICTs in African agriculture, especially when reaching out to farmers. After all, basic mobile phones were technically sophisticated enough to spark the mobile revolution in Africa. Thus, the development of agricultural services that can be delivered through simpler phones is far more useful for most farmers than more advanced technology that is dependent on the internet.

Simple phones lend themselves to diverse types of services. They can be used, for instance, to collect information, such as customer details when registering for a service or crowdsourcing information about crop disease outbreaks. They can also be used for the dissemination of standardized information such as weather forecasts, price information or specific instructions for crop management and postharvest handling. In particular, ICT-enabled financial technology (FinTech) for smallholder farmers can be key in reforming the financing of the agriculture value chain and ecosystem in Southern Africa. Using the previously mentioned emerging technologies, FinTech can increase efficiency and create new financial models for smallholder farmers, for instance to allow for better and simpler ways for payments and to facilitate access to investments, financing or insurance.

Smallholder farmers will not need to understand the sophisticated technologies under the hood of digital services to adopt, access and benefit from their power. The growth of cloud computing, also in Africa (Goldstuck, 2018), is making it possible to run sophisticated systems on centralized computers while using simple phones as interfaces to access the services (Baumüller, 2017). Intermediaries can also play an important role in facilitating access, such as village kiosks, agro-dealers or mobile money agents. Even the data that is incidentally collected by mobile phones can assist in the provision of services. For instance, the U.S. company Branch[19] uses artificial intelligence to analyze digital data, including handset details, SMS logs, repayment history, GPS data, call logs and contact lists, to make lending decisions in Kenya, Nigeria and Tanzania. However, such services should only be rolled out once adequate data protection laws are in place.

Conclusions

It may still be early days for the digital revolution in Southern Africa's food and agriculture sector, but it is essential that policymakers adopt a forward-looking vision to create an innovation environment that supports the use of tomorrow's digital solutions, not only today's. The emerging technologies presented here, when combined with quality data, talent, Africa's youth and their creativity, have the potential to unleash unprecedented advances in the transformation of Africa's agricultural ecosystems and value chains. They can help to establish a modern food sector in Africa that can produce a wide range of food products using a reliable supply of locally available, high quality agricultural products.

With this shift in focus towards the entire value chain and food sector, we do not want to leave smallholder farmers behind in the digitalization age. Dedicated apps for smallholders will continue to have their place in the portfolio of digital solutions. Any improvements in the innovation environment will also benefit providers of such services. At the same time, many of the advanced technologies applied higher up in the value chain can benefit farmers, even if at times indirectly, for instance through better market linkages, data availability or access to financial services. In the longer run, more sophisticated digital solutions will also trickle down to farmers as their digital skills and technologies improve.

Notes

1 Digital in Africa 2018, Hootsuite and we are social, www.slideshare.net/wearesocial (accessed 10 Feb 2019). SADC countries include Angola, Botswana, Comoros, Democratic Republic of the Congo, Lesotho, Madagascar, Malawi, Mauritius, Mozambique, Namibia, Swaziland, Seychelles, South Africa, Tanzania, Zambia and Zimbabwe.
2 http://beatdrone.co
3 www.dronescan.co/
4 www.aerobotics.com/
5 e.g., FarmDrive in Kenya, https://farmdrive.co.ke/credit-scoring
6 e.g., ACRE operational in Kenya, Rwanda and Tanzania, http://acreafrica.com/
7 e.g., the farmer-to-farmer network WeFarm, https://wefarm.org/
8 e.g., Farmcrowdy in Nigeria, www.farmcrowdy.com/
9 www.hellotractor.com
10 http://qualitracegh.com
11 https://twiga.ke/
12 www.sap.com/products/leonardo/blockchain.html
13 www.agriledger.io/
14 www.mediciland.com
15 e.g., the weeding robot Dino by Naïo Technologies which mechanically weeds vegetable rows, www.naio-technologies.com/en/category/dino-robot/
16 e.g., BeatDrone which uses drones to spray on the farms,

17 www.ictregulationtoolkit.org/
18 The gender gap can be attributed, e.g., to lower education levels and income among women as well as cultural and social norms. It is interesting to note that internet access by urban women was higher than that of rural men. This shows that gender is only one aspect influencing internet adoption along e.g. with access to electricity or mobile coverage (RIA et al. 2018).
19 https://branch.co

References

Baumüller, H. (2016) 'Agricultural service delivery through mobile phones: Local innovation and technological opportunities in Kenya', in F. Gatzweiler and J. von Braun (eds) *Technological and Institutional Innovations for Marginalized Smallholders in Agricultural Development*. Springer, Heidelberg, pp. 143–162.

Baumüller, H. (2017) 'Towards smart farming? Mobile technology trends and their potential for developing country agriculture', in K.E. Skouby, I. Williams and A. Gyamfi (eds) *Handbook on ICT in Developing Countries*. River Publishers, Delft, pp. 191–201.

Dinham, J. (2017) 'Could this be Africa's revolution?', in *Annual Report 2017*. Southern African-German Chamber of Commerce and Industry, Johannesburg, pp. 40–44.

Goldstuck, A. (2018) *Cloud Africa 2018*. World Wide Worx and F5 Network, Pinegowrie and Seattle.

Graham, M., Lehdonvirta, V., Wood, A., Barnard, H., Hjorth, I. and Simon, D.P. (2017) *The Risks and Rewards of Online Gig Work at the Global Margins*. Oxford Internet Institute, Oxford.

Kinai, A. (2018) IBM and Twiga foods introduce blockchain-based micro-financing. *IBM Research Blog*, 18 April.

Milano, A. (2018) Sap launches blockchain supply chain initiative. *CoinDesk*, 14 May.

Mothobi, O., Chair, C. and Rademan, B. (2017) SADC not bridging digital divide. *Policy Brief No. 6*. researchICTafrica.net, Cape Town.

Onkokame, M., Schoentgen, A. and Gillwald, A. (2017) *What is the State of Microwork in Africa? A View from Seven Countries*. researchICTafrica.net, Cape Town.

Overstock.com (2018) Medici land governance, an overstock subsidiary, signs MOU with government of Rwanda to implement paperless blockchain land governance and property rights management. *Overstock.com Investor*, 1 November.

Pollock, D. (2018) Africa's blockchain potential untapped, but how can it be implemented? *Forbes*, 23 October.

RIA, LirnEasia and DIRSI (2018) *After Access: Understanding the Gender Gap in the Global South*. RIA, LirnEasia and DIRSI, Cape Town, Colombo and Lima.

Shadab, H. (2014) What are smart contracts, and what can we do with them? *Coin Center*, 15 December.

Siemens (2017) *African Digitalization Maturity Report 2017*. Siemens Southern & Eastern Africa, Midrand.

Tripoli, M. and Schmidhuber, J. (2018) *Emerging Opportunities for the Application of Blockchain in the Agri-food Industry*. International Centre for Trade and Sustainable Development, Geneva.

21 Reducing the impact of mycotoxigenic fungi on food safety and security in Southern Africa

Juliet Akello, George Mahuku, Lindy Rose,
Altus Viljoen, David Chikoye D.,
Pheneas Ntawuruhunga, Katati Bwalya
and Ranajit Bandyopadhyay

Introduction

The spoilage of food by filamentous fungi is a substantial problem in sub-Saharan Africa, with pre- and postharvest losses estimated at 15–100% and amounting to about US$4 billion per annum (Hodges et al., 2011; Affognon et al., 2015). Some fungi may produce toxic secondary metabolites in food and feed, called mycotoxins (Hoopwood and Sherman, 1990), thereby rendering them unsafe for humans and livestock. To date, over 500 types of mycotoxins have been reported, with some fungi being able to produce more than one mycotoxin and some mycotoxins being produced by more than one fungal species.

Worldwide, aflatoxins, trichothecenes, fumonisins, zearalenone, ochratoxin, patulin and ergot alkaloids are considered the most economically and toxicologically important mycotoxins (Bennett and Klich, 2003; Richard, 2007). These mycotoxins are produced by toxigenic fungal species. In the Southern African Development Community (SADC), toxigenic fungal species belonging to *Fusarium*, *Aspergillus*, *Alternaria* and *Penicillium* genera are common in farmland soils, crop debris and foodstuffs (Probst et al., 2014; Kachapulula et al., 2017).

Cereal grains (maize, wheat, sorghum and millet), legume grains (peanuts), tree nuts, wild fruits and insects that are traded in the SADC region are not only hosts of toxigenic fungi, but can also be contaminated with multiple mycotoxins (Probst et al., 2014; Udomkun et al., 2017; Nleya et al., 2018; Kachapulula et al., 2018; Misihairabgwi et al., 2019) (Table 21.1). Conducive environmental conditions, poor agronomic practices, improper drying before storage, insect damage, poor storage facilities, unhygienic handling of produce and informal markets worsen the mycotoxin problem in food and feed value chains (Affognon et al., 2015). Consequently, mycotoxin exposure through eating contaminated food is high in the SADC region (Lauer et al., 2019), posing a significant health hazard to consumers.

Table 21.1 Summary of major mycotoxins reported in cereals and peanuts that were produced and traded in Southern Africa

Mycotoxin type	Food commodity	SADC Country	Maximum level (μg/kg)	Reference
Total aflatoxin	Maize and maize products	Malawi	140	Mwalwayo and Thole, 2016
		Zambia/ Zimbabwe	108/123	Probst et al., 2014
	Sorghum beer	Malawi	1139	Matumba et al., 2011
	Peanuts and peanut butter	Malawi	500	Matumba et al., 2014
	peanut butter	Zimbabwe	622	Mupunga et al., 2014
	Peanut grain	South Africa	131	Ncube et al., 2010
Aflatoxin B1	Maize and maize products	Mozambique	636	Warth et al., 2012
		Zimbabwe	26.6	Hove et al., 2016; Murashiki et al., 2017
		South Africa	741	Mngqawa et al., 2016
		Lesotho	3500	Mohale et al., 2013
	peanut butter	Zambia	10000	Njoroge et al., 2016
	peanut grain	South Africa	35.39	Kamika et al., 2014
Fumonisin (total)	Maize and maize products	Malawi/ Mozambique	9000/10000	Probst et al., 2014
		Zambia	192000	Mukanga et al., 2010
		Zimbabwe	159000	Probst et al., 2014
		South Africa	142800	Mngqawa et al., 2016; Rheeder et al., 2016
	Sorghum	Zimbabwe	187	Mupunga et al., 2014
FB1	Maize and maize products	Mozambique	7615	Warth et al., 2012
		Zimbabwe	1106	Hove et al., 2016; Murashiki et al., 2017
		South Africa	14990	Shephard et al., 2013
		Lesotho	936	Mohale et al., 2013
Zeralenone	Maize	Zimbabwe	369	Hove et al., 2016
Deoxynivalenol (DON)	Maize	Zimbabwe/ South Africa	492/960	Hove et al., 2016
	Wheat flour	South Africa	100	Shephard et al., 2010

Potential negative impacts of mycotoxins in the SADC region

Chronic exposure to mycotoxins (especially aflatoxins, ochratoxin A, fumonisins and ergot alkaloids) may result in health problems such as liver, renal and esophagus cancers, retarded child growth, weakened immunity, reduced fertility, damaged nervous system, egotism and neural tube effects in humans (Wu and Khlangwiset, 2010; Hove et al., 2016; Eze et al., 2018; Lauer et al., 2019)

(Figure 21.1A, 21.1B). Mycotoxins can also be fatal if high quantities are ingested (Richard, 2007; Shephard, 2008; Kamala et al., 2018). The consumption of aflatoxin-contaminated foods in Southern Africa, for example, was noted to result in child undernutrition, increased morbidity and mortality due to suppressed

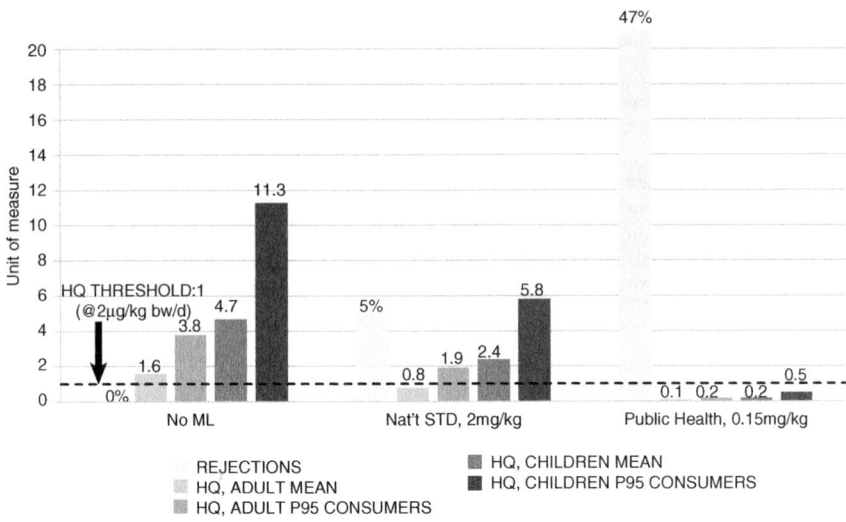

Figure 21.1a Potential impact of fumonisin B1 contamination of maize grain on the health and trade in Zambia; HQ represents "Hazard Quotient"

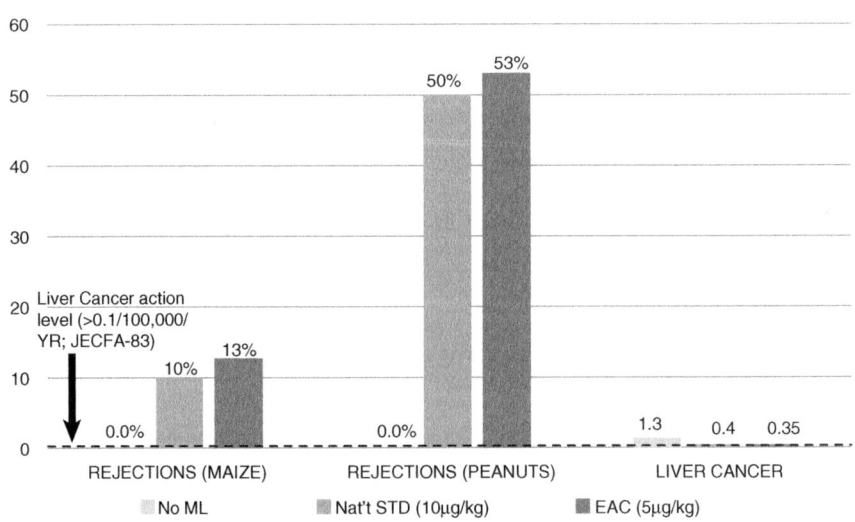

Figure 21.1b Potential impact of aflatoxin B1 contamination of maize and peanuts on health and trade in Zambia

immunity and micronutrient malabsorption (Katerere et al., 2008). The stringent mycotoxin standards set internationally also affects international trade as it prevents access of African products to premium markets in Europe and the U.S. The decrease in productivity, and investments made to mitigate aflatoxins, fumonisins and trichothecenes exposure in Africa, is estimated at US$1 billion and US$500 million per annum, respectively (Wu, 2015).

Mitigation strategies for mycotoxins and challenges

Two review papers that appeared in 2011 alluded to the fact that the global food crisis, including hunger, malnutrition and poverty, can be minimized by addressing pre- and postharvest losses (Foresight, 2011; World Bank, 2011). In Southern Africa, this could include a reduction in mycotoxin-related losses by using appropriate and cost-effective technologies, investments in proper policies, access to finance, pest and disease management, extension and infrastructure.

Prevention and control of mycotoxins in foods

Periodic surveillance, creating awareness and advocacy for safer foods remain a top measure for facilitating behavioural, attitude and mindset changes for reduced mycotoxin contamination and exposure risks (Gelayee et al., 2017). Throughout SSA, limited information is available on the extent of mycotoxins contamination in major staple foods.

Prevention and management of mycotoxins in foods

Information on the extent of mycotoxin contamination in major staple food crops in sub-Saharan Africa is limited with testing crops for mycotoxin contamination mainly limited to maize and peanuts, while the contamination of sorghum and millet has been neglected. SADC governments, thus, need to invest in affordable mycotoxin testing services and the routine analysis of mycotoxin-susceptible or -affected commodities. People should be made aware of the contamination of food with toxins, and the consumption of safe foods should be advocated for. Mycotoxin-prone areas need to be identified using climatic data, and information needs to be provided for the appropriate management of mycotoxigenic fungi. Developing and harmonizing community sensitization messages and platforms will remain key to effectively communicate the harmful effects of mycotoxins and demonstrate the benefits of available solutions to stakeholders involved in the food and feed value chains.

Good agronomic practices and detoxification of contaminated crops constitute some of the mitigation measures used to restrict mycotoxin contamination of food and feed. Simple agronomic practices like timely planting and harvest at the correct maturity, avoidance of water and nutrient stress through proper irrigation and fertigation schedules, crop rotation and minimizing insect damage can lower fungal infestation and subsequent mycotoxin contamination in farm produce (Matumba et al., 2018; Ojiambo et al., 2018; Mukanga et al., 2019). The use of quality seed and mycotoxin-resistant crop varieties may also reduce contamination (Fox and Howlett, 2008). Enormous efforts are currently being made to breed mycotoxin-resistant crop varieties; however, most maize or peanut varieties grown in sub-Saharan Africa are susceptible to fumonisins and aflatoxins, respectively (Fox and Howlett, 2008). The high costs associated with mycotoxin management may hinder adoption of interventions by subsistence farmers. Nonetheless, the planting of drought-tolerant and insect-resistant varieties, in combination with good agronomic and post-harvest handling practices, may be valuable for the integrated management of mycotoxin contamination in grains.

The use of appropriate drying techniques and proper storage facilities, along with proper handling of harvested produce during transportation, can minimize mycotoxin contamination in grains. In developed countries, mechanical driers are used to reduce grain moisture content to safe levels ($< 12\%$ for maize and 9% for peanuts), which prevents fungal growth and mycotoxin contamination in food. In the developing countries including sub-Saharan Africa, smallholder farmers use natural sun-drying in the field, a method that may increase myco-toxin levels depending on the technique and drying duration (Matumba et al., 2018). Storage technologies that limit pest or pathogen multiplication in stored grains, such as hermetic facilities, the Purdue Improved Crop Storage (PICS) bags, cocoons and metallic/ plastic silos – instead of dirty polyethylene bags and rudimentary granaries – help to reduce mycotoxin contamination in stored grain.

The reduction of fungal inoculum in environments where agricultural commodities are produced, transported, processed and stored can prevent pre- and postharvest crop losses due to harmful microbes. The use of antagonistic microorganisms with fungicidal or bactericidal properties to suppress and out-compete harmful microbes are often used to mitigate crop losses (Sharma et al., 2009; Velmourougane et al. 2011). In this respect, non-toxin producing strains of *Aspergillus flavus* has been used with great success to reduce aflatoxins in maize and peanuts (Dorner, 2004, 2009; Bandyopadhyay et al., 2016). Atoxigenic *A. flavus* strains are now widely used in sub-Saharan Africa, including SADC countries, to reduce aflatoxin contamination in maize and peanuts (Bandyopadhyay et al., 2016). In Zambia, countrywide field efficacy trials in peanut fields, for example, demonstrated that the aflasafe biocontrol product can reduce aflatoxin contamination in all agro-ecological zones including areas that receive low amounts of rainfall leading to reduced losses (Table 21.2).

Removal of mycotoxins in foods

Hand-sorting or the use of mechanical sorters to remove insect-infested, or shriveled, rotten and discoloured grains from healthy kernels as well as washing and dehulling are effective means of reducing mycotoxins from agricultural commodities (Matumba et al., 2015). Mycotoxigenic fungi and their toxins can also be decontaminated, degraded, deactivated and detoxified by using physical,

Table 21.2 The effect of aflasafe biocontrol product on the health of peanut consumers, trade and potential food losses in Zambia

Category	Untreated	Aflasafe treated	% Reduction
Samples size	250	250	–
Min (ppb)	< 2	< 2	–
Max (ppb)	7,205	3,168	–
Overall mean (ppb)	220	35	84
% samples > 4 ppb★	38.4	20	48
% samples >10 ppb★	34	16.4	52
% samples > 100 ppb★	20.8	6.4	69
% samples > 1000 ppb★	6	1.2	80

% samples > ★ ppb – Reflects effect on trade or health of consumers or potential food losses and not the effect of aflasafe in the category (row)

chemical or biological treatments. These include modified atmosphere packaging, irradiation, thermal treatment, nixtamalization, ammoniation and the addition of binders to destroy toxigenic microbes and deactivate mycotoxins (Kabak et al., 2006; Fox and Howlett, 2008; Schmidt et al., 2018). Some of these interventions, however, are costly and may only be suitable (e.g., binders) for animal feeds, as they are unsafe for food treatment (Fox and Howlett, 2008).

Policies, legislations and regulations on mycotoxins in foods

Legislation exists in most developed countries to regulate mycotoxin levels in food (Streit et al., 2012). In sub-Saharan Africa, however, food safety policies, legislations and regulations are either nonexistent or, when available, limited to a few mycotoxins such as aflatoxins, ergot alkaloids, deoxynivalenol and ochratoxins only. Even if existing this legislation is seldom enforced, thereby discouraging the trade of safe foods nationally and internationally and, as a result, undermining economic growth on the continent (FAO, 2016). SADC countries, therefore, should sincerely review their food safety and nutrition programmes if they are sincere in reducing the contamination of food with mycotoxins and committed to trading African crops on international markets.

Conclusion

Mycotoxin contamination is an invisible problem, and in many cases, there are no visible symptoms. Often, farmers, traders, processors and consumers are unaware of mycotoxin problem nor potential health risks. Where people are aware of the problem, the lack of easily accessible and cheap tools to measure the level of contamination and the lack of market or legislative incentives to produce aflatoxin-free commodities means that the problem is largely ignored. There is an urgent need to develop and make available to farmers technological solutions that can help them produce mycotoxin-free commodities and processors produce mycotoxin-free products. Low-cost, robust mycotoxin assaying tools are required to monitor mycotoxin contamination of important staples and use the data for risk mapping and targeting of intervention strategies. Several approaches are available for the integrated management of mycotoxins. These range from regulation, good agricultural practices, use of resistant cultivars, biological control through competitive exclusion and good storage practices. Despite the benefits, uptake of those measures is low due to inadequate capacity in terms of knowledge, facilities and financial resources. Strict regulation works well in developed countries; however, this approach is not practical for developing countries that lack the tools, resources and expertise to enforce it. Agronomic practices that require extra time to manage from the farmers are not practical. Creating awareness of mycotoxins and health impact is required to catalyze and stimulate behavioural change to stimulate stakeholders demand for mycotoxin-free products and contribute to nutritional and food safety and reduced health burden from mycotoxin exposure. Incorporating aspects of mycotoxin control into the curriculum of relevant educational/training institutions, capacity building on mycotoxin surveillance and testing, as well as enforcing policies that support production and trade of mycotoxin-safe commodities will offer access to safe food and premium markets.

References

Affognon, H., Mutungi, C., Sanginga, P. and Borgemeister, C. (2015) Unpacking postharvest losses in Sub-Saharan Africa – A meta-analysis. *World Development* vol 66, pp. 49–68.

Bandyopadhyay, R., Ortega-Beltran, A., Akande, A., Mutegi, C., Atehnkeng, J., Kaptoge, L., Senghor, A.L., Adhikari, B.N. and Cotty, P.J. (2016) Biological control of aflatoxins in Africa: current status and potential challenges in the face of climate change. *World Mycotoxin Journal*, 9, pp. 771–789.

Bennett, J.W. and Klich, M. (2003) Mycotoxins. *Clinical Microbiology Reviews*, vol 16, no 5, pp. 497–516.

Dorner, J.W. (2004) Biological control of aflatoxin contamination of crops. *Toxin Reviews*, vol 3, nos 2–3, pp. 425–450.

Dorner, J.W. (2009) Biological control of aflatoxin contamination in corn using a nontoxigenic strain of Aspergillus flavus. *Journal of Food Protection*, vol 72, no 4, pp. 801–804.

Eze, U.A., Routledge, M.N., Okonofua, F.E., Huntriss, J. and Gong, Y.Y. (2018) Mycotoxin exposure and adverse reproductive health outcomes in Africa: A review. *World Mycotoxin Journal*, vol 11, no 3, pp. 321–339.

FAO (2016) *Report of the '2015 Series of International Conferences on Food Loss and Waste Reduction' – Recommendations in Improving Policies and Strategies for Food Loss and Waste Reduction*. FAO Rome, IT. www.fao.org/3/a-bc345e.pdf.

Foresight (2011) *The Future of Food and Farming-Final Project Report*. The Government Office for Science, London, 208pg.

Fox, E.M. and Howlett, B.J. (2008) Secondary metabolism: Regulation and role in fungal biology. *Current Opinion in Microbiology*, vol 11, pp. 481–487.

Gelayee, D.A., Mekonnen, G. B. and Atnafe, S.A. (2017) Practice and barriers towards provision of health promotion services among community pharmacists in Gondar, Northwest Ethiopia. *BioMed Research International*, doi:10.1155/2017/7873951.

Hodges, R.J., Buzby, J.C. and Bennett, B. (2011) Postharvest losses and waste in developed and less developed countries – Opportunities to improve resource use. *Journal of Agricultural Science*, vol 149, pp. 37–45.

Hoopwood, D.A. and Sherman, D.H. (1990) Molecular genetics of polyketides and its comparison to fatty acid biosynthesis. *Annual Review of Genetics, Palo Alto*, vol 34, pp. 37–62.

Hove, M., Van Poucke, C., Njumbe-Ediage, E., Nyanga, L. and De Saeger, S. (2016) Review on the natural cooccurrence of afb1 and fb1 in maize and the combined toxicity of AFB1 and FB1. *Food Control*, vol 59, pp. 675–682.

Kabak, B., Dobson, A.D. and Var, I. (2006) Strategies to prevent mycotoxin contamination of food and animal feed: A review. *Critical Reviews in Food Science and Nutrition*, vol 46, pp. 593–619.

Kachapulula, W.P., Akello, J., Bandyopadhyay, R. and Cotty, P.J. (2017) Aspergillus section Flavi community structure in Zambia influences aflatoxin contamination of maize and groundnut. *International Journal of Food Microbiology*, vol 261, pp. 49–56.

Kachapulula, W.P., Akello, J., Bandyopadhyay, R. and Cotty, P.J. (2018). Aflatoxin contamination of dried insects and fish in Zambia. *Journal of Food Protection*, vol 81, pp. 1508–1518.

Kamala, A., Shirima, C., Jani, B., Bakari, M., Sillo, H., Rusibamayila, N., De Saeger, S., Kimanya, M., Gong, Y.Y. and Simba, A. (2018) Outbreak of an acute aflatoxicosis in Tanzania during 2016. *World Mycotoxin Journal*, 11, no 3, pp. 311–320.

Kamika, I., Mngqawa, P., Rheeder, J.P., Teffo, S.L. and Katerere, D.R. (2014) Mycological and aflatoxin contamination of peanuts sold at markets in Kinshasa, Democratic Republic of Congo, and Pretoria, South Africa. *Food Additives and Contaminants: Part B*, vol 7, no 2, pp. 120–126.

Katerere, D.R., Shephard G.S. and Faber, M. (2008) Infant malnutrition and chronic aflatoxicosis in Southern Africa: Is there a link? *International Journal of Food Safety Nutrition and Public Health*, 1, no 2, pp. 127–136.

Lauer, J.M., Duggan, C.P., Ausman, L.M., Griffiths, J.K., Webb, P., Wang, J.S., Xue, K.S., Agaba, E., Nshakira, N. and Ghosh, S. (2019). Association between maternal aflatoxin exposure during pregnancy and adverse birth outcomes in Uganda. *Maternal and Child Nutrition*, vol 15, no 2, e12701.

Matumba, L., Monjerezi, M., Biswick, T., Mwatseteza, J., Makumba, W., Kamangira, D. and Mtukuso, A. (2014) A survey of the incidence and level of aflatoxin contamination in a range of locally and imported processed foods on Malawian retail market. *Food Control*, vol 39, pp. 87–91.

Matumba, L., Monjerezi, M., Khonga, E.B. and Lakudzala, D. (2011) Aflatoxins in sorghum, sorghum malt and traditional opaque beer in southern Malawi. *Food Control*, vol 22 no 2, pp. 266–68.

Matumba, L., Singano, L., Tran, B., Mukanga, M., Makwenda, B., Kumwenda, W., Mgwira, S., Phiri, S., Mataya, F., Mthunzi, T., Alfred, S., Madzivhandila, T., Mugabe, J., Bennett, B., Chancellor, T. (2018) Managing aflatoxin in smallholder groundnut production in Southern Africa: Paired comparison of the windrow and Mandela cock techniques. *Crop Protection*, vol 112, pp. 18–23.

Matumba, L., Van Poucke, C., Ediage, E.N., Jacobs, B. and De Saeger, S. (2015) Effectiveness of hand sorting, flotation/washing, dehulling and combinations thereof on the decontamination of mycotoxin-contaminated white maize. *Food Additives & Contaminants: Part A*, vol 32, no 6, pp. 960–969.

Misihairabgwi J.M., Ezekiel, C.N., Sulyok, M., Shephard, G.S. and Krska, R. (2019) Mycotoxin contamination of foods in Southern Africa: A 10-year review (2007–2016), *Critical Reviews in Food Science and Nutrition*, vol 59, no 1, pp. 43–58.

Mngqawa, P., Shephard, G.S., Green, I.R., Ngobeni, S.H., de Rijk, T.C. and Katerere, D.R. (2016) Mycotoxin contamination of home-grown maize in rural northern South Africa (Limpopo and Mpumalanga Provinces). *Food Additives and Contaminants: Part B*, vol 9, no 1, pp. 38–45.

Mohale, S., Medina, A., Rodriguez, A., Sulyok, M. and Magen, N. (2013). Mycotoxigenic fungi and mycotoxins associated with stored maize from different regions of Lesotho. *Mycotoxin Research*, vol 29, no 4, pp. 209–219.

Mukanga, M., Derera, J., Tongoona, P. and Laing, M.D. (2010) A survey of pre-harvest ear rot diseases of maize and associated mycotoxins in south and central Zambia. *International Journal of Food Microbiology*, vol 141, pp. 213–221.

Mukanga, M., Matumba, L., Makwenda, B., Alfred, S., Sakala, W., Kanenga, K., Chancellor, T., Mugabe, J. and Bennett, B. (2019) Participatory evaluation of groundnut planting methods for pre-harvest aflatoxin management in Eastern Province of Zambia. *Cahiers Agricultures*. https://doi.org/10.1051/cagri/2019002.

Mupunga, I., Lebelo, S.L. Mngqawa, P., Rheeder, J.P. and Katerere, D.R. (2014) Natural occurrence of aflatoxins in peanuts and peanut butter from Bulawayo, Zimbabwe. *Journal of Food Protection*, vol 77, pp. 1814–1818.

Murashiki, T.C., Chidewe, C., Benhura, M.A., Maringe, D.T., Dembedza, M.P., Manema, L.R., Mvumi, B.M. and Nyanga, L.K. (2017) Levels and daily intake estimates of aflatoxin B1 and fumonisin B1 in maize consumed by rural households in Shamva and Makoni districts of Zimbabwe. *Food Control*, vol 72, pp. 105–109.

Mwalwayo, D.S. and Thole, B. (2016) Prevalence of aflatoxin and fumonisins (B1 C B2) in maize consumed in rural Malawi. *Toxicology Reports*, vol 3, pp. 173–179.

Ncube, E., Flett, B.C.C., Waalwijk, C. and Viljoen, A. (2010) Occurrence of aflatoxins and aflatoxin producing Aspergillus spp. associated with groundnut production in subsistence farming systems in South Africa. *South African Journal of Plant and Soil*, vol 27, no 2, pp. 195–198.

Njoroge, S.M.C., Kanenga, K., Siambi, M., Waliyar, F. and Monyo, E.S. (2016). Identification and toxigenicity of Aspergillus spp. from soils planted to groundnut in Eastern Zambia. *Peanut Science*, vol 43, pp. 148–156.

Nleya, N., Adetunji, M.C. and Mwanza, M. (2018) Current status of mycotoxin contamination of food commodities in Zimbabwe. *Toxins*, vol 10, no 5, pp. 89.

Ojiambo, P.S., Battilani, P., Cary, J.W., Blum, B.H, Carbone, I. (2018) Cultural and genetic approaches to manage aflatoxin contamination: recent insights provide opportunities for improved control. *Phytopathology*, vol 108, no 9, pp. 1024–1037.

Probst, C., Bandyopadhyay, R. and Cotty, P. (2014) Diversity of aflatoxin-producing fungi and their impact on food safety in sub-Saharan Africa. *International Journal of Food Microbiology*, vol 174, pp. 113–122.

Rheeder, J.P., Van der Westhuizen, L., Imrie, G. and Shephard, G.S. (2016) Fusarium species and fumonisins in subsistence maize in the former Transkei region, South Africa: a multi-year study in rural villages. *Food Additives and Contaminants: Part B*, vol 9 no 3, pp. 176–184.

Richard, J.L. (2007) Some major mycotoxins and their mycotoxicosis – An overview. *International Journal of Food Microbiology*, vol 119, nos 1–2, pp. 3–10.

Schmidt, M., Zannini, E. and Arendt, E.K. (2018) Recent advances in physical post-harvest treatments for shelf-life extension of cereal crops. *Foods*, vol 7, no 4, p. 45.

Sharma, R.R., Singh, D. and Singh, R. (2009) Biological control of postharvest diseases of fruits and vegetables by microbial antagonists: A review. *Biological Control*, vol 50, no 3, pp. 205–221.

Shephard, G.S. (2008) Impact of mycotoxins on human health in developing countries. *Food Additives & Contaminants: Part A*, vol 25, no 2, pp. 146–151.

Shephard, G.S., Burger, H.M., Gambacorta, L., Krska, R., Powers, S.P., Rheeder, J.P., Solfrizzo, M., Sulyok, M., Visconti, A. Warth, B. and van der Westhuizen, L. (2013) Mycological analysis and multimycotoxins in maize from rural subsistence farmers in the former Transkei, South Africa. *Journal of Agricultural and Food Chemistry,* vol 61, no 34, pp. 8232–8240.

Shephard, G.S., Van der Westhuizen, L., Katerere, D.R., Herbst, M. and Pineiro, M. (2010) Preliminary exposure assessment of deoxynivalenol and patulin in South Africa. *Mycotoxin Research*, vol 26, pp. 181–185.

Streit, E., Schatzmayr, G., Tassis, P., Tzika, E., Marin, D., Taranu, I., Tabuc, C., Nicolau, A., Aprodu, I., Puel, O., Oswald, I.P. (2012) Current situation of mycotoxin contamination and co-occurrence in animal feed – focus on Europe. *Toxins*, vol 4, no 10, pp. 788–809.

Udomkun, P., Wiredu, A.N., Nagle, M., Bandyopadhyay, R., Müller, J. and Vanlauwe, B. (2017) Mycotoxins in Sub-Saharan Africa: Present situation, socio-economic impact, awareness, and outlook. *Food Control*, vol 72, pp. 110–122.

Velmourougane, K. Bhat, R., Gopinandhan, T.N., Panneerselvam, P. (2011) Management of *Aspergillus ochraceus* and Ochratoxin-A contamination in coffee during on-farm processing through commercial yeast inoculation. *Biological Control*, vol 57, no 3, pp. 215–221.

Warth, B. Parich, A. Atehnkeng, J., Bandyopadhyay, R., Schuhmacher, R., Sulyok, M. and Krska, R. (2012) Quantitation of mycotoxins in food and feed from Burkina Faso and Mozambique using a modern LC-MS/MS multitoxin method. *Journal of Agricultural and Food Chemistry*, vol 60, pp. 9352 – 9363.

World Bank. (2011) Missing food – The case of postharvest grain losses in Sub-Saharan Africa. World Bank – FAO – Natural Resources Institute. World Bank Report 60371-AFR, 2011.

Wu, F. (2015) Global impacts of aflatoxin in maize: Trade and human health. *World Mycotoxin Journal*, vol 8, no 2, pp. 137–142.

Wu, F. and Khlangwiset, P. (2010) Health economic impacts and cost-effectiveness of aflatoxin reduction strategies in Africa: Case studies in biocontrol and postharvest interventions. *Food Additives and Contamination A*, vol 27, pp. 496–509.

22 Small-scale renewable energy as a catalyst for advancing agriculture and food security in Southern Africa

Jon Padgham and Mariama Camara

Introduction

Energy poverty is a key determinant of under development and constrained economic performance across Africa. Approximately 620 million people lack access to electricity and 790 million cook with traditional biomass on unimproved cookstoves, which contributes substantially to land degradation (REN, 2018). With the exception of South Africa, less than 30% of Southern Africa's population has access to electricity and there is strong reliance on traditional biomass fuel for cooking and heating (Mabhaudhi et al., 2018). Given the current situation of low energy access, Africa will not meet the UN 2030 Sustainable Development Goal (SDG) target of universal access to energy until 2080 and access to clean energy for cooking until the middle of the next century (Africa Progress Report, 2015). Meeting that SDG 7 goal of universal energy access would require Africa to spend between US$41 and US$55 billion per year on energy and infrastructure investment; by contrast current annual expenditures are about US$8 billion (Chirambo, 2018).

Sole reliance on a conventional "hard path" to energy development, characterized by high fossil fuel use and a centralized grid, is becoming increasingly untenable as extreme climatic events intensify and technical and market innovations render renewable energy sources delivered on a decentralized grid, or off grid, much more attainable (Africa Progress Report, 2015). Renewable energy in its various forms – solar, hydro, wind and bioenergy – represent an increasingly viable option for at least partially leapfrogging traditional centralized, fossil fuel energy development on the continent (van der Zwann et al., 2018). The scale of these renewable energy forms can vary from macro, grid-connected enterprises to micro, off-grid local applications.

Increasing energy access and rural electrification in Africa is essential for triggering broader gains in agricultural productivity and food security. In remote rural areas, small-scale distributed renewable energy systems offer a more feasible pathway to electrification than large-scale, grid-connection energy sources (Diaz-Maurin et al., 2018). Because of the applicability to low-income rural environments, micro off-grid applications will form the basis for this paper's review and analysis. The paper will focus on solar photovoltaic energy and bioenergy, and both of these energy sources will be examined in the context of rural development and its linkages to food production and food security.

Renewable energy types

Solar photovoltaics

Over the past several years, there has been a steady increase in renewable energy investments in Africa, with this trend mostly attributable to favourable shifts in access to solar energy, driven by falling prices for photovoltaics and the spread of technological and market innovations. These include large investments in solar arrays as well as uptake of micro-grid, nano-grid and off-grid solar power sources. Globally, solar photovoltaic module prices have fallen more than 75% since 2009 (Chirambo, 2018).

Africa has the highest photovoltaic solar energy potential of all the world's regions (Figure 22.1); in Southern Africa solar power potential is approximately 20,000 terawatt-hours per year. There is significant potential for expanding solar energy in the region, given that current installed solar capacity in Southern Africa is less than 1% of that potential (Jadhave et al., 2017). Approaching these potentials would require massive infrastructure investments and policy shifts, strengthening of private and public institutions and changes in incentive structures that would make solar more accessible to low-income consumers.

While Southern Africa is far off from achieving kind of production potential, positive signs of change are increasingly evident across the sub-region and

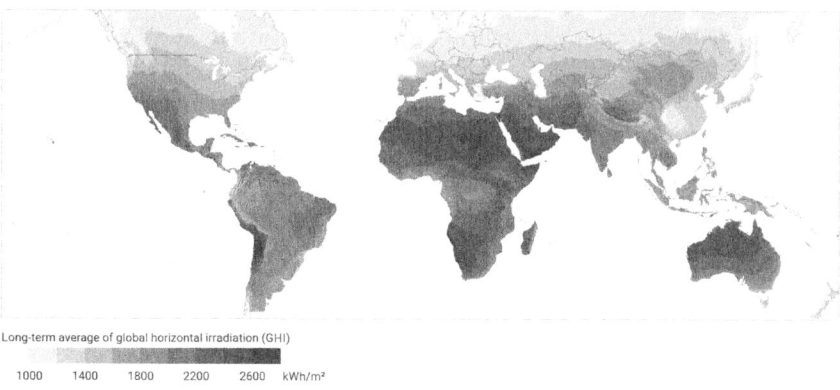

Long-term average of global horizontal irradiation (GHI)

1000   1400   1800   2200   2600   kWh/m²

Figure 22.1 Photovoltaic power potential

Source: Solargis

elsewhere across Africa, ranging from foreign and domestic investments in large-scale solar arrays to new technical and business models that have allowed the emergence of pico-scale solar schemes (REN, 2018). There is strong growth potential in solar PV to serve decentralized energy needs, particularly in rural areas that lack access to the grid and where infrastructure for power transmission is costly and unreliable (Ikejemba et al., 2017). Market innovations for off-grid applications, such as pay-as-you-go (PAYG), based on mobile payment schemes and "plug and play" for solar household systems, have accounted for robust growth in Africa's renewable energy sector over the last several years (Ulsrud et al., 2018). Growth of off-grid solar has been faster in East and West Africa compared with Southern Africa; the exception is South Africa, which, together with Kenya, has had the most rapid growth in cumulative non-hydropower renewable energies. East Africa, and in particular Kenya, has seen strong capital inflows into the off-grid market.

Bioenergy

Crop biofuels

First-generation biofuels in the form of bioethanol from starch sources (e.g., sugarcane, molasses, cassava and cereal crops) and biodiesel production from oilseeds (e.g., soya, oil palm and Jatropha) offer potential for wide-scale bioenergy production in Africa. However, while biofuel production can generate positive energy security and economic growth outcomes, biofuel crop production in Africa also carries significant downside risks related to environmental and social sustainability, including competition for land and water between fuel and food crops, contractual and regulatory obligations that expose farmers to legal risks, changes in land tenure security and extension of agriculture into forested areas to compensate for land lost to biofuel crops (Goetz et al., 2017; Pradhan and Mbohwa, 2014). Recent research in Southern Africa indicates that the type and production method of biofuel can influence whether biofuel production increases or decreases terrestrial carbon stocks (Romeu-Dalmau et al., 2018). Efforts to address crop-biofuel sustainability challenges have been lackluster due to lack of policy responses that go beyond the production potential of biofuels to adequately consider the social and environmental costs of biofuel production (Goetz et al., 2017).

Waste biogas

Biogas offers a promising addition to the biofuels portfolio, and one that is unencumbered by the social and environmental downsides associated with biofuel crops. Biogas has gained a foothold in Asia and India and is viewed as having potential for dealing with environmental and sanitation issues presented by excessive livestock, agricultural and municipal waste in sub-Saharan Africa, while providing energy to rural environments. Domestic biogas digesters (floating drum and fixed dome types) offer multiple benefits for lighting and cooking, which reduces workloads of women and girls and their exposure to indoor pollution, and residues can be used as biofertilizers to increase soils fertility (IRENA, 2017a, 2017b). However, uptake and upscaling of biogas production and use in Sub-Saharan Africa has been uneven. Key barriers to uptake include a lack of 1) socioeconomic and cultural considerations in project development; 2) technicians for maintaining technology; 3) literacy and awareness of users; and 4) funding and subsidies for widespread dissemination of technologies. In South Africa, there is the additional constraint of low efficiency and cost competitiveness of biogas compared with fossil fuels (Mukumba et al., 2016).

Applications of renewable energy to food systems and rural development

Higher rates of energy access and rural electrification in Africa is essential for triggering broader gains in agricultural productivity and food and nutritional security, primarily through increasing access to irrigation, improving food storage and safety and value-added processing. Access to energy can also begin to reduce tightly locked seasonal dependencies associated with smallholder rainfed agriculture. In addition to improving food security, expanding energy access produces a wide array of development benefits linked to stimulating economic growth, improving educational performance and producing better health and gender equality outcomes (Chirambo, 2018).

Use of off-grid and micro-grid solar photovoltaics offers a scale-appropriate technology to expand irrigation in rural areas not connected to grid energy (Hartung and Pluschke, 2018; Wazed et al., 2018). The technology generally consists of photovoltaic panels connected to a pump, set up to an elevated tank or reservoir and adapted for different irrigation types, such as flood, sprinkler or drip irrigation. (The system is analogous to many diesel-powered systems commonly encountered in smallholder farming in Africa.) Increasing access to solar irrigation technologies can spur diversification of cropping systems towards higher-value crops, bring greater stabilization of yields and extend the agricultural season beyond the bounds of the rainy season. Furthermore, solar pumps improve access to household water where economic water scarcity and frequent droughts are high, and the continued decline in the price of photovoltaic components, combined with innovations in finance, has meant that clean cooking systems powered by solar (as well as biogas) are becoming increasingly price competitive with dirty conventional cooking fuel sources (Batchelor et al., 2018). All of these factors are important for rural development, particularly in the context of accelerating warming trends and increasing prevalence of drought in Southern Africa.

However, as with any technology that enlarges water access, widespread use of solar irrigation systems runs the risk of unsustainable water use given that there is no financial signal via a cost per unit of power as with the case with petrol or grid-based energy sources (Hartung and Pluschke, 2018). An integrated approach to solar powered irrigation is required, one that considers access to appropriate financing schemes, the economic viability of solar irrigation systems (and payback time) related to potential for high-value horticultural crops and access to vibrant markets, as well as agronomic practices that promote water conservation and use efficiency.

Increasing food security requires a systems approach that goes beyond a focus on higher crop yields. In particular, postharvest food losses in Africa, estimated at over one-third of its agricultural production, need greater attention. Postharvest technologies require energy to operate, and in much of rural Africa traditional fuels such as firewood and dung cakes are used, resulting in only rudimentary forms of storing and processing. Having access to modern and reliable sources of energy can speed up the adoption of improved storage and processing technologies and foster growth of agro-enterprises. For example, in East Africa appropriate technology pilot efforts are underway, utilizing biogas to provide cold storage for dairy producers, infrared technologies to speed up coffee drying and biogas-powered steam engines to power ancillary food processing for fruit and vegetable drying, grain mills, irrigation and refrigeration (IRENA, 2018). This effort has reduced food losses by 80% for high-value but highly perishable foods. Pilots are also underway in East Africa to use biogas for clean cooking, an important contribution to reducing chronic respiratory and other health concerns associated with use of traditional solid biomass for cooking.

Identifying and addressing constraints to small-scale renewable energy in Africa

While energy access can usher in food productivity and security gains, processes for broader rural transformation depend on structural changes shaped by linkages between agriculture, the rural nonfarm economy, manufacturing, infrastructure, finance and services. Realizing the potential of solar and other forms of renewable energy in Africa will require overcoming important obstacles and bottlenecks that extend well beyond technological issues. These include: 1) poor governance (incoherent and inconsistent policies, inadequate planning and implementation, political agendas, lack of transparency in how energy projects are awarded, etc.); 2) lack of technical skills to support maintenance and management of implemented projects; and 3) economic and social disincentives related to lack of community involvement, informational and financial barriers and economic interests linked to carbon-intensive energy sources (Baptista and Plananska, 2017; Ikejemba et al., 2017; Morrissey, 2017; Ulsrud et al., 2018).

Table 22.1 Key attributes of solar and biogas renewable energy forms

Applications to agriculture and food security	Development outcomes	Key enabling factors	Needs for addressing constraining factors
Irrigation expansion Cold storage Value-added processing Agro-diversification to high-value crops	Reduced sensitivity to dry spells Reduced postharvest storage losses Increased food safety Clean fuel source for cooking Increased income generation	Increased cost effectiveness Flexible and scale-appropriate technology Potential to tap into climate finance schemes for funding Strong potential for growth in the private sector	Appropriate financing schemes to address affordability gap Bottom-up needs and priorities to inform a more enabling policy environment for renewable energy in rural areas Capacity building in finance, private sector and technological skills for energy installation and use More research to understand sociocultural and economic incentives and disincentives Integration of renewable energy frameworks with investments in market creation and rural infrastructure needs

In conclusion, the potential for small-scale renewable energy is quite good, particularly in the case of solar photovoltaics. Falling unit prices for solar, emergence of innovative financing schemes involving the private sector, availability of large-scale financial mechanisms, such as the Green Climate Fund, and the inherent flexibility of the technology are all key factors in increasing the viability of solar power in off-grid and micro-grid systems. However, there are important constraining factors as well related to technical and financial capacities, institutions and governance and policy levers that need greater understanding in order to realize the potential of solar and other forms of renewable energy.

References

Africa Progress Report (2015) *Power, People, Planet. Seizing Africa's Energy and Climate Opportunities,* Africa Progress Panel. ISBN 978-2-9700821-6-3.

Baptista, I. and Plananska, J. (2017) 'The landscape of energy initiatives in sub-Saharan Africa: Going for systemic change or reinforcing the status quo?', *Energy Policy*, vol 110, pp. 1–8.

Batchelor, S., Brown, E., Leary, J., Scott, N., Alsop, A. and Leach, M. (2018) 'Solar electric cooking in Africa: Where will the transition happen first?' *Energy Research and Social Science*, vol 40, pp. 257–272.

Chirambo, D. (2018) 'Towards the achievement of SDG 7 in sub-Saharan Africa: Creating synergies between Power Africa, Sustainable Energy for All and climate finance in-order to achieve universal energy access before 2030', *Renewable and Sustainable Energy Reviews*, vol 94, pp. 600–608.

Diaz-Maurin, F, Chiguvare, Z. and Gope, G. (2018) 'Scarcity in abundance: The challenges of promoting energy access in the Southern African region', *Energy Policy*, vol 120, pp. 110–120.

Goetz, A., German, L., Hunsberger, C., Schmidt, O. (2017) 'Do no harm? Risk perceptions in national bioenergy policies and actual mitigation performance', *Energy Policy*, vol 108, pp. 776–790.

Hartung, H. and Pluschke, L. (2018) 'The benefits and risks of solar-powered irrigation – A global overview', United Nations Food and Agriculture Organization.

Ikejemba, E.C., Schuur, P.C., Van Hillegersberg, J. and Mpuan, P. B. (2017) 'Failures and generic recommendations towards the sustainable management of renewable energy projects in Sub-Saharan Africa', *Renewable Energy*, vol 113, pp. 639–647.

IRENA (2017a) *Biofuel Potential in Sub-Saharan Africa: Raising Food Yields, Reducing Food Waste and Utilizing Residues*, International Renewable Energy Agency, Abu Dhabi.

IRENA (2017b) *Biogas for Domestic Cooking: Technology Brief*, International Renewable Energy Agency, Abu Dhabi.

IRENA (2018) *Sustainable Rural Bioenergy Solutions in Sub-Saharan Africa: A Collection of Good Practices*, International Renewable Energy Agency, Abu Dhabi.

Jadhave, A.S., Chembe, D. K., Strauss., J. M. and Van Niekerk, J. L. (2017) 'Status of solar technology implementation in the southern African developing community (SADC) region', *Renewable and Sustainable Energy Reviews*, vol 73, pp. 622–631.

Mabhaudhi, T., Mpandeli, S., Nhamo, L., Chimonyo, V. G. P, Nhemachena, C., Senzanje, A., Naidoo, D., Modi, A. T. (2018) 'Prospects for improving irrigated agriculture in Southern Africa: Linking water, energy and food', *Water*, vol 10, 1881. doi:10.3390/w10121881

Morrissey, J. (2017) 'The energy challenge in sub-Saharan Africa: A guide for advocates and policy makers Part 2: Addressing energy poverty', Oxfam Research Backgrounder.

Mukumba, P., Makaka, G. and Mamphweli, S. (2016) 'Biogas technology in South Africa, problems, challenges and solutions', *International Journal of Sustainable Energy and Environmental Research*, vol 5, pp. 58–69.

Pradhan, A., Mbohwa, C. (2014) 'Development of biofuels in South Africa: challenges and opportunities', *Renewable and Sustainable Energy Review*, vol 39, pp. 1089–1100.

REN21. (2018) *Renewables 2018 Global Status Report*, Paris: REN21 Secretariat, ISBN 978-3-9818911-3-3.

Romeu-Dalmau, C., Gasparatos, A., von Maltitz, G., Graham, A., Almagro-Garcia, J., Wilebore, B. and Willis, K. J. (2018) 'Impacts of land use change due to biofuel crops on climate regulation services: Five case studies in Malawi, Mozambique and Swaziland', *Agriculture, Ecosystems, and Environment*, vol 246, pp. 314–324.

Ulsrud., K., Rohracher, H., Winther, T., Muchunku, C. and Palit, D. (2018) 'Pathways to electricity for all: What makes village-scale solar power successful?' *Energy Research & Social Science*, vol 44, pp. 32–40.

van der Zwann, B., Kober, T., Longa, F. D., van der Laan, A. and Kramer, G. J. (2018) 'An integrated assessment of pathways for low-carbon development in Africa', *Energy Policy*, vol 117, pp. 387–395.

Wazed, S. M., Hughes, B. R., O'Connor, D. and Calautit, J. K. (2018) 'A review of sustainable solar irrigation systems for Sub-Saharan Africa', *Renewable and Sustainable Energy Reviews*, vol 81, pp. 1206–1225.

23 Trees on farms and farmers in the forest

Good practices and a need for policy

Ingrid Öborn, Rhett D. Harrison
and Sileshi G. Weldesemayat

Introduction

Sustainable intensification of smallholder farming that includes linking production to markets has been a recent focus of research, innovation and development activities in sub-Saharan Africa (e.g., Öborn et al., 2017). Agroforestry – or the integration of trees with crops and/or livestock and diversification of farm production and income sources with tree derived products – is one strategy to sustainably intensify livelihoods' systems and make them more resilient to climate change and other shocks (e.g., Minang et al., 2015). Introducing woody perennials (shrubs and trees) on farmland has been shown to improve food and nutrition security and increase household income (Ajayi et al., 2009, 2011). In addition, planting trees enhances the delivery of products (timber, firewood, fruits, etc.) and increases ecosystem services benefits such as improved soil fertility, microclimate (shade or wind break) and water infiltration capacity (Kuyah et al., 2016). Although, the choice of tree species and agroforestry practices, i.e., how trees are integrated and managed on a farm, differs widely depending on: farmers' resources and needs, climate, soil type, institutional arrangements, knowledge and incentives, e.g., extension, farmers' groups, loan schemes (van Noordwijk, 2019). Trees on farms also mitigate climate change by locking up atmospheric carbon in biomass and soils and improve the environment for biodiversity. Unsurprisingly, therefore agroforestry is often viewed as a win–win option for improving livelihoods and delivering environmental benefits.

In sub-Saharan Africa, wood-fuel accounts for more than 80% of primary energy supply; the vast majority of the population relies on firewood and charcoal for energy, especially for cooking (Iiyama et al., 2014). In rural areas in Southern Africa, firewood is the main sources of energy making access to trees of vital importance, while most urban populations depend on charcoal. Moreover, much of the wood for energy comes from forests resulting in forest degradation. However, there is a huge potential for agroforestry and farm forestry to produce firewood for domestic use and sale and support sustainable charcoal production, which can generate cash income (Iiyama et al., 2014).

This chapter will explore the roles and potentials of trees on farms and farmers in forests for rural transformation in Southern Africa. We describe experiences from on farm integration of nitrogen fixing trees (fertilizer trees) and illustrate how that can affect productivity of maize, vegetables and dairy cattle as well as provide firewood and other ecosystem services and increase farmers' incomes. We demonstrate the importance of community forestry initiatives for smallholders and suggest incentives for communities to protect and sustainably use forest resources.

Integration of fertilizer trees in crop and livestock systems in Southern Africa

Fertilizer tree systems refer to agroforestry practices involving nitrogen-fixing leguminous perennials in crop production, pastures, rangelands and rehabilitation of degraded land (Sileshi et al., 2014). Most fertilizer tree systems have been designed to address problems with soil health and raise crop and livestock productivity in an integrated manner. The contribution of fertilizer trees to soil improvement mainly comes from nitrogen (N) inputs via biological nitrogen fixation (BNF) and the capture of nutrients by tree roots from soil depths beyond the reach of crop roots and their transfer to the soil surface through litter fall, tree pruning and their biological decomposition. There are many agroforestry practices that capitalize on BNF from fertilizer trees (Sileshi et al., 2014); those relevant to the Southern African context will be described here.

Agroforestry practices for maize mixed farming systems

A maize mixed farming system is a crop and livestock integrated system where the dominant crop is maize but which also includes pulses, vegetables, oilseeds and root crops. The livestock component includes cattle, small ruminants and poultry. This system covers a large portion of Southern Africa including Zambia, Malawi, Mozambique, Zimbabwe, Botswana, South Africa, Swaziland, Lesotho and Madagascar. In this system, fertilizer trees are integrated to improve soil fertility and provide animal fodder and other tree products including firewood, poles and timber. In this farming system, improved fallows, relay cropping and intercropping with fertilizer trees have been used to improve soil health and increase crop productivity.

Research conducted in the last two decades in Malawi, Mozambique, Zambia and Zimbabwe demonstrated that fertilizer trees can double or triple yields of maize contributing to food security (Ajayi et al., 2011; Akinnifesi et al., 2010; Sileshi et al., 2014) and reducing production risks (Sirrine et al., 2010). A meta-analysis of published studies found that maize yields can be doubled or tripled relative to unfertilized maize in 45–67% of cases with fertilizer trees (Sileshi et al., 2008, 2014). An additional benefit that is usually underreported in the literature is the stover yield, which is a critical input as a soil cover and livestock feed. Estimates show that 0.2–2.0 t ha^{-1} yr^{-1} of stover can be produced using fertilizer tree systems (Sileshi et al., 2014). In addition, studies on research stations in Zambia have demonstrated that fertilizer trees can reduce weed problems (e.g., *Striga*) and insect pests of maize, especially termites (Pumariño et al., 2015; Sileshi et al., 2005, 2006).

There is substantial evidence showing that fertilizer trees are profitable in terms of returns to land and labour. For example, the net present values (NPV) and benefit cost ratios (BCR) show that fertilizer trees are either comparable or better than the application of inorganic fertilizer in maize cropping (Ajayi et al., 2009). Over a five-year cycle in eastern Zambia, the discounted net benefit of maize grown with *Gliricidia* (US$327 ha^{-1}), *Sesbania* (US$309 ha^{-1}) and *Tephrosia* (US$233 ha^{-1}) compared favourably with maize grown with a recommended inorganic fertilizer (US$349 ha^{-1}). With respect to returns per investment, fertilizer trees performed even better (BCR: 2.8–3.1) than the recommended fertilizer purchased at market price (BCR: 1.8) or at 50% government subsidy of fertilizer (BCR: 2.6) (e.g., Ajayi et al., 2009). Similarly, in central Malawi, intercropping maize with pigeon pea consistently had positive returns across the farmers' resource groups indicating its suitability for a wide range of environments and for poorer farmers (Kamanga et al., 2010).

Biomass transfer for vegetable production in wetlands

Biomass transfer using green manure from fertilizer trees has shown promise in sustainable vegetable production in wetland and nutrition gardens. Biomass transfer is essentially moving green leaves and twigs from one location to another to be used as green manure. Although wetlands are considered extremely vulnerable to poor agricultural practices, rising population pressures have caused their agricultural use to become increasingly important. Wetlands – called *dambo* in Zambia and Malawi, *vlei* in Zimbabwe and South Africa, *molapo* in Botswana, Namibia and Lesotho and *naka* in Angola – are extremely important for dry season agriculture, grazing and water supply (Kuntashula et al., 2004). Traditionally, dry season production of vegetables has been widely practiced in wetlands, particularly in communal areas. With the recent emphasis on nutrition gardens, their utilization is increasing.

Nutrition gardens are often promoted by NGOs and church organizations, targeting the poor and the sick, especially HIV patients, with the aim of improving their standards of living with more emphasis on nutrition and income generation. Nutrition garden sites are often located close to water sources, which often tend to be wetlands. Biomass transfer has been demonstrated to be a sustainable means for maintaining soil-nutrient balances in vegetable production systems in the wetlands in Southern Africa (e.g., Kuntashula et al., 2004, 2006). The trees can be planted on cropland, degraded land or in silvopastoral systems and pruned or lopped. Depending on the nutrient requirement of the vegetable crop, 4–12 t ha^{-1} of leafy biomass (on dry matter basis) may be applied for increased productivity. This method has been demonstrated to be highly profitable in the production of cabbage, rape, onion, garlic and tomato in eastern Zambia (e.g., Kuntashula et al., 2004, 2006).

Fodder trees to improve dairy production

Fertilizer trees have been integrated with livestock production systems, called silvopastoral systems, in Southern Africa with varying degrees of success. Silvopastoral systems may be divided into two broad categories: fodder banks (also called protein banks) and grazing systems. In the protein bank approach, the animals are stall-fed with fodder collected from fertilizer trees grown in blocks on farmland. In the more extensive grazing areas, fertilizer trees are increasingly being planted in association with improved grasses to increase the carrying capacity of pastures or enhance the productivity of grazing cattle. A review of work carried out in Tanzania, Malawi and Zimbabwe has provided substantial evidence of improvement in smallholders' dairy productivity using fodder banks (Chakeredza et al., 2007).

Trees for providing other ecosystem services on farms

Substantial amounts of firewood can be produced through planting trees in various niches (Kamanga et al., 1999; Sileshi et al., 2007; Figure 23.1). According

Figure 23.1 *Sesbania* planted as a relay crop in a maize field on a smallholder farm to improve soil fertility and provide firewood

Source: Photo Ingrid Öborn

to Kamanga et al. (1999), 92–101% of domestic fuelwood needs were met from a hectare of 2–3-year-old *Sesbania* fallows in Malawi. Fertilizer trees also provide other ecosystem services in Southern Africa (Kuyah et al., 2016; Sileshi et al., 2007, 2014). Planting or managing naturally regenerating fertilizer trees can also tighten N cycling in the cropping system, increase carbon sequestration, reduce the need for fertilizer N inputs and lower greenhouse gas emissions (Sileshi, 2016; Sileshi et al., 2014). For example, the wider use of these trees in mixed crop and livestock systems can reduce methane emissions and increase carbon sequestration (Kim et al., 2016; Sileshi et al., 2014). Inclusion of tannin rich fodder in animal feeds can reduce enteric CH_4 production due to the anti-methogenic activity of tannins (Patra and Saxene, 2011). Net soil CH_4 emissions were also reduced when agriculture was shifted to improved fallow (Kim et al., 2016).

Rural households in Southern Africa use trees extensively within their farming systems. The trees can be planted or remnant trees. Trees supply wood fuel for household energy needs and charcoal for sale to urban centres. Trees supply timber for construction, including for example poles for building drying sheds for tobacco. People extensively harvest forest fruits and also grow fruit trees on their own land. Lastly, trees provide ecosystem services to farmers, such as shade and shelter for crops, livestock and people and enhance soil conditions. For example, a recent survey among vulnerable farmers (median hh income = US$200; 80% reported going hungry for one month or more) in Eastern Province, Zambia found that most households identified fuel and food as the main uses of trees, with approximately 20% also identifying medicines and "land benefits" (aka soil improvement) (Figure 23.2, left).

Farmers in the forest

Community forestry

Even where farmers practice more intensified farming, they use forests extensively. They depend on forests for a large part of their fuel needs and also harvest food from the forest (Figure 23.2, top graph). In the past, this dependence on forests for their livelihoods has often set them at odds with local and national authorities who control forest resources. In Zambia a recent change to the forest law (Government of Zambia, 2015, 2018) enabled the establishment of community forests. Under forest management agreements, local communities can gain access to tree resources, such as wood for timber and charcoal production and can collect forest products, like honey, mushrooms and fruit (Figure 23.2, bottom graph). That is, the law provides the right to harvest forest resources. Under the agreement, there are management responsibilities, which would normally include protection of the forest and limiting harvesting of tree resources to agreed-upon sustainable rates. Because of the relative novelty of the concept of community forests in Zambia, and a lack of capacity among district forest offices to negotiate community forest agreements, the uptake to date has been limited

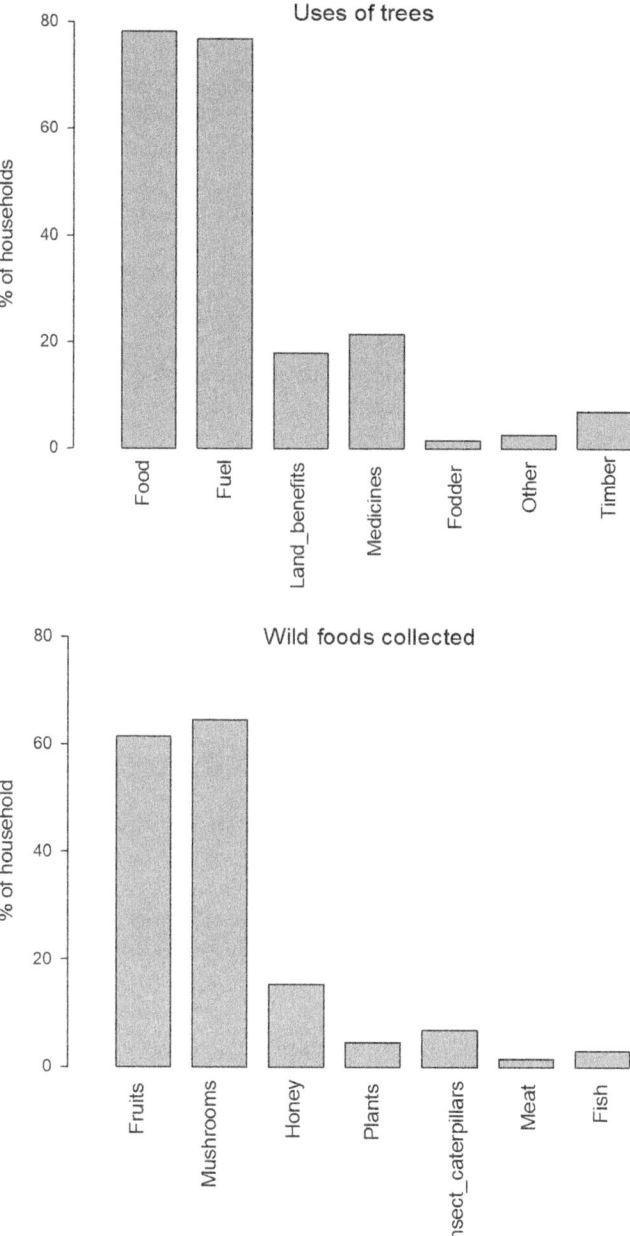

Figure 23.2 Farmers in Zambia are using trees and forest products as part of their livelihoods
   Top: a recent survey in the eastern province of Zambia showed that a majority
   of farmers are using trees for fuel and food, 20% also identified medicines and
   "land benefits" (aka soil improvement)
   Bottom: farmers are collecting wild food, in particular fruits, mushrooms and
   honey, in the forest

Source: Own representation

(33,190 ha as of July 2018). In addition, the government taxes the products, and this has been identified as a disincentive for rural communities to change from the current open access situation. Nonetheless, community forests could be a useful policy tool for delivering more sustainable forest management in a country where forests are widely used by rural communities, but the national government is unable to govern natural resource use. In particular, community forests (Gilmour, 2016) could offer an option for making the charcoal industry more sustainable.

Currently, a large proportion of the charcoal in Zambia is produced by migrant woodcutters who do not bear any of the externalities generated by their activities. Local communities are dis-incentivized to control access to their forests because their resource rights are not guaranteed. Hence, designation of community forests could incentivize communities to protect their forest resources and push the charcoal industry towards a more sustainable path.

Concluding remarks and ways forward

While the scientific foundation for promoting trees on farms has been solid, the challenge has been scaling-up to benefit the millions of smallholders seeking their livelihoods on marginal lands in Southern Africa (Stevenson and Vlek, 2018). The bottlenecks need to be identified and options for creating enabling environments and incentives generated.

There is possibility for farmers and local communities to engage in forestry in Southern Africa in a sustainable manner. The designation of community forests (Gilmour, 2016) is one option.

Over the past five decades, agroforestry science has gone from research on tree species and agroforestry practices on farms, through to studies of the roles of trees in multifunctional landscapes, to more recently paying attention to understanding the agroforestry policy domain and how to bridge the policy gap between agriculture and forestry policies (van Noordwijk, 2019).

To facilitate agroforestry development in Southern African countries and help bridge the gap between policy spheres, the Southern African Development Community (SADC) could develop Guidelines for Agroforestry Development, following the example from Southeast Asia (ASEAN, 2018). Similar developments could guide community forestry in the SADC countries. Adoption of national agroforestry, as for example in Zambia (ZARI and ICRAF, 2013), and community forestry strategies and guidelines can bridge the agriculture and forestry policy divide and enable and incentivize farmers and rural communities to engage in sustainable management of trees and tree products as means to transform their lives and livelihoods.

References

Ajayi, O. C., Akinnifesi, F. K., Sileshi, G. and Kanjipite, W. (2009) Labor input and financial profitability of conventional and agroforestry-based soil fertility management practices in Zambia. *Agrekon* 48: 276–292.

Ajayi, O., Place, F., Akinnifesi, F. and Sileshi, G. (2011) Agricultural success from Africa: the case of fertilizer tree systems in southern Africa (Malawi, Tanzania, Mozambique, Zambia and Zimbabwe). *International Journal of Agricultural Sustainability* 9: 129–136.

Akinnifesi, F. K., Ajayi, O. C., Sileshi, G., Chirwa, P. W. and Chianu, J. (2010) Fertiliser trees for sustainable food security in the maize-based production systems of East and Southern Africa. A review. *Agronomy for Sustainable Development* 30: 615–619.

[ASEAN] Association of Southeast Asian Nations (2018) *ASEAN Guidelines for Agroforestry Development.* Authors: Catacutan, D. C., Finlayson, R. F., Gassner, A., Perdana, A., Lusiana, B., Leimona, B., Simelton, E., Öborn, I., Galudra, G., Roshetko, J. M., Vaast, P., Mulia, R., Lasco, R. L., Dewi, S., Borelli, S., Yasmi, Y. Jakarta, Indonesia: ASEAN Secretariat.

Chakeredza, S., Hove, L., Akinnifesi, F. A., Franzel, S., Ajayi, O. and Sileshi, G. (2007) Managing fodder trees as a solution to human-livestock food conflicts and their contribution to income generation for smallholder farmers in Southern Africa. *Natural Resources Forum* 31: 286–291.

Gilmour, D. (2016) Forty years of community-based forestry. A review of its extent and effectiveness. *FAO Forestry Paper 176.* Rome, Italy; Food and Agriculture Organization of the United Nations.

Government of Zambia (2015) *The Forests Act, 2015 (No. 4 of 2015)*: 81–139.

Government of Zambia (2018) *The Forests Act, 2015 (Act No. 4 of 2015). The Forests (Community Forest Management) Regulations, 2018.* STATUTORY INSTRUMENT NO. 11 of 2018, Supplement to the Republic of Zambia Government Gazette dated Friday, 23rd February, 2018: 77–98.

Iiyama M., Neufeldt, H., Dobie, P., Njenga, M., Ndegwa, G. and Jamnadass, R. (2014) The potential of agroforestry in the provision of sustainable woodfuel in sub-Saharan Africa. *Current Opinion in Environmental Sustainability* 6: 138–147.

Kamanga, B. C. G., Kanyama-Phiri, G. Y. and Minae, S. (1999). Intercropping perennial legumes for green manure additions to maize in southern Malawi. *African Crop Science Journal,* 7: 355–363.

Kamanga, B. C. G., Waddington, S. R., Robertson M. J. and Giller K. E. (2010) Risk analysis of maize-legume crop combinations with smallholder farmers varying in resource endowment in central Malawi. *Experimental Agriculture* 46: 1–21.

Kim D-G., Kirschbaum, M. U. F. and Beedy, T. L. (2016) Carbon sequestration and net emissions of CH_4 and N_2O under agroforestry: Synthesizing available data and suggestions for future studies. *Agriculture, Ecosystems and Environment* 226: 65–78.

Kuntashula, E., Mafongoya, P. L., Sileshi, G. and Lungu, S. (2004) Potential of biomass transfer technologies in sustaining vegetable production in the wetlands (dambos) of eastern Zambia. *Experimental Agriculture* 40: 37–51.

Kuntashula, E., Sileshi, G., Mafongoya, P. L. and Banda, J. (2006) Farmer participatory evaluation of the potential for organic vegetable production in the wetlands of Zambia. *Outlook on Agriculture* 35: 299–305.

Kuyah, S., Öborn, I., Jonsson, M., Dahlin, A. S., Barrios, E., Muthuri, C., Malmer, M., Nyaga, J., Magaju, C., Namirembe, S., Nyberg, Y. and Sinclair, F. L. (2016) Trees in agricultural landscapes enhance provision of ecosystem services in Sub-Saharan Africa, *International Journal of Biodiversity Science, Ecosystem Services & Management* 12: 4, 255–273.

Minang, P. A., van Noordwijk, M., Freeman, O. E., Mbow, C., de Leeuw, J. and Catacutan, D., eds. (2015) *Climate-Smart Landscapes: Multifunctionality in Practice.* Nairobi, Kenya: World Agroforestry Centre (ICRAF) (http://asb.cgiar.org/climate-smart-landscapes/index.html).

van Noordwijk, M., ed. (2019) *Sustainable Development Through Trees on Farms: Agroforestry in Its Fifth Decade.* Bogor, Indonesia: World Agroforestry (ICRAF), pp 420 (www.worldagro forestry.org/trees-on-farms).

Öborn, I., Vanlauwe, B., Philips, M., Thomas, R., Brooijmans, W., Atta-Krah, K., eds. (2017) *Sustainable Intensification in Smallholder Agriculture. An Integrated Systems Research Approach.* London, UK: Earthscan Food and Agriculture, Routledge. 381 pp. https://cgspace.cgiar. org/handle/10568/89642.

Patra, A. K. and Saxene, J. (2011) Exploitation of dietary tannins to improve rumen metabolism and ruminant nutrition. *Journal of the Science of Food and Agriculture* 91: 24–37.

Pumariño, L., Sileshi, G, W., Gripenberg, S., Kaartinen, R., Barrios, E., Muchane, M. N., Midega, C., Jonsson, M. (2015) Effects of agroforestry on pest, disease and weed control: a meta-analysis. *Basic and Applied Ecology* 16: 573–582.

Sileshi, G. W. (2016) The magnitude and spatial extent of *Faidherbia albida* influence on soil properties and primary productivity in drylands. *Journal of Arid Environments* 132: 1–14.

Sileshi, G. W., Akinnifesi, F. K., Ajayi, O. C., Chakeredza, S., Kaonga, M., Matakala, P. (2007) Contribution of agroforestry to ecosystem services in the Miombo eco-region of eastern and southern African. *African Journal of Environmental Science and Technology* 1(4): 68–80.

Sileshi, G. W., Akinnifesi, F., Ajayi, O. C. and Place, F. (2008) Meta-analysis of maize yield response to woody and herbaceous legumes in sub-Saharan Africa. *Plant and Soil* 307 (1–2): 1–19.

Sileshi, G. W., Mafongoya, P. L., Akinnifesi, F. K., Phiri, E., Chirwa, P., Beedy, T., Makumba, W., Nyamadzawo, G., Njoloma, J., Wuta, M., Nyamugafata, P. and Jiri, O. (2014) Fertilizer trees. In: van Alfen, N. K. *Encyclopedia of Agriculture and Food Systems*, Vol. 1, San Diego: Elsevier, pp. 222–234.

Sileshi, G. W., Mafongoya, P. L., Kwesiga, F., Nkunika, P. (2005) Termite damage to maize grown in agroforestry systems, traditional fallows and monoculture on nitrogen-limited soils in eastern Zambia. *Agricultural and Forest Entomology* 7: 61–69.

Sileshi, G. W., Kuntashula, E. and Mafongoya, P. L. (2006) Effect of improved fallows on weed infestation in maize in eastern Zambia. *Zambia Journal of Agricultural Science* 8(2): 6–12.

Sirrine, D., Shennan, C., Snapp, S., Kanyama-Phiri, G., Kamanga, B. and Sirrine, J. R. (2010) Improving recommendations resulting from on-farm research: Agroforestry, risk, profitability and vulnerability in southern Malawi. *International Journal of Agricultural Sustainability* 8: 290–304.

Stevenson, J. R. and Vlek, P. (2018) *Assessing the Adoption and Diffusion of Natural Resource Management Practices: Synthesis of a New Set of Empirical Studies.* Rome: Independent Science and Partnership Council (ISPC).

ZARI and ICRAF (2013) *National Agroforestry Strategy 2013–2020*, November 2013, Lusaka, Zambia.

Part V

Improving policies and processes

This Part provides content on a broad array of policy and process related topics that are underdeveloped. These chapters provide a foundation for the region's policymakers to formulate appropriate policies for strategic investments in market development, infrastructure and institutional support for capacity, among others. It should be noted that there are a large number of policy and process issues handled in Parts III and IV that are also concerned with the need for support from decision makers.

24 Land reform and land tenure for agricultural transformation in Southern Africa

Thulasizwe Mkhabela

Introduction

There is general consensus that land reform is an indispensable yet complicated process that is often loaded with multiple objectives, including economic, social and political underpinnings. Land reform in Southern Africa has been seized with the need to balance the tension between governments' tendencies to rationalize land acquisition and redistribution on the basis of historical injustices and political demands, on one hand, and the basis of valid economic and technical reasons of land reform, on the other hand (Bernstein, 2003). It should be stated that although the former reason for land reform is legitimate, it needs to be counterbalanced with the latter.

The land reform in Southern Africa can be construed broadly as having three distinct, yet related, pillars, as captured in Figure 24.1.

This chapter is particularly concerned with land reform as it pertains to agriculture and agricultural transformation, while recognizing that the land reform process is broader than just agriculture. The first pillar of land redistribution deals with the skewed access to and ownership of land, thus seeking more equitable access to land and ownership. The second pillar of land restitution seeks to return land ownership to people who were previously dispossessed of their land, particularly during colonization and displacement by settler communities. The third pillar includes reforms of existing land tenure systems to facilitate increased security of tenure and an increased sense of land ownership for self-determination of the people.

Despite a declining share in gross domestic product (GDP), agriculture remains a key sector in the economies of Southern African countries. Agriculture is a major employer and foreign exchange earner for most of countries in Southern Africa. Many people in Southern African countries still reside in rural areas, and their main economic activity is farming, predominantly smallholder agriculture.

In Southern Africa, land tenure reform must address a suite of problems brought about by colonization and dispossession. Moreover, pressures of high population growth, deteriorating natural resource base due to degradation, specific food shortages (increasing localized food insecurity incidences) and

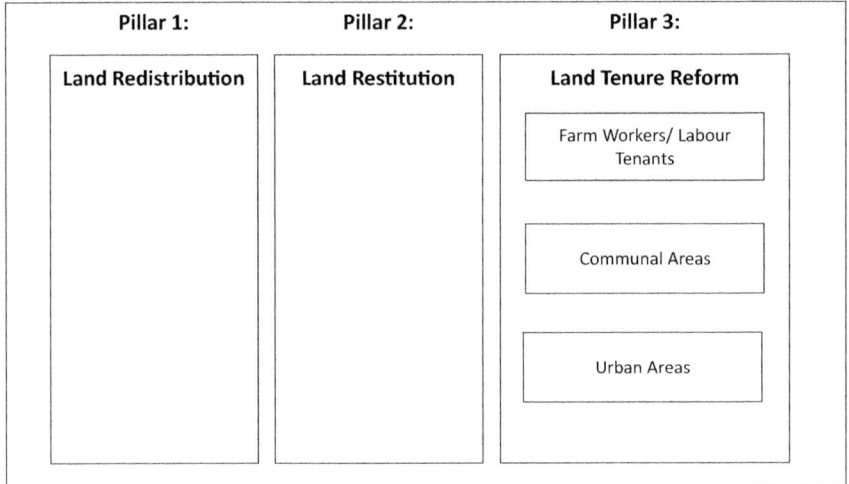

Figure 24.1 Land reform in Southern Africa

conflict over land use have given impetus to the raging debate and contestation on land reform in Southern Africa (Mkhabela, 2006).

The issue of tenure, the ownership or access to a piece of land has long been recognized as a critical factor for security, agricultural transformation, investment and conservation because it determines the linkages between responsibilities and authority over land and natural resources and also incentive structures for sustainable use (Murphree, 1996).

A brief explanation on what land tenure is will be presented before delving deep into a discussion on current land tenure systems in the Southern Africa region. Land tenure can be defined as the "terms and conditions on which land is held, used and transacted" (Adams et al., 1999), whereas, land tenure reform refers to a planned and deliberate change in the terms and conditions, such as the adjustment of the terms of contracts between land owners and tenants or the conversion of more informal tenancy into formal property rights. The fundamental goal of land tenure reform is to enhance and secure people's rights. Such changes are essential to avoid arbitrary evictions and the resultant landlessness. Furthermore, enhanced security of tenure is necessary for land rights holders to invest in the land and use it sustainably.

Prevalent land tenure systems in Southern Africa

There can be no doubt that tenure security matters and it has long been posited that tenure insecurity adversely affects agricultural productivity thereby constraining transformation in this sector of the economy. Land tenure in Southern Africa can be broadly divided into two categories, namely: 1) communal land tenure and 2) private land ownership, often referred to as freehold or title deed.

An examination of land tenure systems needs to take into consideration the complete physical, socioeconomic, political and cultural background and context of the society as a whole. Suffice to say that the present land tenure systems of Southern Africa are a product of historical forces (Mushala et al., 1998). Figure 24.2 summarizes the complexity of land tenure priorities for Southern African households by showing that land rights, and land tenure reform by extension, are a continuum rather than a discontinuity process. Moreover, there is a prevalent coexistence with the formal system of a customary system of land holding and statutory land rights across Southern Africa (Bah et al., 2018). This coexistence has partially given rise to multiple, parallel and, often, even overlapping and contradictory land tenure regimes.

Communal land tenure in Southern Africa is premised on traditional leaders administering land in trust for the benefit of traditional communities residing on such land and for the purpose of promoting the economic and social development of the people. Generally, the communal land tenure system is biased against women and is perceived to carry substantial patronage; thus, it is subject to abuse by the traditional leadership and people that are well-connected – the rural elite. Furthermore, rural folk find it difficult to raise capital for investment into agricultural production on the land that they do not own.

There have been several attempts to remedy the phenomenon of land tenure insecurity in the region. Several traditional leaderships in Southern African countries that have been administering land in trust, have been issuing permission to occupy (PTO) certificates. PTOs are legal documents used to regulate business establishments in communal areas, but these are now being phased out in most Southern African countries. Permission to occupy certificates are being converted into Rights of Leasehold, which are more secure and can be

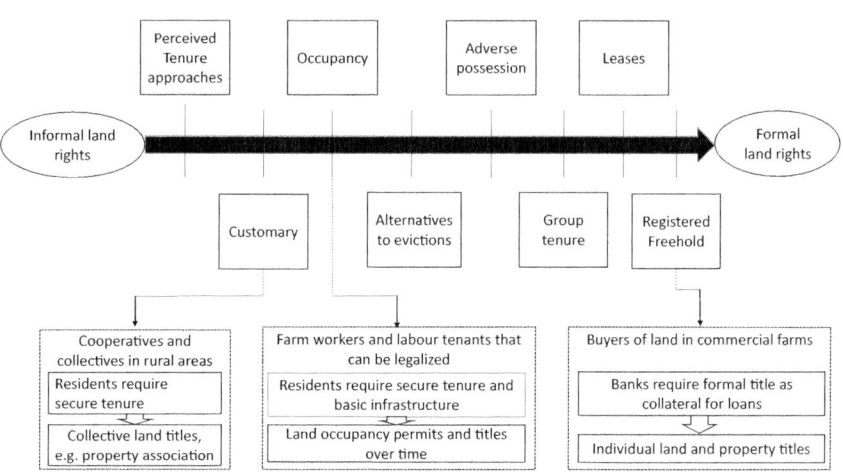

Figure 24.2 Land titling priorities for selected households

Source: Adapted from UN-Habitat's continuum of land rights (UN-Habitat, 2012)

used as collateral by holders to acquire loans if the lease is for periods of ten years or more.

Policies, processes and strategies for agricultural transformation through land reform

Land reform, including tenure reform, can be used as a policy instrument to engender inclusive agribusiness value chains by changing the ownership structure of land, thereby affecting the access to land and produce. This dynamic land reform policy environment creates a dependency for land and produce of large agribusinesses on smallholder indigenous communities as new landowners, indirectly stimulating the establishment of joint ventures, sometimes referred to as inclusive businesses (Mkhabela, 2018; Chamberlain and Anseeuw, 2018).

Land reform and tenure policies should include:

• Security of tenure rights over individual and public lands
• Redistribution of land possession, to include the poor and deprived majority
• Improved land governance
• Enhanced transparency.

Southern African governments and policymakers should look at innovative approaches to land tenure reforms in order to respond to contemporary issues related to security of tenure and agricultural transformation. One such approach is the flexible land tenure system (FLTS). The FLTS is an innovative concept to provide affordable security of tenure to inhabitants in rural areas and informal settlements, which is being piloted in Namibia. The basic tenets of the FLTS is to establish an interchangeable[1] tenure registration system parallel and complimentary to the current formal system of freehold tenure. This approach is somewhat similar to the PTO system used by traditional leaders in rural Southern Africa, although the latter is less formal, as there is no registration of title deed. This approach has already had some successes in Southern Africa. For instance, Botswana has had considerable progress in tenure reform through the integration of traditional tenure with a modern system of land administration for both customary and commercial forms of land use (Adams et al., 1999).

In some Southern African countries, such as the rural KwaZulu-Natal province of South Africa, local rural people are finding other innovative ways of dealing with land tenure issues. There is an emerging approach whereby people exchange communal tenure rights with each other at a fee, with the traditional leadership's endorsement. This approach frees agricultural land to be used by alternative users other than the households that were initially allocated the piece of land.

Conclusions

It has been shown that countries with historically more equitable land distribution achieved growth rates two to three times higher than their counterparts with less equitable land distribution (World Bank, 2006, 2007). Furthermore, there is no doubt that successful land reform which does not result in deleterious economic disruption creates rapid economic growth, as has been demonstrated elsewhere in the world and accordingly documented (Knying, 2009; Laurie, 2016; Lodge, 2018). Against the foregoing assertions, it follows that a redistributive land reform would help reduce rural, and even urban, poverty.

However, there is a caveat. A poorly designed and implemented land reform process intended to unlock the economic assets of communal land by activating dead capital should be demand driven. An indiscriminate supply of tradable tenure in the communal areas may result in destitution of many rural people as an unintended consequence.

It can be concluded that one of the fundamental challenges facing the issue of land reform is ownership or more accurately the best form(s) of tenure systems, arrangements and structure that should apply in Southern Africa. The often misunderstood and even misrepresented notion that the best and only recognizable form of land tenure systems that should pervade is that of outright ownership, especially as it relates to agricultural transformation. This ownership can and does take different forms but is best represented by the fact that the owner (in whatever form or shape, i.e., individual, private company, public company, the government, trust, Community Property Association (CPA), etc.) is recognized by having a registered title to the land. The main problem with this theory, that all land tenure arrangements should be in the form of some direct or indirect ownership, is not shared or even appreciated – as fact – by the vast majority of mainly rural based indigenous south Africans. Most majority of these mainly rural, peri-urban and increasingly urban (townships and sprawling squatter camps) citizens are crying out for some form of enfranchisement and recognizing access to and the rights to land as being at the centre of their most basic of rights, and the lack of it is a serious impediment to agricultural transformation.

Note

1 *Flexible* in the sense that it is at the discretion of the land authorities to choose the appropriate type of tenure for the formalization of an informal settlement; *interchangeable* in the sense that the different tenure types provided for in the parallel registries can be upgraded; and *parallel* to the existing freehold registration system in the sense that parallel institutions will be responsible for the registration of different tenure types.

References

Adams, M., Sibanda, S. and Turner, S. (1999) Land tenure reform and rural livelihoods in southern Africa. *Natural Resource Perspectives,* p. 39. London: An Overseas Development Institute (ODI) publication. ODI.

Bah, El-hadj M.B., Faye, I. and Geh, Z.F. (2018) Unlocking land markets and infrastructure provision. *Housing Dynamics in Africa*, pp. 109–158. London: Palgrave Macmillan.

Bernstein, H. (2003) Land reform in Southern Africa in world – historical perspective. *Review of African Political Economy* 30(96): 203–226.

Chamberlain, W.O. and Anseeuw, W. (2018) Inclusive business and land reform: Corporatization or transform? *Land* 7(1): 18–35.

Knying, K. (2009) The legacy of white highlands: Land rights, ethnicity and the post-2007 election violence in Kenya. *Journal of Contemporary African Studies* 27(3): 325–344.

Laurie, C. (2016) *The Land Reform Deception: Political Opportunism in Zimbabwe's Land Seizure Era*. Oxford: Oxford University Press.

Lodge, T. (2018) Thinking about South Africa land reform. *Focus* 83: 4–13.

Mkhabela, T. (2006) Impact of land tenure systems on land conflicts: Swaziland- a country case study. *Africanus* 36(1): 50–74.

Mkhabela, T. (2018) Dual moral hazard and adverse selection in South African agribusiness: It takes two to tango. *International Food and Agribusiness Management Review* 21(3): 391–406.

Murphree, M.W. (1996) *Wildlife in Sustainable Development: Approaches to Community Participation*. Paper presented to the ODA African Wildlife Policy Consultation. Sunningdale, UK.

Mushala, H.M., Kanduza, A.M., Simelane, N.O., Rwelamira, J.K. and Dlamini, N.F. (1998) Dual tenure systems and multiple livelihoods: A comparison of communal and private land tenure in Swaziland. *Land Reform* 2: 100–110.

UN-Habitat. (2012) *Handling Land: Innovative Tools for Land Governance and Secure Tenure*. United Nations Human Settlements Program. Nairobi: Kenya.

World Bank. (2006) *The Rural Investment Climate: It Differs and It Matters*. Washington, DC: World Bank.

World Bank. (2007) *World Bank Development Report 2008: Agriculture for Development*. Washington, DC: World Bank.

25 Engendering agricultural transformation

Joyce Chitja and Gabisile Mkhize

Conceptualizing gender and the engendering of agriculture in Southern Africa

The definition and meaning of gender are often misunderstood. Some equate gender to only women's issues and view it as a binary category of sex, thus undermining gender as a social construct (Mkhize, 2015). Embedded in our everyday practices and activities, gender is socially constructed and performed and exists in continuum (Lorber and Farrell, 1991). Gender is part of peoples' and communities' daily existence and reflects the ways the world is organized. Further, Krieger (2003) views gender as a social construct that is enforced through culture-bound conventions, roles and behaviours for relations between women and men and boys and girls. In Africa, socially constructed gender roles and patterns often vary across a continuum within and across societies in relation to social divisions premised on unequal gendered power and authority (Mkhize, 2015; Krieger, 2003; Lorber and Farrell, 1991). Fixed gender roles located within feminine and masculine gender schemas produce gender inequalities. For instance, sexist and patriarchal beliefs and cultural practices attach subordinate gender roles to women such as domestication, nurturing and dependency while attaching superior gender roles to men exemplified by leadership, ownership and independence (Mkhize and Njawala, 2016). Such sexist and patriarchal normative gender role assignment reinforces major gender inequalities in societal structures and practices. In gendered agricultural contexts, control and access to agricultural production resources are therefore by default men's forte, despite women being the majority in agricultural production (FAO, 2009).

Rolleri (2013) further elaborates that in many communities and/or societies throughout the world, including Southern Africa, a condition of gender inequality exists where women and the feminine are often devalued, and men and masculine traits are favoured. Here men often possess more power in decision making in personal, community and political arenas, while women's needs and interests are undermined. In most African societies, the learned behaviour leaves women more vulnerable to food insecurity, health and wellbeing (Thamaga-Chitja, 2012). Consequently, these socially constructed realities foster inequalities prevalent in socioeconomic systems, especially that of agriculture. Debatably in Africa and the Southern African Development Community (SADC), women are said to be at the centre of agricultural while in fact they are at the periphery where men mostly own the land and the majority of agricultural socioeconomic assets (Doss et al., 2015). Women remain the invisible stakeholders in agricultural platforms (Whatmore, 2016; Ogunlela and Mukhtar, 2009). Africa is thus seen to be at the bottom-end of the gender equality spectrum of gender inequality practices. Researchers who have worked in African societies have reported on the disenfranchisement and unequal treatment of women in agriculture and development (Olatunji, 2013; Meinzen-Dick et al, 2011).

In the 1970s, feminist scholars, activists and professionals concerned about women's interests decided to design women-centred development approaches such as Women in Development (WID), Women and Development (WAD) and Gender and Development (GAD). These approaches were aimed at pressuring international agencies and developing countries' governments, including those of Southern Africa, to incorporate women in development policy, planning and practice (Bhavnani et al., 2003; Rai, 2002; Moser, 1993). Thus, WID, WAD and GAD seek to bring women of developing countries closer into socioeconomic development processes including in agriculture, in their communities and societies. GAD focuses on empowering women as leaders. WID, WAD and GAD assume that women are important socioeconomic contributors in many societies of the South, and they seek a more sustainable and people-focused approach (Sen and Grown, 1987; Rai, 2002). In trying to address such gender-based inequalities, major actions and plans were developed, and many African countries committed to implementing those plans. For instance, Millennium Development Goal 3 and now Sustainable Development Goal 5 emphasize promotion of gender equality and women's empowerment. In addition to international gender debates, attempts to eradicate gender inequalities, ending gender-based discriminations is integral in the SADC's policies and agendas. In reinforcing a transformative agenda, the SADC (2018) gender policy promotes gender equity in all regions' developmental strategies and practices.

However, despite SADC's intervention, most women in Southern Africa continue to lag behind men in the development process including in agriculture and thus delaying development outcomes that can be achieved through gender inclusivity and collaboration. Just like in most African societies, in Southern African communities land is key to community development and the agriculture for farming. It is clear that women remain invisible in agriculture and land-based livelihood development. Such gender imbalances are not adequately addressed; hence, this chapter envisions engendering of agriculture as a viable and important tenet in agricultural transformation in SADC. Doss et al. (2015) view narratives of women's land ownership in Africa as purely a myth, since most men, irrespective of racial-ethnicities, own and control the land and its means of production and profit. It is within this context that the engendering agriculture and its related value chains is critical. This is no different in countries in Southern Africa, as attested by Honorable Gertrude Mongella of Tanzania when she said, "most men in Africa went to school because their mothers were farmers, [who] are the very same women that maintain the continent" (Hall and Osario, 2004).

Non-transformative gendered agricultural contexts

Currently and historically smallholder farmers, particularly women who dominate rural agriculture, reside in rural Southern Africa within communities that are socioeconomically less endowed. Rural communities in Southern Africa are resource-limited from a macro- and micro-structural point of view reflected in financial, physical, human and social resources. The transformation of these resources into sustainable livelihoods is only possible if there is gender parity, harmony, partnership and an environment that yields a conducive environment to power sharing. Constraints such as low production and productivity, poor access to inputs, credit and poor access to land characterize smallholder farming in Southern Africa (van Schalkwyk et al., 2012). Land is central in the construction of gender and power relations in such communities where rights to land access should be protected (Hall and Osorio, 2014). In spite of women dominating production, they face practice and policy-induced barriers to land access and ownership.

Women farmers' lived experiences of navigating laws and practices that negatively affect their efforts to access land in order to improve their livelihoods and household food security are important in policymaking, yet such women or their voices rarely are present when these policies are designed. As observed by many researchers, including Thamaga-Chitja et al. (2010), Chitja et al. (2016), Hall and Osorio (2014), women often have weaker customary and statutory rights and have barriers to participating in decision-making processes. In most of Southern Africa, including South Africa, the dichotomy of legislative framework and the cultural practices yields a contradictory policy and practice environment where land rights are concerned. In South Africa, for example, despite the clear policy on equal rights of access to all people in South Africa in the constitution, women in rural areas are negatively affected by the reality of this dichotomy, where traditional courts and the communities actuate biased and socially embedded gendered laws and practices that wield a different reality (Odeny, 2013; Ntsebeza, 2004). In most Traditional Authorities in South Africa and similar cultural contexts in Africa and SADC countries (e.g., Malawi, Zimbabwe, Lesotho and Tanzania), women are often viewed as minors (Thamaga-Chitja et al., 2010; Tripp et al., 2008), lacking access to land because of patriarchal, cultural and religious embedded norms, even where progressive statutory law exists. Evidently, cultural practices that are steeped in patriarchy tend to recognize males as heads of households and land access rights holders more than women regardless of women's dominance in the production of food in rural South Africa and as household heads in over 60% of households (Schatz, 2011). Mechanisms to mitigate this are key to protect the most vulnerable women. Agriculture acts as a key survival strategy for the majority of underdeveloped and developing countries globally to conquer food and nutrition insecurities. A study by (van Arendonk, 2015) found that the position of agriculture within the economy seems more important in developing countries and less important in developed countries, however different gender associated challenges have shown to be among the factors limiting maximum potential and capabilities of production.

Agriculture and agriculture related land rights have been mostly linked with the masculine practices in relation to land rights where women are mostly associated with in-house related activities. Such beliefs have limited women's access to agricultural land, knowledge and skills, which could potentially increase food production for household food and nutrition security. In Southern Africa, women commonly own less land, and what they own is most often of lower quality compared to the land possessed by men (Adeniyi, 2010). Women in Africa, including Southern Africa, only own 1% of the land and have to cope with limited access to supporting inputs such as financial and mechanical assets (Adeniyi, 2010). Given the point that women are key players towards alleviating poverty and food insecurity at household level, gender inequality and limited access to agricultural inputs in Southern Africa limits their potential. Addressing gender inequality in agriculture can possibly improve food production, reducing household food and nutrition insecurity. However, apparent inadequate appreciation and understanding of the negative role of related "custom" and "culture" among political leaders and civil servants, including extension services and TAs, impede progress and investment.

A study by Diiro et al., 2018 showed that between male managed and female managed agricultural plots, there is a significant improvement in productivity when the women managed the plots. These findings provide evidence that women's empowerment in agriculture positively contributes to reducing the gender gap in agricultural productivity and improving, specifically, productivity from farms managed by women (Diiro et al., 2018). Efforts to increase women's participation in poverty alleviation projects should encourage women's capacity building by providing incentives and opportunities that enable them to increase their competitive potential in their enterprise and policymaking. Policies that protect rights of and access for women to education and encourage them to complete at least a high school degree will help increase opportunities for women to participate in poverty alleviation (Nkemnyi et al., 2017). Given the beliefs and cultural practices, and understanding that the majority of rural African communities possess them, changing gender inequality in agriculture is challenging. To address this issue, the involvement of government and nongovernmental organizations (NGO) focused on farmer development using empowerment and agency building approaches is important. However, government extension and NGOs will need to be well endowed in participatory approaches that aim and bring a paradigm shift. Periodic feedback to policymakers is critical during this process. Agricultural funding should be directed equally be directed towards resources for women farmers and such actions are likely to increase their productivity and household food security and welfare. Some of these funds could be used to train both men and women with technical and "soft" skills and knowledge required to adapt and thrive in a changing business, sociocultural and climate environment. Formulation of gendered initiatives that enable women with improved access to land and markets is critical for real change.

Gender equity policies, processes and transformation

As discussed, in many African and SADC countries, including South Africa, statutory and cultural laws and practices clash spectacularly, leaving women negatively impacted with limited access to agriculturally productive land (Agarwal, 1994; Thamaga-Chitja and Morojele, 2014). The authors therefore argue for gender equity policies and processes in agricultural development context to be revisited to be more gender inclusive. Moreover, cultural and religious practices that prevails despite progressive statutory law require capacity and capability of extension agents and communities empowered in change management and creating new, yet positive, realities.

Gender policy can be strengthened by promoting gender mainstreaming and gender machineries in all SADC societies to promote the implementation of gender equality in farming. SADC promotes gender equity as a fundamental human right pivotal to the regional development initiatives (SADC Gender Policy, 2005). Thus, promoting gender centred development policies is significant for poverty eradication and family sustainability not only in SADC but in Africa at large. In addition, Technical Centre for Agricultural and Rural Cooperation (CTA) argues that women's transformation in agriculture needs to be tracked through and index, the agribusiness access Index (Adedeji et al., 2019). There should be a creation of a gender machinery, being correctly tooled personnel in extension research policy and community to facilitate and monitor gender equal participation in agricultural development projects, including farming.

Prior to democracy, Southern African legislation access to human and land rights only for the minority. Achieving equality was among the main priorities aimed to be achieved after obtaining democracy. The democratic South African government after 1994 began to make laws and implement programmes aiming to dismantle inequality. Similarly, countries in the southern Africa region countries have had attempts at a policy level to correct this, but evidence of change at a ground level is thin. For example, the 2008 SADC policy on gender and development prioritizes some substantive targets for achieving gender equality by 2015. Since agriculture is the main domain for fighting food insecurity, land reform and redistribution was within the primary of those programmes. Although the set of plans and acts to dismantle inequality were put into ideas and plans, it appears that the majority of actions addressed male related inequalities, resulting in woman getting least of economic activities, infrastructure and skill straining. SADC (2018) notes that various gender inclusive legislations are in existence in its regions, mostly evident in their gender policies that promote gender machinery, gender mainstreaming and women's empowerment. However, despite this, gender-orientated strategies and attempts, representation and participation of women in farming land ownership, decision making and profit making remain marginalized and inferior to that of men who continue to fully control and own the land, its usage and its profits. The chapter argues for the engendering of agricultural transformation policies and programmes, and the disabling of socially imbedded roles of women which need to be disrupted to bring about equity yielding transformation.

The following framework, in Table 25.1, puts forward key issues, policy and desired change ignitors. It includes the Centre for Agricultural and

Table 25.1 Key factors in gender policy change

Wabi-indicator	Policy action	Ignitors	Who is responsible
Access to land	Evaluate existing policies and programmes for gender equity, rewrite policy and establish effective programmes	Implement agriculture and gender inclusive policy; train, equip and skill all personnel	Government, cultural/ traditional authorities, farmers
Finance and Investment	Evaluate the level of access to finance and tailor make policy	Facilitate equal access and appropriate finance and investment portfolios	Government, NGOs, farmers
Skills Support	Gender inclusive stakeholders to identify the needs	Continual appropriate training and retraining programmes	Extension services and NGOs,
Networking and Capacity Development	Assess the availability of educational and technical extension resources for women	Establish skills- based programmes for women	Extension services, NGOs, farmers
Knowledge and Technology	Analyze existence and access of efficient and labour-saving tools and equipment	Implement updated technical training, using accessible state-of-the-art technology	Extension services, NGOs, private sector
Overcoming Embedded Sociocultural Practices and Belief	Assess the belief and practices in each context and align to equality and fairness	Capacitate extension, lead farmers and NGO personnel as change agents; raise awareness and campaigns on gender inequalities and inclusivity	Government, extension services, cultural/social bodies, farmers
Appropriate Recognition of Women in Society	Assess and highlight women's contributions to agriculture and publicize at platforms	Implement practical and proportional participation of women in agricultural programmes, practices and policies	Government, extension services, cultural/social bodies, farmers, media

Rural Cooperation (CTA) developed index named *Women Agribusiness Index Indicator (WABI)*

Land access, finance and investment, skills, networking and capacity development are all elements that are critical for addressing gender equity in transforming agriculture. However, the correct environment that is natured by skilled and capacitated extension personnel, NGO and lead farmers to confront backward cultural practices that impede transformation based on biased gender practices is critical.

Conclusions

Transformation is a process. A solution for engendered change may lie in a process where power between men and women is addressed through engagement, negotiation, learning and change. This should result in agency for both men and women. Stakeholders also need gender empowerment in this paradigm shift, e.g., extension services, NGOs, traditional leaders, lead farmers, community structures, input suppliers, etc., because of their role in farmer development. Empowering women for greater access to resources, knowledge and support is important. This process requires capacitated personnel functioning as a gender machinery alongside legislation, policies and practices to bring forth the desired change. However, changing the environment that houses the social and cultural constructs of women's position in communities by overcoming female stereotypes and traditional gender roles in rural communities is key and will have a positive result. This should yield better access and a more equal position in society for women.

References

Adedeji, O., Sudarkasa, M., Campbell, D.R. and Reynier, A. (2019) *CTA Technical Brief: Women's Agribusiness Access Index*. Technical Center for Agricultural and Rural Cooperation, Wageningen: CTA. https://cgspace.cgiar.org/bitstream/handle/10568/99706/2066_PDF.pdf.

Adeniyi, L. (2010) *Women Farmer's and Agriculture Growth: Challenge and Perspective for Africa Face the Economic Crisis*. Poster presented at the Joint 3rd African Association of Agricultural Economists (AAAE) and 48th Agricultural Economists Association of South Africa (AEASA) Conference, Cape Town, South Africa.

Agarwal, B. (1994) "Gender and command over property: A critical gap in economic analysis and policy in South Asia." *World Development* 22(10): 1455–1478.

Agarwal, B. 2011. "Food Crises and Gender Inequality," *Working Papers* 107, United Nations, Department of Economics and Social Affairs.

Bhavnani, K., Foran, J. and Kurian, P. A. (2003) *Feminist Futures: Re-imagining Women, Culture and Development*. London: Zed Books Ltd.

Chitja, J.M., Mthiyane, C.C.N., Mariga, I.K., Shimelis, H., Murugani, V.G., Morojele, P.J., Naidoo, K. and Aphane, O.D. (2016) Empowerment of women through water use security, land use security and knowledge generation for improved household food security and sustainable rural livelihood in selected areas in Limpopo. Water Research Commission. *WRC Report* No. 2082/1/15. ISBN: 978-1-4312-0744-2.

Diiro, G.M., Seymour, G., Kassie, M., Muricho, G. and Muriithi, B.W. (2018) "Women's empowerment in agriculture and agricultural productivity: Evidence from rural maize farmer households in western Kenya." *PLoS One* 13(5): e0197995.

Doss, C.R., Kovarik, C., Peterman, A., Quisumbing, A.R. and van den Bold, M. (2015) *Gender Inequalities in Ownership and Control of Land in Africa: Myths Versus Reality.* Washington, DC: International Food Policy Research Institute.

FAO (2009) *State of Food Insecurity in the World, 2009.* Rome: Food and Agriculture Organization of the United Nations.

Hall, R. and Osorio, M. (2014) *Agricultural Investment, Gender, and Land in Africa: Towards Inclusive Equitable and Socially Responsible Investment. Conference Report 2014.* Multistakeholder Conference on Agricultural Investment, Gender and Land in Africa Towards inclusive, equitable and socially responsible investment. Protea Hotel, Stellenbosch (Cape Town), South Africa, 5-7 March 2014 www.plaas.org.za/sites/default/files/publications-pdf/AIGLIA%20Report_Web.pdf

Krieger, N. (2003) "Genders, sexes, and health: What are the connections – and why does it matter?" *International Journal of Epidemiology* 32(4): 652–657.

Lorber, J. and Farrell, S.A. eds., (1991) *The social construction of gender* (pp. 309–321). Newbury Park, CA: Sage.

Meinzen-Dick, R., Quisumbing, A., Behrman, J., Biermayr-Jenzano, P., Wilde, V., Noordeloos, M., Ragasa, C. and Beintema, N. (2011) *Engendering Agricultural Research, Development and Extension.* International Food Policy Research Institute

Mkhize, G. (2015) "Problematising rhetorical representations of individuals with disability – disabled or living with disability?" *Agenda* 29(2): 133–140.

Mkhize, G. and Njawala, L. (2016) "Rethinking hegemonic masculinity and patriarchal privilege within heterosexual interpersonal relationships." *Oriental Anthropologists* 16(2).

Moser, C. (1993) *Gender Planning and Development: Theory, Practice and Training.* London: Routledge.

Nkemnyi, F. M., Fombu, C. Y. N., Kwobenyi, N. G., & Mumbang, C. (2017). *An Assessment of the Role of Women in Development and Poverty Alleviation in Cameroon.* In I. Piot-Lepetit (Ed.), Cameroon in the 21st Century: Challenges and Prospects. Volume 2: Environment and People (pp. 219–234). New York: Nova Science Publishers, Inc.

Ntsebeza, L. (2004) "Democratic decentralisation and traditional authority: Dilemmas of land administration in rural South Africa." *The European Journal of Development Research* 16: 1, 71–89. doi:10.1080/09578810410001688743.

Odeny, M. (2013) *Improving Access to Land and Strengthening Women's Land Rights in Africa.* Paper presented at the World Bank Conference on Land and poverty, Washington, DC.

Ogunlela, Y.I. and Mukhtar, A.A., (2009) Gender issues in agriculture and rural development in Nigeria: The role of women. *Humanity & Social Sciences Journal* 4(1): 19–30.

Olatunji, C.-M. P. (2013) "An argument for gender equality in Africa." *CLCWeb: Comparative Literature and Culture* 15(1): 9.

Rai, S. (2002) *Gender and the Political Economy of Development.* Cambridge, UK: Polity Press.

Schatz, E. (2011) "Female headed households contending with HIV/AIDS related hardship in rural South Africa." *Health & Place* 17(2):598–605.

Sen, G. and Grown, C. (1987) *Development, Crises and Alternative Visions.* London. Earthscan.

South African Development Community (SADC) (2018) "SADC gender policy (English)." https://www.sadc.int/files/8414/0558/5105/SADC_GENDER_POLICY_-_ENGLISH.pdf. Accessed: 03 April 2019.

Thamaga-Chitja, J.M. (2012) "How has the rural farming woman progressed since setting of Millennium Development Goals for eradication of poverty and hunger?" *Agenda* 26(1): 67–80.

Thamaga-Chitja, J.M., Kolanisi, U. and Murugani, V. (2010) "Is the South African land reform programme, gender sensitive to women's food security and livelihoods effort?" *Agenda* 86: 122–134.

Thamaga-Chitja, J.M. and Morojele. P. (2014) "The context of smallholder farming in South Africa: Towards livelihood asset-building framework." *Journal of Human Ecology* 45(2):147–155.

Tripp, A., Casimiro, I., Kwesiga, J. and Mungwa, A. (2008) "Contents." In *African Women's Movements: Transforming Political Landscapes* (p.V). Cambridge: Cambridge University Press.

Van Schalkwyk, J.A., Groenewald, G.C.G., Fraser, A., Obi, A. and van Tilburg, A. (2012) *Unlocking Markets for Smallholder in South Africa*. Wageningen: Wageningen Academic Publishers.

van Arendonk, A. (2015). *The development of the share of agriculture in GDP and employment*. Masters Dissertation. Wageningen University, Netherlands.

Whatmore, S. (2016) *Farming Women: Gender, Work and Family Enterprise*. Springer.

26 Building human capacity to transform agriculture in Southern Africa

Anusuya Rangarajan and Joyce Chitja

Introduction

Recent estimates indicate that 90% of the world's farms are family owned and operated, and 84% of those families farm on less than 2 ha (Lowder et al., 2016). Small farms (less than 20 ha) supply diverse and nutritious products to local food systems while advancing household food security (Herrero et al., 2017). Among the Southern African Development Countries (SADC), smallholder farmers represent the majority of producers, farming an average of 1 to 5 ha. The government of South Africa recognizes small-scale agriculture as a potentially sustainable contributor to food security (Altman et al., 2009) and as a development strategy in the 2010 National Development Plan (Wiggins and Keats, 2013) because it increases household income and nutrition with minimal investment.

To continue to thrive these farmers constantly change their practices and leverage their limited assets and resources in creative ways. Developing this type of nonmaterial ability or agency refers to the ability of smallholders and their organizations to position themselves in a market, to make effective choices to advance their interests and to be able to act on those choices (Sonia, 2012). This type of creative adaptation is underpinned by social networks, education and training and personal motivation and planning and risk management. This type of agency is not static; in fact the dynamic nature of learning, sharing information, reflection, innovation and anticipation or forward planning more completely describes the adaptation process that is fundamental to building resilient farming systems (Tschakert and Dietrich, 2010). The key policy and practice question is how can research, extension and policy support the development of human adaptive capacity or agency of smallholder farmers and thereby support continuous transformation of agriculture?

Smallholder farmers in Southern Africa, and around the world, face several critical challenges to their long-term viability and sustainability. These challenges can be briefly summarized within production strategies, access to land and inputs, access to markets, relevant training, farmer organizational capacity and capability, appropriate technology and entrepreneurship and business development.

Perhaps the most pressing production challenge being faced by smallholders is adaptation to climate change. Ecologically and economically resilient and relevant farming strategies are needed to overcome the drought, flood and temperature perturbations associated with climate change. Solutions need to be scale and resource relevant, quickly adaptable by women farmers and have some immediate measurable benefit to farm productivity to secure implementation long-term. Many technologies are available that meet these criteria- but outreach is limited by labour hours and strategies used by educators.

In South Africa and to an extent in Zimbabwe, there is another fundamental challenge particularly for black smallholder farmers- new farm startup. While repatriation is celebrated, the reality is that many returning to the land find they lack knowledge on starting and growing a farm business. Traditional knowledge has been lost given generations out of farming; new farming technologies are not readily accessible if one is unaware of the supportive networks for information flows. In other parts of the world, beginning farmer training has become a priority for federal programmes given the general aging of the current farmer population. Loss of farmers and their businesses represents a risk to national food security as well as rural community viability.

Limited access to production knowledge, inputs and financial credit can curtail growth and adaptability of any farmer to climate and other challenges, regardless of scale. Access to inputs (e.g., soil amendments, pesticides, fertilizers and packaging) at price points, in appropriate quantities and at locations nearby smallholders' farms, is the first challenge. Understanding the wise and best use of these inputs is yet another hurdle, given language and literacy barriers. Agricultural credit from appropriate and reputable sources, or via community shared savings, can support growth and development of farm enterprises. However, access to such financing may be out of reach for many smallholder farmers.

For many women farmers, secure access to land, whether through secure rights of access, title or long-term lease agreements, is of particular concern. Women's access to agricultural land or any other land is precarious, often accessed through secondary means through natal or marital relationship. Without secure access to land, smallholder farmers and in particular women farmers may have limited feasible or affordable options for restructuring production strategies for climate resilience. Investing in soil quality, for example, requires a long-term view; incentives are few if tenure is questionable.

Many smallholder farmers rely upon direct sales to neighbours or through local farmers' markets. Access to wholesale markets represents a diversification and growth strategy, particularly for situations in which a farmer may be very skilled at growing select crops or raising livestock. Access to wholesale markets is limited by farmer awareness of terms and quality required, economical transportation options to move product to markets and concerns with brokers that siphon profits of sales. While working cooperatively or collaboratively with other farmers to achieve the economies of scale for product volume is an attractive approach, these types of cooperatives are riddled with social and equity challenges that can quickly lead to their undoing.

Entrepreneurship was defined by Dr Howard Stevenson (Harvard University) as "the pursuit of opportunity beyond resources controlled" (Eisenmann, 2013). Pursuit characterizes the focused attention to an opportunity, which might include new products, markets, models or cost saving strategies. The resources that an entrepreneurial farmer has to leverage are their human, social, natural and financial capitals. They must be creative and take risks to grow their businesses. While central to entrepreneurship is personal agency, or the capacity to act independently and make individual choices, there are many structures (e.g., class, gender, ability) that can influence or limit a smallholder's decisions. Scalable examples of resilient, thriving smallholder farming in SADC are needed to inspire farmers to transition from subsistence to growth.

How can educators build capacity and capability among smallholder farmers to anticipate and adapt to challenges? What are potent strategies in education and engagement that feed the fire of optimism and entrepreneurship? How do training programmes foster dynamic and future-focused "anticipatory" planning required to truly prepare for climate change and other agricultural risks?

Participatory and popular education techniques

While there are many farming challenges that can be addressed through straightforward technology transfer, these strategies may fall short in securing transformation of smallholder farmer practice. Resilience to climate change and other environmental stresses requires more agro-ecological, integrative farming strategies that include local knowledge, social networks and political economies to build sustainable and equitable systems (Méndez et al., 2013) (Mistry and Berardi, 2016).

Conveying the complexity of agro-ecological system design without assuming advanced education and understanding of biological processes requires a participatory or action-research approach. Indigenous knowledge of a locale's biodiversity and natural resources can provide insights to support adaptation of appropriate solutions (Snapp et al., 2010). Participatory action research is well established as a strategy to support cogeneration of knowledge; the process empowers individual and group action by weaving local knowledge with scientific findings and strengthens trust with outside educators (Snapp and Heong, 2003). The approach allows marginalized communities to give voice to their own knowledge, draw from lived experiences and problem-solve towards their own solutions.

The educator requires skills and careful planning to fully implement these types of participatory approaches. Facilitation strategies for success include how to structure group process, establish working agreements and ensure participation from all members. Collaborative problem identification coupled with cogeneration of solutions by researchers/educators/extensionists strengthens the appropriate application of strategy within the farmers' context.

The Smallholder Horticultural Empowerment and Promotion (SHEP) project provides an example where farmers and implementers co-identify and co-implement solutions (JICA, 2019). SHEP has assisted farmers to double their income in Kenya (JICA, 2019) and has been implemented in Lesotho, Malawi, South Africa and Zimbabwe since 2018. Similar to the SHEP model in Kenya, models and case studies of inclusive, empowering educational approaches have been described for agricultural enterprises in South Africa (Mabaya et al., 2011).

Building capacity of service providers

Most extension agents or other educators come from formal education programmes with specific disciplinary focus. Continuing professional development ensures these front-line individuals' technical knowledge and skills are in sync with current research (e.g., climate risks) or best practices. Prioritizing their engagement directly with farmers is critical to deploying this valuable development resource. Public sector extension work is necessary for the growth of the agriculture in southern Africa but limitations in resources, excessive bureaucratic work and the diversity of extension goals limits transformational success (Raidimi and Kabiti, 2017). Therefore, investments in extension services must be deeply connected with communities to ensure both educator and farmer satisfaction.

Extension personnel are rarely trained in participatory action research strategies, which include facilitation strategies, group or leadership development or nontraditional modes of engagement. Yet, these "soft" skills are essential "tools" in the trainer's toolbox to support transformative change among smallholder farmers. These skills require the educator to constantly be shifting roles, from teacher, to learner, to networker. Dogmatic adherence to a prescribed educational agenda will only serve to document numbers "trained" but none transformed. Policy and strategy must support the mix of skills needed to cultivate dynamic educators with both the foresight and humility to plan thoughtfully before entering communities with "solutions". Participatory engagement, with full commitment and mutual respect among researchers, educators, students and citizens serves to leverage the assets of the whole while also strengthening democracy (Post et al., 2016).

Several of these types of skills are briefly described in the following sections. An example of a comprehensive approach to engaged curriculum focused on agro-ecology and climate change is presented by Bezner Kerr et al. (2019).

Mapping the knowledge ecosystem

If the goal is to strengthen the human capacity and agency of farmers, the initial focus must be on supporting professional development of those who connect with them regularly. While this will obviously centre on funded agriculture extension officers/educators, the pool of service providers extends to all who impact the viability of these farmers, including NGO agents, agriculture input suppliers, product buyers, financiers, environmental and regional planners, etc.

Collectively, these individuals represent the knowledge "ecosystem" that supports farm viability. Mapping and understanding the influence of these current and historic actors in a region is essential prior to investing in new training. Engaging this diversity of perspectives will support design of programmes that are reinforced by multiple individuals and farmer contact points. Supporting professional development across these groups, together, could foster greater collaboration towards training goals.

Extension investment in SADC remains sparse, uncoordinated and varied in quality (Zwane and Davis, 2017). In South Africa, extension renewal and evaluation exercises have taken place and identified gaps in policy environment, professionalism, capacity and technical competencies (Davis and Terblanché, 2016). Recent policy investments point to a renewal in agricultural extension through the development of new coordinating bodies that promise to promote farmer development (e.g., New Partnership for Africa's Development's Comprehensive African Agricultural Development Programme (CAADP, 2003) and the Southern Africa Regional Forum for Agricultural Advisory Services (SARFAAS) (Zwane and Davis, 2017).

Mapping of community infrastructure

Ubuntu is a South African concept that refers to compassion, solidarity and the interdependence of the individual with a community. It speaks to values of caring, sharing, cooperation and compassion. It recognizes that inextricable link among us, summarized as "I am because we are". This principle is at the core of these participatory and asset-based approaches that can build human capacity to support the transformation of agriculture.

Participatory methods build upon this shared South African idea of Ubuntu to foster group action. Several methods can help farmer groups clarify resource assets and constraints. Examples of group activities include participatory mapping, time use surveys, participatory photography and seasonal calendars and focus groups. The participatory activities can uncover shared knowledge, attitudes and perceptions, as complements to individual interviews and case studies. In all cases, the educator/facilitator must consider their role in supporting discussion, having community agreement on that role and setting aside their own knowledge or opinions. These skills require training and coaching to effectively build trust within the community group and maintain open, engaged dialogue and debate.

For example, participatory mapping helps a community and advisors understand the community's relationships to their land, natural resources, community resources and other assets (Corbett, 2009). A set of guiding questions helps participants create their own visual map by first asking them to locate built landmarks (e.g., churches, community centres, roads,) followed by natural resources (water sources, forests), their own homesteads and then agricultural assets (e.g., input suppliers, fences, markets, best soils). Discussion can then focus on uncovering challenges, such as flood zones, best or degraded soils, poor roads or transport, social isolation and opportunities for change. Through this activity, communities can be more empowered to share priorities for training from educators or investment from local governments.

Participatory photography is another strategy in which community members can identify, record and then critically discuss opportunities and challenges using the photos as prompts (Wang and Burns, 1997). Given a set of prompts and a camera (cell phones now make this easy), the community members capture images to provide evidence as to their understanding of ecological, production or marketing concepts, their access to resources or the challenges they face individually or as a group. For example, asking farmers if they have a problem with access to water may elicit are vocal response of "no". Yet when they take photos of their water source, it can become clear that hand carrying water from a small stream could in no way support expanding agricultural production. The photos become a focus point for discussion around challenges and solutions. This strategy also allows for full engagement of those who do not read or write in the dominant language used in the community. Images can support much more rich discussion and memorable conclusions, compared to sharing survey or research results.

Mapping the challenges

Surveys, storytelling, focus groups and other small group discussions can support community inquiry into shared challenges. In addition, dramas can help bring humour into the process, particularly when exploring sensitive issues, such as inequities in access to resources or power differentials (Bezner Kerr et al., 2019). Drama done by facilitators can be used to illustrate complex problems, supporting group discussion and analysis. Subtle shifts in understanding can pave the way for greater farmer confidence in their own knowledge and more openness to collaboration (Bezner Kerr et al., 2019).

Role plays or participatory theatre create active ways for a community to explore issues, analyze possible changes and explore power relations that may impact solutions (Sloman, 2012). The educator/facilitator is essential to "set the stage" with some scenarios that can guide but do not control the creative direction of the role play. The risk is that the outcome can be unpredictable. Yet, these moments provide opportunities for new dialogue. The process can energize a community, strengthen cohesion and be entertaining for all. As a facilitation strategy, role plays and drama can enliven educational programmes and create memorable stories.

Timely knowledge sharing

Farmer field schools are a popular educational strategy in which groups of farmers gather to inquire and experiment about various agricultural issues of importance to their farms (Sustainet, 2010). The strategy combines traditional education with hands-on or experiential activities, to deepen understanding of the problem and context. The farmer groups repeatedly gather at a farm to work together to analyze and solve particular challenges. Critical to the success of this engaging process is a skilled facilitator who has the confidence and ability to guide but not direct, supporting the farmers in sharing their own knowledge (Godrick et al., 2014). Extension personnel need to be trained to facilitate this strategy.

Farmer field schools can build or strengthen farmer networks and collaboration. Unlike field demonstrations, the process uses repeated observations and inquiry by the group. Structuring small experiments or trials together supports peer-to-peer learning, trust building and can lay groundwork for more complicated group activities such as joint marketing of products.

Digital platforms offer new strategies to quickly provide relevant information to farmers or facilitate their sharing information with each other via texting on cell phones (e.g., https://wefarm.co/). Educators have an opportunity to provide advice on planning, inputs, production, harvest, pests and their control, postharvest handling and market pricing – just a few areas in which educators could leverage a digital platform. While the most efficient strategy currently may be texting, short videos or photo sharing will allow for those with limited literacy to participate. Photos can also support rapid diagnosis of pest problems. Data could be shared in a "programmed" manner, according to seasonal calendars (e.g., what diseases/insect and at what growth stages along with available pesticide/fungicide to effectively treat them) or in response to weather anomalies and reach more farmers than possible individual or community field visits.

Another novel example of a small-farm relevant cell phone application is bringing the sharing economy to booking tractors (e.g., www.trotrotractor. com/ or www.hellotractor.com/home). These platforms facilitate tractor services particularly for smallholder and women farmers as well as encourage entrepreneurship and community collaboration.

Evaluation benchmarks

Investing in strategies to assess and benchmark access to essential agricultural resources can support ongoing development of appropriate and effective programmes aimed at increasing human capacity for African farmers. Of particular interest is tracking progress of women farmers, a significant portion of SADC farmers. The Women Empowerment in Agriculture Index (IFPRI, 2012) outlines five domains for empowerment- production, resources, income, time and leadership. The recent brief by the Technical Center for Agricultural and Rural Cooperation argues for more comprehensive benchmarks or indices to assess women farmers' access to agriculture resources that support entrepreneurship (Adedeji et al., 2019). Included in this index are 1) access to land; 2) access to finance; 3) access to markets; 4) access to infrastructure, technology and equipment; 5) access to education and capacity-development resources; 6) access to business information and networks; and 7) access to policymakers/ policy dialogue. These types of indices provide strategies for understanding the full ecosystem of factors influencing farmer decision making towards agriculture advancement.

Conclusions and policy recommendations

Building human capacity of farmers first requires building capacity among extension educators. Learning to do engaged work is not easy. Professional development must start in post-secondary training of extension educators, outfitting them with the facilitation skills and confidence to deploy participatory strategies as easily as technical solutions.

To move from technical service provider to transformative agent will require new skills, confidence and programme design from the educators. Hosting training events and sharing technical information will certainly provide smallholder farmers with choices. Creating an environment to support the personal realizations that precede transformative change and action requires more than traditional technology transfer. Grounding farmer education in participatory or asset-based approaches can foster this progression. The transforming extension agent fosters resilient social networks, guides – not prescribes – and encourages peer-to-peer learning and challenges farmers to grow "beyond resources controlled".

The work of transformation does not lie only with the field extension educator. It includes the network of publicly engaged scholars and university academics, school educators, practitioners, citizens and nonprofit leaders whose knowledge and commitment are needed for long-term capacity building among the target farmers (Uta et al., 2018). A revived and invigourated sense of Ubuntu, or interdependence, among these various actors will support innovation and creative problem solving. Participatory engagement, with full commitment from all stakeholders to yield social learning for all stakeholders, including facilitators, extension researchers, educators and farmers, will best leverage the assets of the whole.

References

Adedeji, O., M. Sudarkasa, D.R. Campbell and A. Reynier (2019) *CTA Technical Brief: Women's Agribusiness Access Index. Technical Center for Agricultural and Rural Cooperation.* Wageningen: CTA. Accessed March 2, 2019. https://cgspace.cgiar.org/bitstream/handle/10568/99706/2066_PDF.pdf.

Altman M., T.G.B. Hart and P.T. Jacobs (2009) Household food security status in South Africa. *Agrekon* 48(4): 345–361.

Bezner Kerr, R., S.L. Young, C. Young, M.V. Santoso, M. Magalasi, M. Entz, E. Lupafya, L. Dakishoni, V. Morrone, D. Wolfe, and S.S. Snapp (2019) Farming for change: Developing a participatory curriculum on agroecology, nutrition, climate change and social equity in Malawi and Tanzania. *Agriculture and Human Values.* Accessed February 20, 2019. https://doi.org/10.1007/s10460-018-09906-x.

Comprehensive Africa Agriculture Development Program (CAADP). (2003) *New Partnerships for Africa's Development.* www.caadp.net/pdf/CAADP. ISBN 0-620-30700-5.

Corbett, J. (2009) *Good Practices in Participatory Mapping: A Review Prepared for the International Fund for Agricultural Development.* Accessed February 22, 2019. www.ifad.org/documents/38714170/39144386/PM_web.pdf/7c1eda69-8205-4c31-8912-3c25d6f90055.

Davis, K.E. and S. Terblanché (2016) Challenges facing the agricultural extension landscape in South Africa. *South African Journal of Agricultural Extension* 44(2): 231–247.

Eisenmann, T.R. (2013) Entrepreneurship: A working definition. *Harvard Business Review.* https://hbr.org/2013/01/what-is-entrepreneurship.

Godrick, S.K., J. Okoth and E. O'Brien (2014) *Farmer Field Schools: Key Practices for DRR Implementers.* FAO. Accessed March 10, 2019. www.fao.org/3/a-i3766e.pdf.

Herrero M., P.K. Thornton, B. Power, J.R. Bogard, R. Remans, S. Fritz, J.S. Gerber, G. Nelson, L. See, K. Waha, R.A. Watson, P.C. West, L.H. Samberg, J. van de Steeg, E. Stephenson, and M. van Wij (2017) Farming and the geography of nutrient production for human use: A transdisciplinary analysis. *The Lancet Planetary Health* 1(1): e33–e42. Accessed February 20, 2019. https://doi.org/10.1016/S2542-5196(17)30007-4.

International Food Policy Research Institute (IFPRI). (2012) *Women's Empowerment in Agriculture Index.* Accessed May 12, 2019. http://ebrary.ifpri.org/utils/getfile/collection/p15738coll2/id/126937/filename/127148.pdf.

Japan International Cooperation Agency (JICA). (2019) *Smallholder Horticulture Empowerment and Promotion (SHEP).* Accessed May 14, 2019. www.jica.go.jp/english/our_work/thematic_issues/agricultural/shep.html.

Lowder, S.K., J. Soet and T. Raney (2016) The number, size, and distribution of farms, smallholder farms, and family farms worldwide. *World Development* 87: 16–29. Accessed February 20, 2019. https://doi.org/10.1016/j.worlddev.2015.10.041.

Mabaya, E., K. Tihanyi, M. Kaaran and J. van Rooyen (2011) *Case Studies of Emerging Farmers and Agribusinesses in South Africa.* Sun Press. ISBN: 978-1-920338-65-7.

Méndez, E.V., C.M. Bacon and R. Cohen (2013) Agroecology as a transdisciplinary, participatory, and action-oriented approach. *Agroecology & Sustainable Food Systems* 37: 3–18.

Mistry, J. and A. Berardi (2016) Bridging indigenous and scientific knowledge. *Science* 352: 1274–1275. Accessed February 12, 2019.

Murphy, S. (2012) *Changing Perspectives: Small-scale Farmers, Markets and Globalisation.* London/The Hague: IIED/Hivos. Accessed March 2, 2019. www.ictsd.org/sites/default/files/downloads/2012/08/changing-perspectives-small-scale-farmers-markets-and-globalisation-murphy-iied.pdf.

Post, M.A., E. Ward, N.V. Longo and J. Saltmarsh (2016) *Publicly Engaged Scholars: Next Generation Engagement and the Future of Higher Education.* Sterling, Virginia: Stylus Publishing. p. 286.

Raidimi, E.N. and H.M. Kabiti (2017) Agricultural extension, research, and development for increased food security: the need for public-private sector partnerships in South Africa. *South African Journal of Agricultural Extension* 45(1): 49–63.

Sloman, A. (2012) Using participatory theatre in international community development. *Community Development Journal* 47(1): 42–57. Accessed February 15, 2019. https://doi.org/10.1093/cdj/bsq059.

Snapp, S.S and K.L. Heong (2003) Scaling up: Participatory research and extension to reach more farmers. In: *Managing Natural Resources for Sustainable Livelihoods: Uniting Science and Participation.* Eds. B. Pound, S, Snapp, C. McDougall and A. Braun. UK: Earthscan. pp. 67–87.

Snapp, S.S., J. Malcolm, B. Blackie, R.A. Gilbert, R. Bezner Kerr and Y.G. Kanyama-Phiri. (2010) Biodiversity can support a greener revolution in Africa. *Proceedings of the National Academy of Sciences* 107: 20840–20845. Accessed February 20, 2019. www.pnas.org/content/107/48/20840.

Sonia, Murphy (2012). *Changing Perspectives: Small-scale farmers, markets and globalisation.* London/ The Hague: IIED/Hivos. Accessed March 2, 2019. https://www.ictsd.org/sites/default/ files/downloads/2012/08/changing-perspectives-small-scale-farmers-markets-and-glob alisation-murphy-iied.pdf.

Sustainet, E.A. (2010) *Technical Manual for Farmers and Field Extension Service Providers: Farmer Field School Approach.* Nairobi, Kenya: Sustainable Agriculture Information Initia- tive. Accessed December 2, 2019. www.sustainetea.org/downloads/file/11-farmer-field- school-approach.html.

Tschakert, P. and K.A. Dietrich (2010) Anticipatory learning for climate change adaptation and resilience. *Ecology and Society.* Accessed February 20, 2019. www.ecologyandsociety. org/vol15/iss2/art11/.

Uta, W.U., K. Collins, K. Anema, L. Basco-Carrera and A. Lerebours (2018) Stakeholder engagement in water governance as social learning: Lessons from practice. *Water Interna- tional* 43(1):34–59.

Wang, C. and M.A. Burns. (1997) Photovoice: Concept, methodology, and use for participa- tory needs assessment. *Health Education and Behavior* 24(3): 369–387.

Wiggins, S. and S. Keats. (2013) *Smallholder Agriculture's Contribution to Better Nutrition.* Over- seas Development Institute. Accessed February 19, 2019. www.odi.org/sites/odi.org.uk/ files/odi-assets/publications-opinion-files/8283.pdf.

Zwane, E.M. and K.E. Davis (2017) Extension and advisory services: The African renais- sance. *South African Journal of Agricultural Extension* 45(1): 78–89.

27 Urban markets and agricultural transformation in Southern Africa

Thulasizwe Mkhabela

Introduction

Agriculture constitutes a significant part of the economies of all African countries; thus, as a sector it can contribute meaningfully towards the eradication of poverty and hunger, increasing intra-Africa trade and investments, accelerating industrialization, job creation and shared prosperity, among other continental priorities (NEPAD, 2013).

In the last 30 years, Africa's population has doubled overall and tripled in urban areas, and by 2030, more than half of Africa's population will reside in urban areas (Crush et al., 2011). Rapid urbanization, on the one hand, has created an "invisible crisis of urban food security" (Crush and Frayne, 2010). On the other hand, rapid urbanization has become one of the major driving forces for agricultural transformation, creating opportunities and demand for huge urban markets for food and other related agricultural product as demand for food increases due to more mouths to feed. Market demands for agricultural and food products, including staple crops (mainly maize for Southern Africa), have been historically one of the most important factors determining the choice of agricultural production systems and choice of enterprises by farmers in Southern Africa. International and national factors have historically dominated the process, but this has recently changed with regional factors such as internal migration, urbanization, a burgeoning middle class with stronger purchasing power and consequently increased demand for both staple and lifestyle foods assuming a greater role (Browder and Godfrey, 1997; Brondízio et al., 2002).

Against the foregoing background, it is imperative for both producers of agricultural products, including farmers, agribusinesses and policymakers, alike, to understand the marketing of agricultural and food products in urban areas. Urban markets for agricultural and food products are complex systems dogged by several challenges that require both understanding and proper governance. Improving the policies and processes for greater impact of urban markets should address the following tenets:

- Eliminating market failures, such as concentration and anti-competitive behaviours by dominant market players
- Absence of services such as credit facilities, insurance, etc.

- The provision of public goods, such as transport infrastructure, bulk storage facilities, access to energy and water, information on prices, etc.
- Market regulations addressing distortions caused by international markets, price volatility and integration with global markets.

It is worth noting that while African agricultural production has increased over the years, it has not kept pace with population; thus, demand for food outstrips supply leading to the importation of substantive quantities of food products from international markets. Furthermore, it should be noted that the increase in African agricultural production has largely been due to increased area under planting rather than productivity growth. Most African countries' agriculture, including the Southern Africa region, is focused primarily on supplying national markets. However, the picture is beginning to change, with regional trade increasing owing to the proliferation of free trade areas that are reducing the cost of doing business among African countries.

Opportunities for the African agriculture sector abound. It is posited that the food market in Africa will triple by 2030, when it is expected to account for over US$1 trillion compared with US$313 billion in 2013 (Byerlee et al., 2013). In order to exploit these opportunities, the markets need to be functioning efficiently. Efficient agricultural markets enable cross-border trade in staple food and contribute to economic growth and food security. Better policies facilitate commodity exchanges (such as the South Africa Future Exchange [SAFEX] in Johannesburg), grain traders and agribusiness and international financial institutions to promote warehouse finance and expansion.

Prevalent urban agricultural markets in Southern Africa

The different agricultural commodities and products are generally marketed differently, largely dependent on their perishability or shelf life. Furthermore, there are nuanced differences between the different countries in terms of available marketing outlets for agricultural products. For commodities, such as maize – the staple crop – the price is formed internationally following the Chicago Board of Trade (CBOT) and other large future exchanges, including SAFEX. A commodity is defined as "an intermediate good with a standard quality, which can be traded on competitive and liquid global international markets" (Clark et al., 2001, p. 3). An important characteristic of commodities is the quality of the good which is verifiable *ex ante*, that is, information can be found easily before purchase. While commodities quality has been characterized by an increasing high degree of standardization, costs to acquire information about the product are generally low (Geman, 2005), which makes these goods suitable to trade.

In Southern Africa, SAFEX serves as the price setting mechanism for grains (maize, wheat) and oilseeds (soybeans, sunflower, etc.) where buyers and sellers trade their commodities. The futures exchange approach to marketing provides stability in terms of price in the sense that traders and buyers can enter into a contract at an agreed price and volumes prior to physical exchange of goods.

Nowadays, the bulk of agricultural commodities are sold on forward contracts with limited quantities sold on the spot market, mainly by sellers who have storage facilities for hoarding their product to release at a later stage when supplies are low. Traders have storage facilities such as silos located strategically close to areas of production and transportation infrastructure to move the goods closer to urban markets where they are sold.

However, fresh produce and livestock are a different kettle of fish. Fresh produce markets are an integral, although diminishing, part of price formation, distribution and marketing of fresh produce in Southern Africa, particularly in South Africa. South Africa has about 18 National Fresh Produce Markets (NFPMs) owned by the municipalities of the cities in which they are located. There are also a number of smaller private fresh produce markets.

In South Africa, as would be expected, the four largest NFPMs are located in the four largest cities, namely, Johannesburg, Tshwane (Pretoria), Cape Town and Durban. These four NFPMs account for more than 74% of turnover and volumes traded through fresh produce markets. The Johannesburg Fresh Produce Market (also known as the Joburg Market) is the largest in the country in terms of volumes of fresh produce traded and income generated. The Joburg Market accounts for 47.7% of revenue generated in 2014–15, while Tshwane Market, being the second largest, accounted for 21.8% during the same period (Lekgau, 2016).

The foregoing discussion notwithstanding, the trend in fresh produce marketing is moving away from traditional wholesale marketing channels such as NFPMs towards supermarket chain stores and other direct marketing schemes, also known as alternative food networks (AFN), for example, farmers' markets. The importance of supermarkets has been discussed extensively in Chapters 2.5 and 4.3 in this book and also in the literature (for example, see Crush, 2019; Louw et al., 2008; D'Haese and Van Huylenbroeck, 2005; Reardon and Hopkins, 2006); thus, this chapter will only provide a synoptic discussion. Figure 27.1 presents a schematic illustration of the urban agri-food supply chain with supermarkets in the epicentre of it.

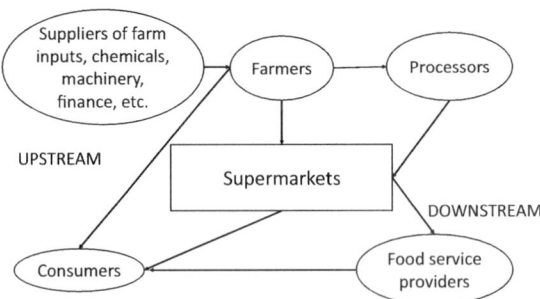

Figure 27.1 The urban agricultural food–supply chain

Source: Own representation

It is now common knowledge that there has been a gradual transformation in agricultural urban agricultural markets marked, to a large extent, by a shift in the distribution of power in the agri-food supply chains away from manufacturers (processors) of branded food products to the global supermarkets chains (Burch and Lawrence, 2005; Burch and Lawrence, 2007). The Southern Africa market is largely dominated by a few large supermarkets of South African origin such as Pick 'n Pay, Shoprite & Checkers, Spar (Dutch-owned but prominent in South Africa), Woolworths and Massmart (now part of the Walmart empire). However, there are a few local supermarkets that are making their presence felt in some Southern African countries. For example, Choppies, a Botswana supermarket, is growing in Botswana, Namibia, South Africa and Tanzania. Nakumatt, a Kenyan supermarket, is another one that is growing into the Southern Africa market.

Traditionally, supermarkets were just distribution channels for branded agrifood products from food manufacturers, but this has changed drastically with the shift in power in favour of supermarkets and supermarkets now have food brands of their own in their endeavours to be vertically integrated along the entire value chain. In the infant stages of advent of supermarkets own brands, popularly known as house brands, these brands were considered inferior to well-known brands owned by food manufacturers, but this perception has changed, with house brands now considered at par with established brands in the market and are often sold at lower prices than equivalent name-brand items (Greenblat, 2013; Wilkinson, 2002; Anderson and Narus, 1990). Furthermore, supermarkets are increasing engaging directly with farmers to source fresh produce bypassing the processors and shortening the value chain.

As African economies continue to develop and advance, it is expected that the dominance of supermarkets as markets for agri-food products in urban areas will also increase, as evidenced in the developed world (Wilkinson, 2002). Next the discussion turns to farmers' markets.

Farmers' markets are a growing trend in Africa, particularly in South Africa. Farmers' markets aim to bring producers and consumers together under direct marketing schemes, also known as alternative food networks for local and sustainable production and consumption of food (Figueroa-Rodriquez et al., 2019; Brown, 2001). The Australian Farmers' Market Association aptly defines a farmers' market as "a predominantly fresh food market that operates regularly within a community, at a focal public location that provides a suitable environment for farmers and food producers to sell farm-origin and associated value-added specialty foods products directly to consumers" (AFMA, 2014, p. 3). Farmers' markets are increasingly becoming commonplace in and around large cities and towns surrounded by farming communities in South Africa, and the phenomenon is spreading into neighbouring countries within the Southern African Development Community (SADC). The proliferation of farmers' markets is fuelled by several factors. Increasingly farmers are recognizing the value of farmers' markets as an alternative marketing channel. The advantages of farmers' markets to farmers include direct selling to consumers, thus cutting

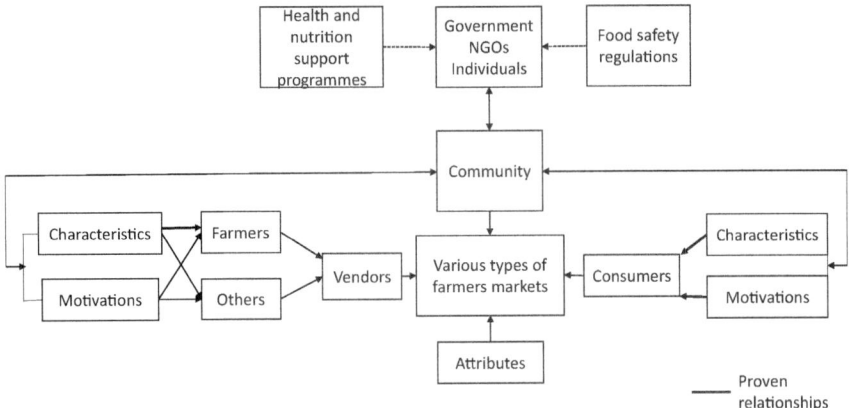

Figure 27.2 Partial framework for understanding farmers' markets

Source: Own representation

out the middleman and increasing revenue and also connecting with the consumers, thus better understanding the consumers' tastes and preferences. From the consumer's point of view, farmers' markets offer fresh food directly from the farm and satisfy the consumer's need to consume locally produced food with a lower carbon footprint. Furthermore, consumers perceive farmers' markets to a "special atmosphere", friendly, personal and smaller places compared to supermarkets (Sommer et al., 1981).

From available literature, it can be seen that the concept of farmers' markets is generally understood under three main aspects, namely the market and consumers, healthy eating promotion and food safety and environmental concerns (Kirwan, 2006; Fendrychová and Jehlička, 2018; Figueroa-Rodriquez et al., 2019). Figure 27.2 provides a partial framework for understanding farmers' markets based on the author's observations.

The important components of the system as depicted in Figure 27.2 represent the starting point: vendors, intrinsic attributes of the market, consumers, support organizations/institutions/individuals and the community, which relates to all the components at different levels.

Policies, processes and strategies to improve and enhance urban markets

There is general consensus that agriculture plays a pivotal role in economies of African countries, and marketing is an integral part of a well-functioning agricultural economy. Although agriculture is generally perceived as a rural activity, there is no denying that for the rural household "the landscape of daily life includes both rural and urban elements" (Douglass, 1998, p. 125). The rural-urban linkages are part and parcel of the local reality for household members carrying out the diverse tasks of generating income on and off-farm, maintaining a living space in the village and going to local and even distant towns and cities for shopping, marketing, work and specialized services. Thus, policies and strategies for enhancing urban agricultural markets should, a priori, also be aimed at linking rural and urban areas to overcome the urban- rural divide by incorporating the foregoing reality into development frameworks and, further, identify policy measures to foster mutual benefits for both town and country households.

At the top of the list of support services for creating an enabling environment for large and urban agricultural markets to thrive and be inclusive is the provision of quality infrastructure linking rural areas to urban areas. This infrastructure includes the construction and maintenance of good quality road network to facilitate an efficient movement of goods and services between the areas of production (often rural) and the marketplace, often in towns and cities. A good road network also reduces transportation costs and time spent travelling between two points, thus ensuring that agricultural goods reach their destination in good quality.

Farmers can take advantage of the rise of supermarkets by aggregating their produce, through secondary marketing cooperatives, to meet the quantities demanded by supermarkets. Supermarkets are reluctant to deal with individual smallholder farmers in sourcing their supplies as dealing with a plethora of suppliers and increases in transection costs, leading to greater variability in the quality of products. There is also the need to for farmers to be better organized in order to meet the regulations and standards in agri-food supply chains. Policymakers in African countries need to be cognizant of the stringent quantity and quality requirements of supermarkets, which often act as barriers to entry into formal and lucrative markets for most smallholder farmers and put in place policies and strategies to aggregate smallholder farmers produce.

Urban agriculture is another avenue that needs to be explored. Increased urban agricultural production, both in the cities and peri-urban areas would ensure sufficient throughput of marketable agricultural produce within shorter distances to large urban markets, thus enhancing their viability and efficiency.

Such policies and strategies aimed at enhancing urban agricultural markets would also invariably improve food security in urban areas through increasing the supply of food and reducing the price of food positively responding to improved availability, accessibility and affordability of food. The rural areas will continue to be the main source of agricultural and food products for the foreseeable future in African countries, including Southern Africa. This assertion is premised on the fact that there is still vast arable and agricultural land available in the rural areas, and considerable investments have been made in the form of infrastructure, such as irrigation and storage facilities.

Conclusions

Large agricultural markets, predominantly located in urban areas, will continue to play a pivotal role in the agricultural transformation process and economic development of African countries. The supply and demand of food will continue for as long as human beings exist, and properly functioning markets provide an ideal mechanism to allocate food and incentivize farmers and agribusinesses to continue producing food. It has been clearly demonstrated by numerous studies that urban markets are a major market outlet for agricultural products in Africa due to the high populations in urban areas and the proximity to export markets through better infrastructure. Large urban agricultural markets play an important function in generating income, employment creation and addressing food security challenges faced by many African urban areas.

It is posited in this chapter that urban agricultural markets will continue to grow and expand throughout the Southern Africa region and integration between neighbouring countries will grow as part of the reality presented by the process of agricultural transformation. Already, several agribusinesses and vendors from countries bordering South Africa, such as Eswatini (formerly known as Swaziland), Botswana and Lesotho source their fresh produce from South Africa, especially from the Joburg Market. A number of agro-processors, such as animal feed manufacturers, also procure their raw materials (Maize, soybeans, etc.) from SAFEX.

References

Anderson, J. and Narus, J. (1990) A model of distribution firms and manufacturer firm working partnerships. *Journal of Marketing* 54: 42–58.

Australian Farmers' Markets Association – AFMA. (2014) AFMA Strategic Plan 2014–2016. www.farmersmarkets.org.au/strategic-plan/. Accessed: June 24, 2019.

Brondízio, E.S., Safar, C.A.M. and Siqueira, A.D. (2002) The urban market for Açaí fruit (Euterpe oleracea Mart.) and rural land use change: ethnographic insights into the role of price and land tenure constraining agricultural choices in the Amazon estuary. *Urban Ecosystems* 6: 67–97.

Browder, J. and Godfrey, B. (1997) *Rainforest Cities: Urbanization, Development and Globalization of the Brazilian Amazon.* Columbia University Press, New York.

Brown, A. (2001) Counting farmers' markets. *Geographical Review* 91(4): 655–674.

Burch, D. and Lawrence, G. (2005) Supermarkets own brands, supply chains and the transformation of the agro-food system. *Journal of Sociology of Agriculture and Food* 13(1): 1–18.

Burch, D. and Lawrence, G. (2007) *Supermarkets and Agri-food Supply Chains: Transformations in the Production and Consumption of Foods.* Edward Elgar Publishing Ltd., Cheltenham, UK.

Byerlee, D., Garcia, A.F., Giertz, A. and Palmande, V. (2013) *Growing Africa-Unlocking the Potential of Agribusiness.* World Bank, Washington, DC.

Clark, E., Lesourd, J-B. and Thieblemont, R. (2001) *International Commodity Trading: Physical and Derivative Markets Wiley Trading/Ephraim Clark, Jean-Baptiste Lesourd and Rene Thieblemont.* Wiley, Chichester, New York.

Crush, J. (2019) *The Growing Role of Supermarkets in Africa's Food Security.* www.businesslive.co.za.

Crush, J. and Frayne, B. (2010) *The Invisible Crisis: Urban Food Insecurity in Southern Africa.* AFSUN Series No.1, Cape Town.

Crush, J., Hovorka, A. and Tevera, D. (2011) Food security in Southern African cities: The place of urban agriculture. *Progress in Development Studies* 11(4): 285–305.

D'Haese, M. and Van Huylenbroeck, G. (2005) The rise of supermarkets and changing expenditure patterns of poor rural households: a case study of the Transkei area, South Africa. *Food Policy* 30(1): 97–113.

Douglass, M. (1998) A regional network strategy for reciprocal rural-urban linkages: An agenda for policy research with reference to Indonesia. *Third World Planning Review* 20(1): 124–154.

Fendrychová, L. and Jehlička, P. (2018). Revealing the hidden geography of alternative food networks: The travelling concept of farmers' markets. *Geoforum* 95: 1–10.

Figueroa-Rodriquez, K.A., Álvarez-Ávila, M del Carmen, Castillo, F.H., Rindermann, R.S. and Figueroa-Sandoval, B. (2019). Farmers' market actors, dynamics, and attributes: A bibliometric study. *Sustainability* 11(3): 745–760.

Geman, H. (2005) *Commodity and Commodity Derivatives: Modelling and Pricing of Agriculturals, Metals, and Energy.* John Wiley & Sons, West Sussex.

Greenblat, E. (2013) Customers before suppliers, says Woolies boss. *Sydney Morning Herald,* 28 February. www.smh.com.au/business/retail/customers-before-suppliers-says-woolies-boss-20130228-2f7z0.html. Accessed: 17 May 2019.

Kirwan, J. (2006) The interpersonal world of direct marketing: Examining conventions of quality at UK farmers' markets. *Journal of Rural Studies* 22: 301–312.

Lekgau, S. (2016) *Participation of Black Farmers in the Joburg Market.* Presented at the 1st Subtropical Fruits Summit. Tzaneen, South Africa.

Louw, A., Jordaaon, D., Ndenga, L. and Kirsten, J.F. (2008) Alternative marketing options for small-scale farmers in the wake of changing agri-food supply chains in South Africa. *Agrekon* 47(3): 237–308.

New Economic Partnership for African Development – NEPAD. (2013) Comprehensive African Agriculture Development Programme (CAADP): Agriculture, Food Security and Nutrition. 2013 Report (CAADP Implementation Support). www.nepad.org. Accessed: June 24, 2019.

Reardon, T. and Hopkins, R. (2006) The supermarkets revolution in developing countries: Policies to address emerging tensions among supermarkets, suppliers and traditional retailers. *European Journal of Development research* 18(4): 522–545.

Sommer, R., Herrick, J. and Sommer, T. (1981). The behavioral ecology of supermarkets and farmers' markets. *Journal of Environmental Psychology* 1: 13–19.

Wilkinson, J. (2002) The final foods industry and the changing face of the global agro-food system. *Sociologia Ruralis* 42(4): 329–346.

28 Importance of small rural markets in the transformation of Southern African agriculture

Simphiwe Ngqangweni, Ndiadivha Tempia and Moses Herbert Lubinga

Introduction

In most African countries, agriculture is still the largest contributor to Gross Domestic Product (GDP), the biggest source of foreign exchange, accounting for about 40% of the continent's hard currency earnings and the main generator of savings and tax revenues (Ruane and Sonnino, 2010). As 70% of the continent's extreme poor depend on agriculture for their livelihood, it is estimated that growth in agriculture is about four times more effective in raising incomes of extremely poor people than growth originating from other sectors (World Bank, 2008). However, growth in agriculture, especially emanating from rural markets, has been limited. This can be attributed to the dualistic system which favour large-scale commercial farmers to the detriment of smallholders (Van Rooyen, 1997) (see also Kherallah et al., 2002).

Christy (2001) posits that smallholder farmers' access to markets is a function of the characteristics and performance of the food marketing system that they find themselves in. Such economic performance can in turn be explained by a range of complex issues, including the structure of the marketing system and social, cultural and behavioral aspects of institutions, human values and goals. The purpose of this chapter is to present an analysis of the constraints and challenges that limit or otherwise enhance the impact of small rural markets on agricultural transformation in Southern Africa. Policy proposals that have the potential to improve the development of rural markets through increased income as well as the long-term impacts on food security will be assessed. As it is the case in the broader African continent, the rural economy of Southern Africa is based on smallholder agriculture (Livingston et al., 2011), whose development is constrained by poor physical infrastructure, weak private markets, widespread information asymmetries and low levels of marketed surplus (World Bank, 2009).

Based on Christy's (2001) argument that a study of the dynamics and performance of one segment of the agro-food marketing system (small rural markets in this case) calls for a look into the entire complex system whose performance is in turn affected by a multiplicity of factors, the next section summarizes these factors as they are relevant in the context and character of small rural markets in Southern Africa.

Factors affecting development of small rural markets

There is an increasing demand for food in developing countries of sub-Saharan Africa. This trend is expected to continue due to increasing population, urbanization, changing lifestyles and consumer preferences associated with increasing disposable incomes (Soji et al., 2015). It is also expected to create opportunities for small-scale producers while consumers are likely to benefit from modern retail structures and an advanced processing sector. With access to modern retail outlets supplied by a sophisticated processing sector, consumers increasingly have access to a greater variety of better-quality food products (Mather, 2005). However, the implications of these changes may be less positive for small-scale farmers and small medium enterprises (SMEs) involved in food processing and production where changing market requirements and strict quality control mechanisms are being put in place along the value chain to ensure that high quality products with long shelf life reach the consumer. Such stringent measures as well as other challenges such as high transport costs, poor health management practices, poor nutrition, lack of infrastructure and poor marketing management, inadequate knowledge and not keeping abreast with current technology (Spies and Cloete, 2013) drive informal livestock producers away from the formal market. Christy (2001) classifies the key factors that drive transformation in small rural markets into two broad categories, namely *structural* and *institutional* factors.

Structural factors

One of the key characters of small rural markets in Southern Africa is that they tend to be informal. Becker (2004) defines the "informal economy" as "the unregulated nonformal portion of the market economy that produces goods and services for sale or for other forms of remuneration". In the context of agriculture in Africa, including Southern Africa, informal markets provide a platform for farmers, traders and consumers to exchange agricultural commodities and money, perform barter deals and share knowledge (Dhewa, 2016).

Although the informal economy in general has been stigmatized in various ways in literature, being associated with such labels as "undeclared labour", "tax evasion", "unregulated enterprises" and "illegal and criminal activity" (Becker, 2004), it plays a vital role in developing economies such as those of Southern Africa. In South Africa, 40% of all beef production takes place in communal areas (Sotsha et al., 2018), which could be considered part of the informal economy. In Namibia, over 50% of all cattle are produced in communal areas. Small-scale farmers often cannot compete in formal markets on the basis of health, safety and quality standards (Kalundu and Meyer, 2017). Moreover, while food retailing in developing countries historically operated in the informal market, implementation of private standards particularly in the supermarket sector have proliferated recently, and these are often out of reach for smallholder farmers in terms of cost (Das Nair and Chisoro, 2015). Smallholder farmers are therefore restricted to operate informally. In some cases, the informal nature of small rural markets creates opportunities for speculators that lead to smallholder farmers selling produce (notably livestock) at unprofitably low prices (Uchezuba et al., 2009).

Delgado (1997, 1999) argues that structural constraints facing smallholder farmers manifest themselves in high and often prohibitive transaction costs, that is, the full costs of carrying out exchange, including marketing costs. A number of studies have analyzed these in detail – see Musara et al. (2018) among others.

Institutional factors

Institutional bias that favoured large-scale, modern, urban and capital-intensive production over small-scale peasant farming in much of the region has left a legacy that has proven difficult to correct (Matungul, 2000). In addition to structural constraints, a large body of literature also mentions institutional constraints as drivers of agricultural transformation in the region – see Ortmann and King (2010) for a comprehensive review. In a study that includes two Southern African countries (Zambia and Mozambique), the World Bank (2009) concluded that institutional bottlenecks and policy constraints have hampered economic growth in general and agricultural growth in particular.

In his study of communities in rural South Africa, Matungul (2000) introduces the notion of "informal constraints" (as opposed to formal rules or institutions) as factors affecting participation in rural markets in the region. He states that informal constraints come from tradition and customs and are social norms of a culture transmitted from one generation to the next. In their study of northern Namibian cattle farmers, Duvel and Stephanus (2000) concluded that farmers tend to be overshadowed by cultural or socioeconomic incentives and considerations in their marketing decisions.

Policies and strategies to improve market access for small farmers

There is rich literature on various strategies that can be implemented to develop rural markets in Africa. There is consensus that structural and institutional factors are important in achieving efficiency and growth in rural or small markets. Hazell et al. (2010) contend that the strategies and policies to develop small farmers will vary by context, implying that policy needs to match the circumstances and change through time. A number of policy proposals are outlined in the following section.

Institutional changes

Successful implementation of policies that promote small farm development will require more than just technical capacity but rather actions from various players. These include government agencies, civil society, farmer organizations and agribusiness firms. Weakness in administrative and technical capacity have been exacerbated by structural adjustments that neglected, rather than reformed, many public institutions serving rural areas. In addition, the demolition of the marketing boards has shifted the control of agricultural activities into the hands of the private sector, where small farmers remain marginalized. Policies that reform and strengthen public agencies that serve rural areas will require political will, especially where there are vested interests. Even in the aftermath of structural adjustments, many African countries still experience heavy presence of government intervention in markets due to the resistance from entrenched political and bureaucratic interests that retain the control of policy level that are useful for patronage and rent-seeking purpose (Hazell et al., 2010).

Increased participation

According to Hazell et al. (2010) innovative ideas that involve partnerships between government, civil and community bodies as well as private firms can be more cost effective and may offer possibilities for involving local people and communities. Increased participation from local people and communities is heavily propagated by many stakeholders as a workable poverty reduction strategy both within the relevant government ministries and outside (Foster et al., 2001).

Studies that advocate numerous benefits that smallholder farmers could gain when they participate in the decision-making process include Aphane (2011) and Alene et al. (2008). Alene et al. (2008) outline the benefits associated with the participation, which vary and range from improved cash-flow, market access and viable farming entities. Thus, participation itself is catalytic to access the markets and sustainability of smallholder farming. According to Smith (1983), participation gives farmers the right to be consulted, involved and informed by other stakeholders. This kind of interaction has potential to enrich the farmers' decision-making processes, build trust and network among the participating farmers as well as share formal and informal knowledge. Table 28.1 that follows outlines the benefits of initiatives based on increased participation from the affected parties as well as the institutional support. The National Red Meat Development Programme (NRMDP) coordinated by the National Agricultural Marketing Council (NAMC) in South Africa, provides anecdotal evidence of the success of such processes.

Future impacts

Food and nutrition security

Due to dysfunctional rural markets, countries may produce enough and readily available food but it may not be accessible to both rural and urban households (Timmer, 2017), thereby compromising a country's/region's food and nutritional status. Developing small rural markets is anticipated to enhance access to nutritious food from the producing areas to areas where it is needed. The

Table 28.1 The NRMDP, South Africa

Institutional Support: NAMC in collaboration with the Department of Rural
    Development and Land Reform in South Africa, provincial governments of the Eastern
    Cape, KwaZulu-Natal and Northwest (with Limpopo and Mpumalanga provinces
    being considered as part of future expansion), district and local municipalities in the
    aforementioned provinces, as well as some municipal development agencies

The purpose: Afford communal livestock farmers an opportunity to increase their
    income through greater participation in the market and better marketing of their cattle

Main drivers:
- To enable farmers to understand the structure, operation and requirements of red meat
  markets
- To develop marketing channels that will increase their participation in the red meat
  markets
- To assist with practical training to align the age, health and breeding of animals more
  closely to market demand

Achieved outcomes:

During the funding period (five years) more than R100 million (US$7 million) was
    generated from sales by communal cattle farmers with jobs in catering, transport and
    construction created along the value chain. Furthermore, during the funding period,
    16 custom feeding facilities were successfully constructed in five provinces, thereby
    providing sustainable solution to the already stressed natural communal grazing pastures,
    enhancing the nutrition and the marketability of the cattle. This has translated into
    improved livelihoods in the beneficiary areas.

---

development of food marketing systems is bound to foster structural, agricultural and dietary transformations.

When small rural markets are developed, food will be easily transformed in
time, from one place and form to another (Ogutu and Qiam, 2018). Transformation in form entails value addition into more nutritious products. A functional food marketing system will also help to address challenges of soaring food
prices, thereby enabling those who cannot produce it on their land to afford it.
Reducing the distance to markets is noted to increase household food security
due to the lower transaction costs involved in distributing food items. Small
rural markets will also counter the dominance of the few big retail food market
groups which force small food markets out of business due to price competition. The small rural markets enable the distribution of food items in relatively
less geographical accessible communities. However, there is a risk that food
market systems may lead to food insecurity due to easy access to less nutritious
but cheaper processed food items. In this case, food insecurity manifests in form
obesity in grown-ups and stunting in children.

Income generation

Studies by Lubinga et al. (2018) revealed that the sale of livestock through
auctions in rural areas in the KwaZulu-Natal and Eastern Cape provinces of
South Africa generate higher incomes. Developing small rural markets is bound

to minimize postharvest losses, increase product shelf life and also bring more income to market participants. Beyond the farm produce, development of rural markets is also anticipated to stimulate growth in the use and sales of agricultural inputs. Furthermore, development of markets will stimulate the purchasing power of the rural communities.

Poverty alleviation

Market development fosters poverty alleviation in its different forms (Ogutu and Qiam, 2018). For instance, through the NRMDP that helps communal livestock farmers to participate in formal markets, farmers were able to acquire more assets. Therefore, further, establishment of marketing systems will be critical in reducing poverty. However, it is important to emphasize the need for special market-linkage support to help farmers to ensure equality in incomes received.

Conclusions

This chapter presented a concise analysis of the constraints and challenges that limit agricultural transformation in Southern Africa region as well as the policies that can unlock the potential in underdeveloped small rural markets. Potential benefits such as improved nutrition, income and poverty alleviation were also presented. While there is vast literature detailing the constraints that limit agricultural transformation in Southern Africa, much still needs to be done in terms of strategies and actual implementation of the policies. Any successful policy will require more than just technical capacity. Thus, engagement from various bodies, both public and private, is crucial for agricultural transformation. The main challenge that rural markets face is access to the vibrant economy where maximum profits can be earned. Structural barriers and institutional barriers are considered the most inhibiting factors towards market access. Smallholder farmers in the rural markets cannot break these barriers alone. Technical and institutional capacity from both the public and private sectors is required. The NRMDP coordinated by the NAMC in South Africa provided anecdotal evidence of the benefits that can be achieved through processes which involve partnerships between government, civil and community bodies as well as private firms. Such processes have the potential to increase income and to address poverty.

References

Aphane, M.M. (2011) *Small-Scale Mango Farmers, Transaction Costs and Changing Agro-Food Markets: Evidence from Vhembe and Mopani Districts, Limpopo Province*, Masters in Economics, University of Western Cape, Cape Town, Republic of South Africa.

Alene, A.D., Manyong, V.M, Omanya, G., Mignouna, H.D., Bokanga, M. and Odhiambo, G. (2008) Smallholder market participation under transactions costs: Maize supply and fertilizer demand in Kenya. *Food Policy*, 33(4), pp. 318–328.

Becker, C. F. (2004) *Fact Finding Study: The Informal Economy*, Stockholm: Swedish International Development Agency (SIDA).

Christy, R. D. (2001) Evaluating the economic performance of alternative market institutions: Implications for the smallholder sector in Southern Africa. *Agrekon*, 40(4), pp. 522–536.

Das Nair, R. and Chisoro, S. (2015) *The Expansion of Regional Supermarket Chains: Changing Models of Retailing and the Implications for Local Supplier Capabilities in South Africa, Botswana, Zambia, and Zimbabwe.* WIDER Working Paper 2015/114, Helsinki: UNU-WIDER.

Delgado, C. (1997) The role of smallholder income generation from agriculture in Sub-Saharan Africa. In: L. Haddad, ed. *Achieving Food Security in Southern Africa: New Challenges, New Opportunities*, Washington, DC: International Food Policy Research Institute (IFPRI), pp. 145–173.

Delgado, C. (1999) Sources of growth in smallholder agriculture in integration of smallholders with processors in Sub-Saharan Africa: The role of vertical integration and marketers of high value-added items. *Agrekon*, 38(S1), pp. 165–189.

Dhewa, C. (2016) *Knowledge in Informal African Markets*, Stellenbosch University: Unpublished MPhil Thesis.

Duvel, G. and Stephanus, A. (2000) A comparison of economic and cultural incentives in the marketing of livestock in some districts of the northern communal areas of Namibia. *Agrekon*, 39(4), pp. 656–664.

Foster, M., Brown, A. and Naschold, F. (2001) Sector programme approaches: Will they work in agriculture? *Development Policy Review*, 19(3), pp. 321–338.

Hazell, P., Poulton, C., Wiggins, S. and Dorward, A. (2010) The future of small farms: Trajectories and policy priorities. *World Development*, 38(10), pp. 1349–1361.

Kalundu, K. and Meyer, F. (2017) The dynamics of price adjustment and relationships in the formal and informal beef markets in Namibia. *Agrekon*, 56(1), pp. 53–66.

Kherallah, M. et al. (2002) *Reforming Agricultural Markets in Africa*, Baltimore: Johns Hopkins University Press.

Livingston, G., Schonberger, S. and Delaney, S. (2011) *Sub-Saharan Africa: The State of Smallholders in Agriculture. Conference on New Directions for Smallholder Agriculture 24–25 January 2011,* Rome, IFAD.

Lubinga, M., Mazibuko, N. and Sotsha, K. (2018) Comparing prices received by participating and non-participating farmers in the custom feeding initiative of the National Red Meat Development Programme: A case of KwaZulu-Natal Province. *Sylwan*, 163(3): 30–43.

Mather, C. (2005) The growth challenges of small and medium enterprises (SMEs) in South Africa's food processing complex. *Development Southern Africa*, 22(5), pp. 607–622.

Matungul, P. M. (2000) Buying patterns for staple foods amongst rural households in KwaZulu-Natal: Informal constraints perspective in Impendle and Swayimana. *Agrekon*, 39(2), pp. 142–150.

Musara, J. et al. (2018) Market participation and marketing channel preferences by small scale sorghum farmers in semi-arid Zimbabwe. *Agrekon*, 57(1), pp. 64–77.

Ogutu, S. and Qiam, M. (2018) *Commercialization of the Small Farm Sector and Multidimensional Poverty. Global Food Discussion Paper 117*, Gottingen: University of Gottingen.

Ortmann, G. F. and King, R. P. (2010) Research on agri-food supply chains in Southern Africa involving small-scale farmers: Current status and future possibilities. *Agrekon*, 49(4), pp. 397–417.

Ruane, J. and Sonnino, A. (2010) Agricultural biotechnologies in developing countries and their possible contribution to food security. *Journal of Biotechnology*, 156, pp. 356–363.

Smith, L.G. (1983) *Impact Assessment and Sustainable Resource Management*, Harlow: Longman.

Soji, Z., Chikwanda, D., Jaja, I.F., Mushonga, B. and Muchenje, V. (2015) Relevance of the formal red meat classification system to the South African informal livestock sector. *South African Journal of Animal Science*, 45(3), pp. 263–277.

Sotsha, K. et al. (2018) Factors influencing participation into the National Red Meat Development Programme (NRMDP) in South Africa: The case of the Eastern Cape Province. *OIDA International Journal of Sustainable Development*, 11(01), pp. 73–80.

Spies, D.C. and Cloete, C.P. (2013) *Performance and Marketing Options for Red Meat in the Formal and Informal Value Chains in the Free State Province.* Conference paper presented at the 19th International Farm Management Congress, SGGW, Warsaw, Poland.

Timmer, P. (2017) Food security, structural transformation, markets and government. *Asia & the Pacific Policy Studies*, 4(1), pp. 4–19.

Uchezuba, I., Moshabele, E. and Digopo, D. (2009) Logistical estimation of the probability of mainstream market participation among small-scale livestock farmers: A case study of the Northern Cape province. *Agrekon*, 48(2), pp. 171–183.

Van Rooyen, J. (1997) Challenges and roles for agriculture in the Southern African region. *Agrekon*, 36(2), pp. 181–205.

World Bank (2008) *World Development Report: Agriculture for Development*, Washington, DC: World Bank.

World Bank (2009) *Awakening Africa's Sleeping Giant: Prospects for Commercial Agriculture in the Guinea Savannah Zone and Beyond*, Washington, DC: World Bank.

29 Agricultural growth corridors in sub-Saharan Africa – new hope for agricultural transformation and rural development?

The case of the Southern agricultural growth corridor of Tanzania

Michael Brüntrup

Introduction

The basic idea of Agricultural Growth Corridors (AGCs) is to directly combine the planning and implementation of large transport infrastructure development (basically roads and railways) with sectoral initiatives in agriculture, in order to create territorially defined rural and economic transformation hubs. Usually this also includes both backward industries (services and input supply) and forward industries (processing and packaging, logistics, etc.). This aims to overcome weaknesses of more narrow approaches to rural development in poor countries (Ashley and Maxwell, 2001). Sectoral approaches get easily stuck by not including bottlenecks in transport and marketing, often induced by weak infrastructure, while infrastructure investments in rural areas yield lower returns than in urban areas due low density of beneficiaries and lack of supply response capacity, thereby structurally disadvantaging public (and private) rural investment.

AGCs are a relatively recent approach in sub-Saharan Africa (SSA) but follow on from older approaches of spatial development initiatives (World Bank, 2009). Over the last decade they have gained prominence and international support, including through: the United Nations recognition of Yara, the world's biggest fertilizer company, in the World Business and Development Awards for its corridor concept; the World Economic Forum since 2008; the African Union's New Partnership for Africa's Development (NEPAD); and the G7's New Alliance for Food Security and Nutrition. AGCs now figure prominently in several national development strategies on the continent (Kaarhus, 2011; Gálvez-Nogales, 2014; Reeg, 2017).

In order to fully understand the particularities, strengths and weaknesses of AGCs in SSA, a few additional elements need to be highlighted: many AGCs are planned as add-ons of other types of corridor, often for mineral transport from mine to port (Weng et al., 2013); they are usually planned and managed as strategic public-private partnerships; and they have large-scale, land-based enterprises in production (plantations) as their backbone. African AGCs often also have regional integration in their target system, in particular between coastal and land-locked countries (Byiers et al., 2014; Gálvez-Nogales, 2014; Reeg, 2017). These particularities are not defining requirements of AGCs but are very important for the priorities of individual initiatives, the way they are executed and their public perception.

While many AGCs are highly ambitious, they are also often strongly contested, in particular due to their linkages to agro-industries and private sector involvement (Paul and Steinbrecher, 2013; Byiers et al., 2014). There is a significant body of literature critiquing large-scale land acquisitions and agro-industry investments in SSA (see Brüntrup et al., 2018). The types of agricultural investments that AGCs in SSA pursue involve a whole range of risks – environmental, social and economic – and in particular they are blamed to exacerbate poverty and food insecurity through land-grabbing and modern farm technology use inappropriate for smallholders.

A stringent evaluation of the impacts of AGCs in SSA is not yet possible, since the approach is young and nowhere yet fully implemented and comprises elements with a very long implementation time. In addition, due to their complexity, a comprehensive evaluation would be extremely difficult even if long-term data was available. Accordingly, this paper cannot provide a rigorous or ex-post impact assessment of any AGCs in SSA, but it can contribute to that debate by providing a mix of evidence and analysis of concepts for one of the oldest AGCs in SSA, the Southern Agricultural Growth Corridor of Tanzania (SAGCOT). The paper brings together four sources of evidence:

- an extensive qualitative study of large agro-investments of the type favoured by SAGCOT and most AGCs – the nucleus – outgrower scheme (NOS) – in three subsectors (sugar, rice and tea), and of the sociopolitical and economic environment, explicitly including the SAGCOT initiative, with more than 280 interviews of a wide array of stakeholders (farmers and farmer groups, farmer and civil society organizations, local and national authorities, investors, researchers, donors) (Brüntrup et al., 2018);
- several short-term site visits and interviews with selected stakeholders, as well as participation in the 2017 SAGCOT field day by the author;
- a systematic analysis of the literature of spatial development initiatives (SDIs) in developing countries regarding the lessons for AGCs by Reeg (2017);
- quantitative impact studies with a large-N sample (>600 farm households) for two older NOSs, in rice and sugar, in the SAGCOT area, often regarded as models (Herrmann, 2017). The available literature on SAGCOT was also reviewed.

The chapter is structured as follows: the first section (*The Southern agricultural growth corridor of Tanzania*) starts with a short overview of SAGCOT and its level of implementation. The section that follows (*Assessing the factors for effective implementation of SAGCOT*) then reviews how the lessons of successful implementation of SDIs in general are relevant for, taken into account, and (if already visible) implemented in SAGCOT. Next, *Impacts of individual investments* summarizes the impacts of older NOSs on local farmers, workers and rural communities. The final section offers conclusions and recommendations for SAGCOT and more broadly for AGCs in SSA.

The Southern agricultural growth corridor of Tanzania

SAGCOT was one of the first, and may be the most prominent, of the African AGCs. According to a World Bank report (2016, p. 6): "The success of the SAGCOT project will not only help to modernize agriculture in Tanzania but also provide lessons for other countries which have large untapped potential to improve their agricultural productivity and lift large populations from living in poverty". The SAGCOT initiative was launched during a World Economic Forum (WEF, 2012) meeting in Dar es Salaam in 2010 (WEF, 2016). The corridor stretches from the Tanzanian port, Dar es Salaam, to Malawi and Zambia along already existing road and rail transport infrastructure (Figure 29.1). The initial, and still promoted, blueprint sets the target of US$2.1 billion of private investment, along with US$1.3 billion of public sector grants within 20 years, to bring 350,000 hectares under "profitable production", creating at least 420,000 new employment opportunities and lifting more than two million people out of poverty, while assuring regional food security (SAGCOT, 2011).

The initial plan foresaw six regional clusters with several NOSs, in each of which large-scale, land-based investments (plantations and processing factories: the nuclei) were to be combined with smallholders (the contract farmers or outgrowers; here both terms are used synonymously) who would receive inputs and services from the nuclei and deliver products to them (see Figure 29.1).

The SAGCOT Centre, the main SAGCOT entity, was created in 2011 as a public-private partnership by the Agricultural Council of Tanzania, the Confederation of Tanzania Industries (CTI) and the Rufiji Basin Development Authority (RUBADA), a government authority tasked with water resource management and multisectoral development in the Rufiji basin, which covers part of SAGCOT. The number of private and public sector partners grew from about 20 initially to more than 50 in 2018 (Bergius et al., 2018). The Centre does not conduct implementation of investments itself; its main tasks are promoting shared vision, sharing information and mobilization. The Tanzanian government launched and supported the initiative by designating high level staff, but they did not directly invest much (less than 5% of US$22 million over five years) (SAGCOT, 2018). Instead, they aimed to support SAGCOT through its regular sector policies, in particular agricultural sector policies and programmes and the presidential initiative "Big Results Now!" of 2013, which aimed, among other things, to fast track 25 NOSs in the SAGCOT region, in the subsectors of sugar, rice and maize (URT, 2015). Several donors supported SAGCOT with smaller amounts and/or specific projects, but it was not until

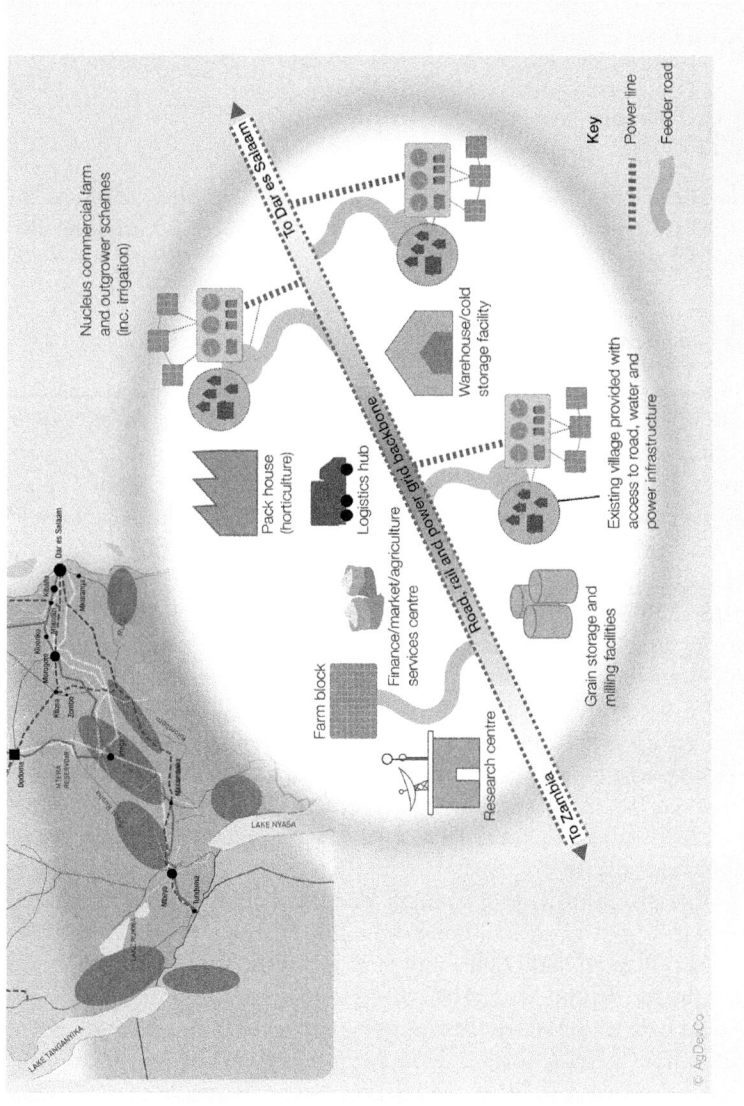

Figure 29.1 Map of the SAGCOT corridor and schematic organization of clusters 2011

Source: Permission granted by AgDevCo 23072019

late 2016 that the major and strongly contested, US$70 million, World Bank SAGCOT Investment Project was approved to support key SAGCOT entities (the SAGCOT Centre and Tanzania Investment Centre) and provide matching grants for the private sector. The project prematurely closed at the end of January 2019, apparently over disputes with the new government about the use of the matching grants for the private sector vs. local communities and administrations (World Bank, 2019a).

Assessing the factors for effective implementation of SAGCOT

This section assesses the implementation of SAGCOT according to the main clusters of factors identified by Reeg (2017) as key for successful implementation of SDIs in general.

a. Geography and natural resources

The choice of SAGCOT and the clusters is not much disputed. The backbone infrastructure exists and is in acceptable shape. Dar es Salaam is an important market for food and has the most important national port, though its congestion is an issue. The international dimension of SAGCOT is not yet of importance in the planning, though landlocked and densely populated Malawi could become an interesting market in the future, while rural parts of Zambia and Democratic Republic of the Congo may be important production hubs.

As to natural resources, the corridor zone has good production potential, with many areas providing few constraints to agricultural intensification (SNAPP, 2016), including various agro-ecological zones, from tropical to temperate. Options for crops and related value chains are manifold. The choice of SAGCOT priority commodities has changed substantially from the initial blueprint (rice, other grains, livestock/beef, sugar, citrus, banana and other horticultural products) to the first phase of implementation, which concentrated on one cluster (Ihemi) (tomatoes, dairy, soya, tea and potatoes), which shows the many options available.

Land is still available in large quantities. For instance in the Ihemi cluster, more land has gone out of agricultural production than has been newly cultivated in the past 20 years, although more land is projected to be brought into cultivation in the future (SNAPP, 2016). But this does not mean that fallow and other land is unused; it is used for various purposes, including firewood collection and pastoralism – sometimes only in stress years and for regeneration of soil fertility. For the favourite NOS model, large unsegmented portions of land are needed for the nucleus farms. This is the real challenge, although this is not a scarcity but a governance problem (see d. Governance).

An additional issue is water. While water reserves are still sufficient in general, there is strong seasonal variation and some erratic fluctuation over longer periods, with significant droughts and irrigation water scarcity. As a result, there is competition between water needs for energy and agriculture (SNAPP, 2016). Some private investors opt for the construction of smaller dams that can attenuate seasonal fluctuation; however, for longer drought periods this is not sufficient. With a massive increase in irrigated area, there would be a need for integrated management of the whole basin to carefully provide water withdrawal licenses that not only respect the competition between human users but also leave a minimal base stream flow. Also, some erosion protection and water infiltration improvement measures would have to be designed or coordinated. SAGCOT is well placed to organize such integrated water resource management, with the responsible government body for that task, RUBADA, being one of its founding members, while SAGCOT should be able to bring in higher, member-based acceptance for coordination efforts. However, until now, not much has been done in this regard, since hardly if any early investments did materialize.

b. Hard infrastructure

Infrastructure is a cornerstone of any SDI and should be one for AGCs. The infrastructural backbone of SAGCOT, the main transport axis by road and rail, already exists – in contrast to other corridors and corridor approaches. While this is a great financial, organizational and time benefit for the initiative, it makes the transport and marketing effect of SAGCOT invisible for farmers and locations near to the axis.

However, additional infrastructure is required to transform the existing transport corridor into an AGC, including: feeder roads to open up the lands further away from the main road, irrigation systems (larger and smaller dams and canals), electricity lines and warehouses for storage and cooling, just to name the most important ones (see Figure 29.1). One of the key ideas of the SSA AGC model is that, while the public sector is short of funds, the private sector can provide these for themselves and for the outgrowers and local communities. Partially, such provision of local infrastructure can already be observed in existing NOSs. They are seen as corporate social responsibility projects or as core investments to compensate and remunerate communities. Sometimes there are legal obligations: for example, local road taxes for tea companies. In some cases (electricity, dams, irrigation), regulatory hurdles impede such business models (see d. Governance). In other cases, larger investments (e.g., longer feeder roads, electricity covering several NOSs or whole subregions within the clusters or improvements to the central axis), are too expensive even for the big private agro-industry partners, in contrast to large mining investments which are much larger in investment volume in total and per unit of land area. Also, if they are not linked to a cost-covering supply scheme, individual investors are unlikely to be willing to fund infrastructure that not only benefits their own business but also that of their competitors and others. Since the regional and central governments are unable to fund such infrastructure development, large donors step in. The World Bank's Tanzania country strategy (World Bank, 2018) lists several initiatives that can support SAGCOT infrastructure needs, but they are outside of the SAGCOT investment project and would need additional negotiations and coordination with other parts of the government, possibly not committed to SAGCOT but with their own priorities.

Social infrastructure investments in SAGCOT are left to negotiations between investors and the local population. Often, investors promise a range of different structures and services and determine that these will be paid for out of profits, not all upfront, making them dependent on the success of the business. This creates high expectations, and since investors do not meet all of them upfront, this often generates mistrust. For the attraction of highly qualified staff, the lack of local quality infrastructure for them and their families is also a problem. Investors in rural areas have to pay high salaries and provide additional benefits for this kind of staff, which is rare in Tanzania. At the level of clusters, possibly common solutions could be found, but no such plans are known.

c. Soft infrastructure

As with hard infrastructure, the provision of soft infrastructure through the government is generally inadequate in Tanzania, as elsewhere in SSA, and one of the key elements of SAGCOT is to have this provided by the private sector. This is particularly true for finance and for the more sophisticated services necessary for the upgraded value chains that are able to serve higher consumer segments and export markets. The tea and sugar subsectors provide evidence that the private sector is willing to provide these specialized services. However, as in the case of hard infrastructure, investors are not ready to fund services that also serve noncontributing, free-riding competitors and may even weaken their own position. Thus, difficult and detailed trade-offs must be negotiated, and what can be provided and funded by whom must be sorted out. Research on the three subsectors of rice, tea and sugar (Brüntrup et al., 2018) shows that complicated arrangements must and can be designed to balance public and private interests, reduce side-selling and free-riding, encourage or oblige enterprises to contribute, support farmers to cooperate, assure quality of services and combine financial and various nonfinancial services. The exact design depends on the particular structure, needs and partners of each subsector; historic path dependency of the site; past experiences; the power of regulation and implementation; the constellation of local and sometimes nonlocal actors; the impact consciousness and vision of the investor; and so on.

This strong specificity of NOS models and the site-specific conditions means that, though SAGCOT can provide a platform for organizing soft infrastructure, specific solutions mostly have to be found in the specific value chains of organizations and often even for individual NOSs. A blueprint is rarely the solution; good practices must be shared and adapted to local conditions.

d. Governance

Binding governance within and outside SAGCOT is almost the same. Only the nonbinding internal rules to which investors subscribe by joining SAGCOT make (in principle) a difference in governance of the investments and the corridor. This is different from most other SDIs, for which special rules and regulations are put in place and are often very far-going (e.g., on tax and tariffs, labour, land and other resources and services). This is because a space of one third of the national territory is simply too big, and borders of the corridor and of the clusters are too permeable to draw boundaries around and create specific regulations for within the corridor. The key governance issues are:

- Access to large tracts of agricultural land, and often also smaller tracts of construction land is limited, particularly for foreign investors. No single large tract of land, including old government farms, could be cleared for an investor under SAGCOT, and not even under the Big Results Now! initiative. The land ownership and transfer regimes and corresponding legal systems are complicated, opaque, partially contradictory, prone to political interference from all sides and time consuming, and the necessary resources for implementation and monitoring are usually not available.
- The general investment climate for private investments is very weak and has worsened in the last three to four years. In 2019, Tanzania was ranked 144 among 190 economies (World Bank, 2019b).
- Trade policy is unreliable. Although several agricultural sectors, including rice, tea and sugar, are in principle protected under Tanzania's or the East African Community's tax regime, explicitly to encourage investments, smuggling, corruption and influential importers weaken that protection and make investments unprofitable and/or risky. Erratic export bans (district or national) for some crops have a similar effect.
- Specific subsector regulations on products (see trade policy) and services – for instance on quality control, local sale of electricity, financial services and entry of international staff – also add to investor reluctance.

The SAGCOT Centre is aware of these problems. It has established many formats and working groups to collect complaints and discuss them with the administration and policymakers. Some of the formats established include a Land Use Dialogue Forum in partnership with the Sustain-Africa programme, a robust policy network, strategic policy partnerships, focal people, parliamentary excursions, national and regional field days and strategic learning journeys (SAGCOT, 2018). SAGCOT claims to have contributed to some changes in governance related to agribusiness – for instance the abolition of value added tax on feed for livestock to improve the competitiveness of homemade feed against imported.

Thus, SAGCOT can play a role in governance, in particular through networking, mutual learning and managing dialogue and conflict (e.g., over land issues), decentralized services (e.g., extension services), local investments and local export bans. These all have to be managed through communication, building trust, organizing debates and finding or facilitating solutions, and this is one of SAGCOT's most important tasks. Many national regulatory issues are negotiated at sector and subsector level; however, other issues of regulation implementation, practice and negotiation of informal rules and use of local resources are negotiated at the local level, particularly in Tanzania, which has a relatively high decentralization of government functions, including agricultural extension. The need to be present at the local level has brought SAGCOT quickly to the limits of its capacities. Some years after SAGCOT's launch, only one cluster, Ihemi, had been permanently staffed, and only in 2017 did a second cluster, Mbarali, follow.

What has long been underestimated, is the resistance not only of some interest groups and actors outside the corridor but also of civil society worried about social and environmental harm of an agricultural development model including the private sector, in particular large-scale industries. SAGCOT tried to address the concerns of critics in the blueprint document by labelling its model "green growth" agriculture, by highlighting its principles, by creating an environmental and a social feeder group and by collaborating with scientific bodies. However, some significant resistance has remained. While some common understanding has developed, it is obvious that the visions of different stakeholders are still far apart, particularly with regard to the nuclei, the role of large agro-industries and conventional agriculture (Various CSO, w.d.).

Impacts of individual investments

SAGCOT very much relies on the concept of NOSs. Thus, the impact of SAG-COT can, to a large extent, be assessed by looking at the impacts of its individual NOS investments and the number of such NOSs additionally created through SAGCOT. Other effects at aggregated level, such as the impacts of regulation, side-effects of infrastructure or improved soft infrastructure, will also have to be taken into account but can be expected to be of minor importance since, as argued earlier, SAGCOT is not very active or powerful with regard to these issues.

The evidence is mixed as to the success and local development impacts of NOSs. A distinction should be made between short- and long-term impacts. For the long term, the two formal studies of a rice and a sugar NOS (Herrmann, 2017) show significant and important average positive impacts for people directly involved. Sugarcane outgrowers gained 120–150% higher per capita incomes than the matching control households (depending on the matching method and specification), sugarcane workers 84–99% more and rice workers (there were no rice outgrowers at that time) around 50%. Poverty rates were about 30–40% lower for all groups. The qualitative surveys of Brüntrup et al. (2018) mostly support these positive findings. In particular, people directly affected by tea and sugar NOSs (workers and outgrowers) were found to be better off, as also indicated by long queues to get jobs and in-migration of workers from other places. The indirect effects on the surrounding populations are also quite large, demonstrated by general in-migration to the small urban centres near larger NOSs and their vibrant development. All investors engage in some form of social infrastructure and services, and in more recently established NOSs they are often contractually agreed upon.

However, some smaller groups of people are also affected negatively – for instance, when in isolated markets local food prices increase due to a strong increase in local demand and/or, to a lesser degree, lower local food production. In the short term, land transfers are the main cause of negative effects, as well as distributional conflicts. In the case studies that feature in Brüntrup et al. (2018), there were a few instances where land grabbing was an issue, where displaced people were not well compensated or used their monetary compensations unwisely or where access to some resources and service points was cut off. This was more of a problem in recent investments, although land grabbing has been an issue in the country since colonial times. The literature provides more case studies where land grabbing was a risk in land-based investments, particularly foreign investments (see Brüntrup et al., 2018 for a discussion of these findings). The high number and complexity of impacts in various dimensions; the different, mostly qualitative, methodologies used; the different valuation set-ups and priorities; as well as a highly emotional and sometimes ideological attitude towards industrial and large-scale agriculture may explain these differences of findings of impact. And without any doubt, a wrongly designed or executed large agro-investment can do a lot of harm in a given rural area.

There are many factors explaining and potentially influencing these positive and negative outcomes of individual NOSs, some of them have been mentioned in the previous chapter on governance. Many influences are strongly driven by local forces: the local monopoly situation; the size and duration of the investment; the role of government as shareholder, mitigator and/or arbitrator; irrigation, the type of crop, its market and trade policy; the quality of seeds available; the sophistication of production and integration into farming systems, as well as the type of interplay between investors, administrations and smallholders; and value addition and the package of support services and inputs available. The individual attitude of the investors and the management are also important factors in terms of local impacts. There is strong evidence that the investments that survive in the long run are those which provide important advantages to local communities (Brüntrup et al., 2018). Since SAGCOT has no direct power over these determinants and only has a narrative, its convening power and a voluntary code of conduct to influence them, it seems plausible to reiterate that the impact of SAGCOT will mainly be determined by the sum of the additional investments it is able to lure into the country and much less by the quality of the individual investment impacts.

Conclusions and recommendations

AGCs are a relatively new phenomenon in SSA, though other corridor and SDI approaches have been known in Africa and in other regions for a long time. One striking particularity of African AGCs, at least of the type represented by SAGCOT, is the strong role of the private sector in the design of the corridors and of NOSs.

SAGCOT is one of the earliest and arguably the most prominent AGC in SSA. It is now in the ninth year after its launch. However, the reported results can hardly be described as a success. SAGCOT has certainly achieved a lot in organizing the private sector, orchestrating its interests and voice and shaping the conception of public-private partnerships in Tanzania. However, not a single new NOS, the key element of the approach and blueprint, has been established under the SAGCOT label. SAGCOT claims a number of other investments, jobs and production increases as successes, but these are mostly based on older agro-investments and other projects – as far as available data allows us to judge. The end of the SAGCOT Investment Project represents a serious setback.

Our impact studies, however, show that good, long-term NOSs can yield very important gains for outgrower farmers, workers and communities; while badly designed, unscrupulous investments can do significant harm. Some especially vulnerable groups particularly have to be taken into account, including local migrants, women and pastoralists. However, the biggest challenge for NOSs seems to be bad governance of the economic and institutional environment, which harms investments, investors and local partners alike by hampering successful implementation of the investments and generation of profits.

Despite the obvious weakness of Tanzania and likely of many other SSA countries in creating an enabling environment, the AGC approach with large private sector participation remains appealing since it shows a solution to the various parallel and interdependent weaknesses of agricultural value chain development and rural infrastructure projects by governments and donors in rural areas of poor countries such as Tanzania: lack of funds, technical and economic know-how, long-term orientation and market orientation. What can be learned from SAGCOT for AGC development?

- A high degree of participation and transparency is needed, in particular the involvement of smallholder farmer organizations and local level actors, to provide credibility, to help find the right partners and solutions and also to involve at least some large (including international) civil society organizations, which are so important for building international support. This is more important when large-scale land acquisitions are envisaged.
- An authoritative coordination and implementation is needed. A private or public-private partnership entity such as the SAGCOT Centre may be an indispensable element, but it is not strong enough to quickly trigger important changes. This can only be achieved by a high-level government authority that accepts being guided by the partnership.
- There is a need for catalytic, flexible funding of a sufficient dimension to kick start projects while organizing the longer-term networking, to build capacities of some stakeholders and around some investments, to create early success stories and to keep up and enhance support and credibility.
- More flexibility in the choice of crop and business models should be allowed. Not all value chains should be forced or planned into one predefined

scheme with regard to responsibilities, contractual arrangements, shares, etc. If NOSs and other models requiring large-scale land investments are the key element of an AGC, land issues must be well managed, with more emphasis placed on good governance. If these conditions are not fulfilled, other models of vertical integration of the private sector and smallholder farmers should be preferred, even if this reduces the number of viable options for subsectors and investors.

- Smaller territorial entities (e.g., clusters) are better to start with.
- Long time horizons, particularly in pioneering projects and partnerships, are needed.

References

Ashley, C. and Maxwell, S. (2001) Rethinking rural development. *Development Policy Review*, 19(4), 395–425.

Bergius, M., Benjaminsen, T. A. and Widgren, M. (2018) Green economy, Scandinavian investments and agricultural modernization in Tanzania. *The Journal of Peasant Studies*, 45(4), 825–852.

Brüntrup, M., Schwarz, F., Absmayr, T., Dylla, J., Eckhard, F., Remke, K. and Sternisko, K. (2018) Nucleus-outgrower schemes as an alternative to traditional smallholder agriculture in Tanzania – strengths, weaknesses and policy requirements. *Food Security*, 10(4), 807–826.

Byiers, B., Molina, B. and Engel, P. (2014) *Agricultural Growth Corridors: Mapping Potential Research Gaps on Impact, Implementation and Institutions*. www.fao.org/3/a-bp142e.pdf, accessed 19 March 2019.

Gálvez-Nogales, E. (2014) *Making Economic Corridors Work for the Agricultural Sector* (Agribusiness and Food Industries Series No. 4). Rome: Food and Agricultural Organisation of the United Nations.

Herrmann, R. (2017) Large-scale agricultural investments and smallholder welfare: A comparison of wage labor and outgrower channels in Tanzania, *World Development*, 90, 294–310.

Kaarhus, R. (2011) Agricultural growth corridors equals land-grabbing? Models, roles and accountabilities in a Mozambican case. In *International Conference on Global Land Grabbing* (pp. 6–8).

Paul, H. and Steinbrecher, R. (2013) *African Agricultural Growth Corridors and the New Alliance for Food Security and Nutrition. Who Benefits, Who Loses?* EcoNexus Report. www.econexus.info/files/African_Agricultural_Growth_Corridors_%26_New_Alliance_-_EcoNexus_June_2013.pdf, accessed 19 March 2019.

Reeg, C. (2017) Spatial development initiatives – potentials, challenges and policy lesson. *The German Development Institute/Deutsches Institut für Entwicklungspolitik (DIE) Studies*, 97. DIE: Bonn. www.die-gdi.de/uploads/media/Study__97.pdf, accessed 25 March 2019.

SAGCOT (2011) *SAGCOT Investment Blueprint*, Dar Es Salaam. http://sagcot.co.tz/index.php/mdocuments-library/#, accessed 19 March 2019.

SAGCOT (2018) *The Journey of the SAGCOT Initiative 2013–2018*. http://sagcot.co.tz/index.php/mdocuments-library/#, accessed 19 March 2019.

SNAPP (Science for Nature and People Partnership) (2016) *Encouraging Green Agricultural Development in the SAGCOT Region of Tanzania*. www.nceas.ucsb.edu/files/snap/SAGCOT_Final_Report.pdf, accessed 19 March 2019.

URT (2015) *Agricultural Sector Development Programme 2 (ASDP-2), Transforming the Agricultural Sector*, DRAFT 0, Dar Es Salaam, Tanzania.

Various CSO (Civil Society Organisations) (w.d.) *Feedback and Recommendations from CSOs for the "Greenprint" Strategy of the SAGCOT Initiative.* www.tnrf.org/Greenprint.pdf, accessed 19 March 2019.

WEF (World Economic Forum) (2012) *Putting the New Vision for Agriculture into Action: A Transformation Is Happening.* http://www3.weforum.org/docs/WEF_FB_NewVision Agriculture_HappeningTransformation_Report_2012.pdf, accessed 19 March 2019.

WEF (World Economic Forum) (2016) *Grow Africa: Partnering to Achieve African Agriculture Transformation.* http://www3.weforum.org/docs/IP/2016/NVA/GrowAfrica_Partner ingtoAchieveAfricanAgricultureTransformation_Jan2016.pdf, accessed 19 March 2019.

Weng, L., Boedhihartono, A. K., Dirks, P. H., Dixon, J., Lubis, M. I. and Sayer, J. A. (2013) Mineral industries, growth corridors and agricultural development in Africa. *Global Food Security*, 2(3), 195–202.

World Bank (2009) *World Development Report 2009: Reshaping Economy Geography.* https:// openknowledge.worldbank.org/bitstream/handle/10986/5991/WDR%202009%20 -%20English.pdf?sequence=3&isAllowed=y, accessed 19 March 2019.

World Bank (2016) Southern Agricultural Growth Corridor of Tanzania Investment Project. http://projects.worldbank.org/P125728/tanzania-southern-agriculture-growth-corridor-investment-project?lang=en, accessed 19 March 2019.

World Bank (2018) *Country Partnership Framework.* http://documents.worldbank.org/curated/ en/669801521338458808/pdf/Tanzania-CPF-FY18-22-SECPO-February-14-0221 2018.pdf, accessed 24 March 2019.

World Bank (2019a) *Southern Agricultural Growth Corridor of Tanzania Investment Project.* http://projects.worldbank.org/P125728/tanzania-southern-agriculture-growth-corridor-investment-project?lang=en, accessed 19 March 2019.

World Bank (2019b) *Ease of Doing Business in Tanzania.* https://tradingeconomics.com/ tanzania/ease-of-doing-business, accessed 19 March 2019.

30 Policy options for cropping systems diversification in Southern Africa

Giuseppe Maggio and Nicholas J. Sitko

Introduction

Crop diversification cuts across the economic development and climate change adaptation policy agenda of many countries in sub-Saharan Africa (SSA). For example, about a third of the SSA countries that have submitted a Nationally Determined Contribution plan to the United Nations Framework Convention on Climate Change (UNFCCC) list diversification among their key climate adaptation objectives (UNFCCC, 2019). Crop diversification is seen as a mechanism to increase agricultural productivity and production value, and thus to sustain food security of smallholders and to build resilience to adverse weather and market price fluctuations.

Despite the recognition diversification's central role for the future of agricultural in SSA, diversification levels are quite limited in many smallholder systems in the region. Moreover, adoption of more diverse cropping systems is often hindered by government policy actions that reduce incentives for appropriate crop diversification pathways. In several southern African countries, for example, input and output subsidies often target staple food crops and may inadvertently push farmers to adopt lowly diversified cropping systems or of cropping systems with poor compatible with prevailing agro-ecological conditions.

The attributes of the crops composing a given cropping system, as well as their interactive effect, are the principal determinants of the productivity, the production value and the resilience of the cropping system to external shocks, such as drought, flood or fluctuation in prices. When farmers evaluate the inclusion of a given crop within their cropping system, they often must weigh the advantages and drawbacks that the crop will bring to the system, conditional on the expected weather and agro-ecological condition, production cost and marketability. For example, the inclusion of staple crops that are tolerant to high temperatures, such as cassava, may help a farmer to ensure their production against the occurrence of droughts. However, due to its limited commercial value the inclusion of this crop may reduce the overall value of a farmer's production (Schlenker and Lobell, 2010). Intercropping with legume crops, such as pigeon peas, beans and groundnuts, can help improve soil quality and nutrient content, thus improving the performance of other crops in the system, but their inclusion may also be conditioned on prevailing market factors for these crops (Kerr et al., 2007; Sileshi et al., 2008). Others, such as cash crops, can bring higher returns from their marketization, but their price suffer of high interannual volatility, impeding a correct prediction of the final profit from their cultivation (Chapoto et al., 2013). Unpacking the potential benefits and risks associated with the inclusion of particular crops into a cropping system is critical for developing appropriate crop diversification policy strategies.

Focusing on the southern African countries of Malawi, Mozambique and Zambia, this chapter empirically examines crop systems diversification drivers and impacts in order to provide policymakers with insights into viable strategies to enhance smallholder productivity and build resilience. This is done using nationally representative household surveys that capture information at an individual-, field- and community-level for multiple agricultural seasons (IAI, 2015; IHPS, 2013; RALS, 2015). The analysis focuses on maize smallholder farmers, as maize remains the dominant staple food in these countries, and measures the effect of diversification in terms of maize productivity and crop income volatility.

This capture moves beyond standard measures of diversification used in economic literature, which rely on abstract indicators such as the Gini coefficient or the Margalef index, which do not provide insights into the underlying crops farmers include in their systems (Arslan et al., 2018). To address this weakness, the chapter considers seven possible cropping systems adopted by the farmers, based on combinations of four categories of crops: dominate staple (maize), alternative staple (e.g., cassava, millet, sweet potato), legumes (e.g., groundnuts, pigeon pea, beans) and cash crops (e.g., cotton, sunflowers, tobacco).

The geography of cropping systems

Figures 30.1 displays the distribution of the dominant cropping systems in Malawi, Mozambique and Zambia, at the district level, defined in terms of the cropping systems with the greatest share of cultivated land for that district. Note that the maize-legume-staple system is most widely adopted in Mozambique and Zambia, while in Malawi the majority of the area is under maize monocropping and maize-legume. In all cases, the dominant cropping systems are primarily subsistence oriented.

In general, spatial clustering of cropping systems is evident within each country, indicating internal spatial spillovers in adoption, which is likely associated with variations in infrastructural and market development. In eastern Zambia, for example, the dominant system is the maize-legume-cash crop system, which is associated with private investments in legume and cotton markets in the region. This system is also dominant across the border in Mozambique and may reflect market spillovers between the two countries. This is likely due to the limited infrastructural connectivity in this region of Mozambique, which therefore favours cross-border trade over domestic markets for farmers close to

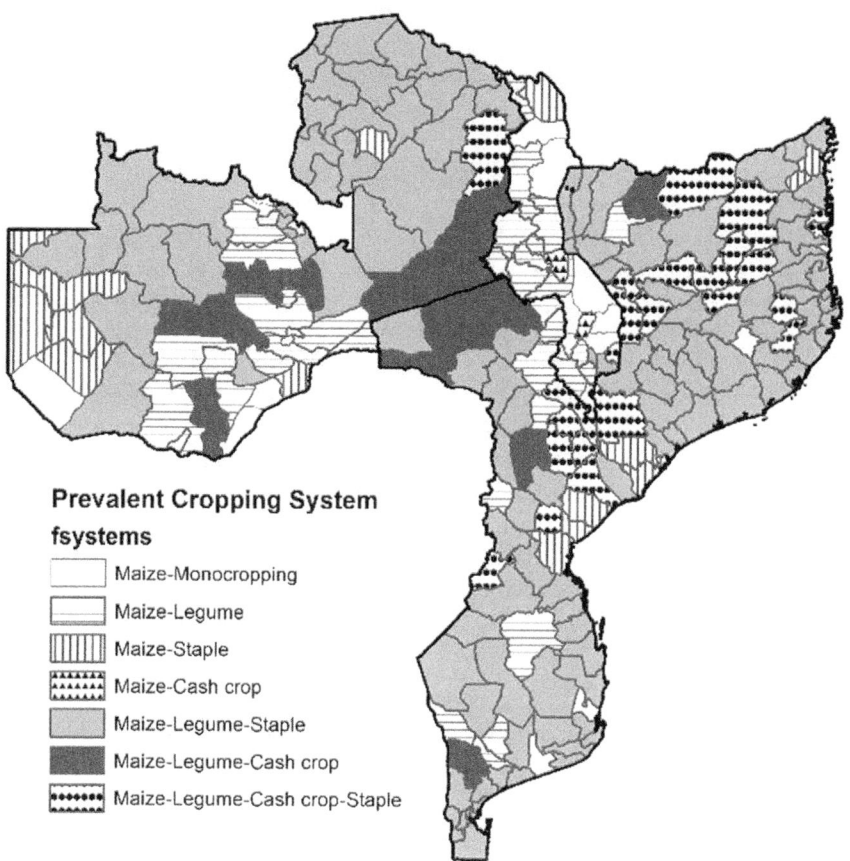

Figure 30.1 Prevalent cropping system at the district level

Note: prevalent cropping system is the one with the highest relative share of land cultivated at the district level

Source: Maggio et al., 2018

the border. This market spillover is not, however, apparent between Malawi and Zambia where cross border trade is more restricted.

Understanding the limitations of dominant cropping systems

Several country-specific limitations are apparent in the dominant cropping systems and help explain the limited adoption of more market-oriented systems. In Malawi, the dominant systems are maize monocropping and maize-legume, suggesting a low general level of diversification. In this country, socioeconomic and institutional factors appear to disincentivize diversification. Diversification in Malawi is most prevalent among larger and wealthier farms, as these farmers have greater physical and financial resources to incorporate additional crops into their systems. For most farmers in Malawi, where average land holdings are less than 1 hectare, maize production is often prioritized to meet household food security requirements. At institutional level, the farmers appear to diversify only they have access to functional, competitive and stable input and output markets. The lack of stability in prices may disincentive diversification towards certain crops. For example, the low adoption of alternative staple and cash crops may be linked to their price volatility compared to legumes. As shown in Table 30.1, between 2010 and 2014 the prices of cassava, cotton and sorghum prices were 14, 15 and 74% more volatile than groundnut prices.

In Mozambique, 47% of the households grows systems of three crops, including maize, with staples such as cassava/sorghum and at least a legume, such as beans and pigeon pea. Despite being highly diversified, this system is associated to low adoption of inorganic fertilizer and low crop income gains compared to monocropping (see Table 30.2). The reason behind the diffusion of this system relies on its resilience to weather fluctuations, especially to drought shocks. However, it would be important to find solutions to help farmers moving

Table 30.1 Groundnut prices are less volatile than alternative staples and cash crops in Malawi

Crop	Price growth rate (2010–2014)	Price volatility (PV)	PV relative to Groundnuts
Cassava	0.29	0.16	14%
Cotton	−0.18	0.16	15%
Groundnuts	0.27	0.14	–
Sorghum	0.22	0.24	74%

Source: Authors own elaboration using FAOSTAT

Table 30.2 Adoption of fertilizer and contribution in income of the cropping systems in Mozambique

	Share of Adopters	Inorganic fertilizer adopters	Crop Income (US$ 2010)	Change in Crop Income compared to MM
Maize monocropping (MM)	6%	4%	220	–
Maize-Legume	19%	4%	397	180%
Maize-Staple	9%	2%	448	113%
Maize-Cash Crops	2%	23%	741	165%
Maize-Legume-Staple	47%	3%	361	49%
Maize-Legume-Cash Crops	5%	28%	1100	305%
Maize-Legume-Cash Crops-Staple	12%	10%	688	63%

Source: Authors own elaboration using the survey data

from this subsistence-oriented system towards the adoption of more market-oriented systems, such as the maize-legume-cash crops system. Indeed, adopters of cropping systems including cash crops show substantial gains in terms of crop income and resilience, compared both to maize monocropping and to maize-legume-alternative staple systems.

In Zambia, about half of the farmer population relies on systems with three or more crops. However, there is an apparent geographic disconnection between levels of cropping system diversification and the frequency of climatic risks that farmers experience. In particular, in the north and northwest of the country the dominant cropping system is a three-crop system comprising legumes and cassava. This system would be particularly well suited for drought prone areas, but the zones where it is adopted most receive substantially more rainfall than the rest of the country. Maize monocropping and two-crop systems, in contrast, prevail in the south of the country, where both land fragmentation and rainfall variability are higher. As shown in Figure 30.2, higher rainfall risks, measured as quartiles in historical volatility of rainfall, is strongly associated with lower level of diversification. In particular, the share of farmers adopting one- or two-crop systems is higher in high risk areas and lower in low risk areas (Figure 30.2). Addressing this spatial disconnect in cropping system diversification may be an important avenue for improving climate resilience among Zambian smallholders.

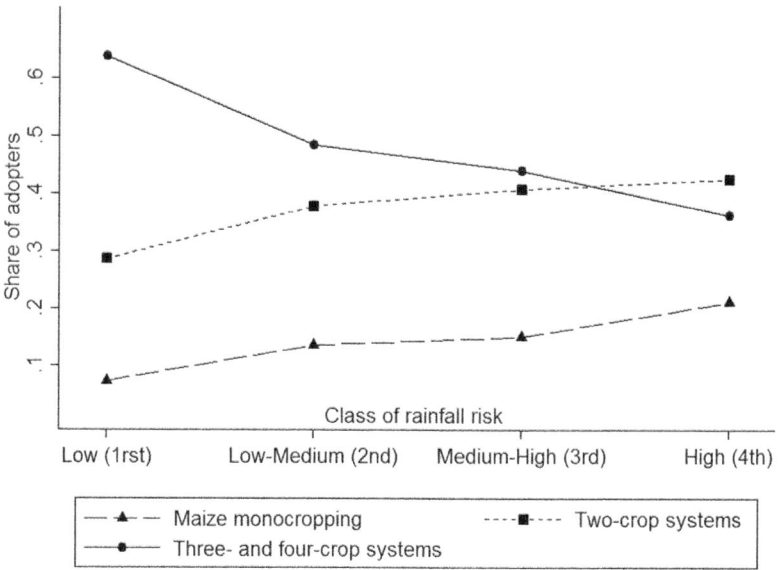

Figure 30.2 Farmers are adopting more vulnerable systems for higher classes of rainfall risk in Zambia

Source: Own representation using RALS 2015 data

Drivers and impact of cropping systems diversification

In general, three factors appear to influence the adoption of more diversified systems across the three countries. First, private output market development is a key driver of adopting more diverse and commercialized cropping systems. This is partially driven by the impact of transactions costs on seed prices, including maize seed prices. Given scarce resources, farmers often prioritize purchasing staple food seeds before investing in other inputs. Contrary to expectations, when maize seed prices increase the probability of diversification often declines. For example, in Malawi, an increase of maize seed prices by 2.1 US$ decreases by the probability of adopting a four-crop system by 2%.

Second, the presence of parastatal marketing boards appears to affect farmers' decisions about diversification. The parastatal marketing boards operate in Malawi and Zambia and typically focus on purchasing maize from individual farmers and cooperatives at pan-territorial price that typically exceeds prevailing market prices. This elevated market price acts as a disincentive to diversification. In Zambia, for example, as the distance between farmers and buying depots for the Food Reserve Agency increase, the probability of adopting more diverse three- and four-crop systems increases significantly.

Finally, larger land holdings and household wealth status are found to be significant drivers of diversification in all cases. These better-off households are able to manage the costs and risks of experimenting with new cropping systems and integrating a larger range of food and nonfood crops into their farm systems.

The magnitude of the impact of cropping system diversification on maize productivity varies across countries and cropping systems, but several commonalities emerge from the analysis. All the cropping systems exhibit higher levels of maize productivity compared to maize monocropping when controlling for all other factors. The only exception of is for systems comprised of maize with an alternative staple crop. This is likely because alternative staple crops do not confer agronomic benefits to maize, as legumes typically do, nor do they typically provide commercialization options, like cash crops, which would support farmers to intensify production with capital inputs. The adoption of three- and four-crop systems in general results in higher maize yields than maize monocropping. In Malawi, for example, farmers adopting three-crop systems based on maize, legumes and a third crop (cash or alternative staple), are likely to increase their maize yield of between 15 and 37 compared to monocropping.

In terms of the resilience of cropping systems, the empirical results highlight important differences between the seven cropping systems across the three countries. This heterogeneity is likely due to variations in the market conditions for the crops composing the systems within each country and differences in agro-ecological characteristics. For example, in Malawi, only the maize-legume system appears to reduce significantly farmers' crop income volatility, while adopters of all the other cropping systems experience levels of income volatility that are similar to those of maize monocropping. In Mozambique, only systems based on three crops, such as maize-legume-cash crops, are effective in reducing crop income volatility. In Zambia, all the cropping systems appear to reduce volatility in crop income compared to maize monocropping, with an increasing effect for more diversified cropping systems.

Policy conclusions for cropping system diversification

What actions can policymakers undertake to support the adoption of more diversified and resilient cropping systems? The results from the analysis suggest that policymakers should prioritize three areas of intervention in their countries.

Policymakers will need to incentivize investments in the private input and output markets

A set of direct and indirect policies may help the development of such markets and improve the level of diversification of the agricultural sector. For example, diversification may be enhanced through the promotion of stable macroeconomic conditions and predictable trade policy, which may induce private investments in agricultural markets. If necessary, trade restrictions on staple foods, such as maize, should be applied in predictable ways and supplemented with import support measures.

Implementing policies that favour competition in the seed sector may help seeds multiplication and, in the medium term, crop system diversification. The three countries differ in their level of seed market development. Zambia, for example, has a relatively more advanced seed market compared to Malawi and Mozambique, and different public and private actors participate to its sector under the umbrella of the Zambia Agriculture Research Institute (ZARI). Malawi and Mozambique observe a low number of seed breeders and foundation seeds producers, with a consequent low number of seeds traded along all the value chain. This is often associated with the low prices set for the foundation seeds produced by the public institutions. It is necessary therefore to create incentives through the creation of public-private partnerships along all the seed value chains. In certain contexts, a stricter enforcement of the laws defending breeders' rights and supportive legislation for small-scale and community-based breeders may incentivize investments in the sector.

Policymakers may need to rethink public intervention in markets to achieve crop diversification objective

It is necessary to reconsider interventions into output markets when these generate distortive effects to diversification, especially in periods of normal or above normal production. For example, floor prices may be relaxed when international prices are higher than internal ones to incentivize trading and investment in cash crops. Public market boards selling large quantities of maize in certain regions may inadvertently reduce the price taken by local producers and thus reduce the possibility of investment into alternative staples, legumes and cash crops during the next agricultural season. Any policy supporting the development of contract farming sectors should include, among its objectives, a minimum level of diversification at the farm level, so to improve smallholders' economic efficiency and economic resilience to unexpected market and weather shocks.

Finally, ongoing reforms to support programmes, such as the Farm Input Subsidy Programme (FISP) in Malawi, should consider crop diversification as one of their key objectives. Amendments may include more flexibility in the types of seeds, improvement in the quality of seeds and promotion of seeds compatible with the local agro-ecological conditions of the zone of distribution.

Policies to improve land tenure security are important for diversification

Land acquisition by urban resident and large agribusiness firms are widespread in the region (Sitko and Jayne, 2014; Jayne et al., 2016).

To guarantee land access to smallholders, policymakers may consider policies regulating the land acquisition from large agribusiness or companies outside the farming sector, while supporting mortgage and micro-credit finance for the most productive smallholders. Land constrains are particularly binding in urban and peri-urban areas, where farmers operate in zones with high population densities and high land fragmentation (Sitko and Chamberlin, 2016). To facilitate land consolidation in those areas, policymakers may consider supporting the network of public transportation to the countryside of the most densely populated cities. An improved transportation network may help farming households in accessing cheaper land and may help decrease the price of land in the peri-urban areas. Finally, policymakers may design simplified land registration procedures and provide support to help land registration of the households still residing on statutory land, who often lack the knowledge of how to secure their land rights.

References

Arslan, A., Cavatassi, R., Alfani, F., Mccarthy, N., Lipper, L. and Kokwe, M. (2018) Diversification under climate variability as part of a CSA strategy in rural Zambia. *The Journal of Development Studies*, 54(3), 457–480.

Chapoto, A., Haggblade, S., Hichaambwa, M., Kabwe, S., Longabaugh, S., Sitko, N. and Tschirley, D. (2013) Institutional models for accelerating agricultural commercialization: Evidence from maize, cotton and horticulture. No. 154940. Michigan State University, Department of Agricultural. *Food, and Resource Economics.*

IAI. (2015) *Inquérito Agricola Integrado. Ministério da Agricultura e Segurança Alimentar.* Maputo, Moçambique: República de Moçambique.

IHPS. (2013) *Malawi Integrated Household Panel Survey.* Lilongwe, Malawi: Government of Malawi.

Jayne, T.S., Chamberlin, J., Sitko, N., Muyanga, M., Yeboah, F.K., Nkonde, C., Aneeuw, W., Traub, L., Chapoto, A. and Kachule, R. (2016) Africa's changing farmland ownership: The rise of the emergent investor farmer. *Agricultural Economics*, 47(S1), 197–214.

Kerr, R. B., Snapp, S., Chirwa, M., Shumba, L. and Msachi, R. (2007) Participatory research on legume diversification with Malawian smallholder farmers for improved human nutrition and soil fertility. *Experimental Agriculture*, 43(4), 437–453.

Maggio, G., Sitko, N. J. and Ignaciuk, A. (2018) *Cropping System Diversification in Eastern and Southern Africa: Identifying Policy Options to Enhance Productivity and Build Resilience* (No. 288953). Food and Agriculture Organization of the United Nations, Agricultural Development Economics Division (ESA).

RALS. (2015) *Agricultural Livelihoods Survey*. Lusaka, Zambia: Government of Zambia.

Schlenker, W. and Lobell, D. B. (2010) Robust negative impacts of climate change on African agriculture. *Environmental Research Letters*, 5(1), 014010.

Sileshi, G., Akinnifesi, F. K., Ajayi, O. C. and Place, F. (2008) Meta-analysis of maize yield response to woody and herbaceous legumes in sub-Saharan Africa. *Plant and Soil, 307* (1–2), 1–19.

Sitko, N. J. and Chamberlin, J. (2016) The geography of Zambia's customary land: Assessing the prospects for smallholder development. *Land Use Policy*, 55, 49–60.

Sitko, N. J. and Jayne, T. S. (2014) Structural transformation or elite land capture? The growth of "emergent" farmers in Zambia. *Food Policy, 48*, 194–202.

United Nations Framework Convention on Climate Change (2019) *Nationally Determined Contribution Registry, Country Pages*. https://www4.unfccc.int/sites/NDCStaging/Pages/All.aspx.

31 Entrepreneurship as an economic development strategy for agricultural transformation in Southern Africa

Ralph D. Christy and Mohammad Karaan

Introduction

Entrepreneurs[1] are recognized globally for their contribution to economic growth. Frequently, entrepreneurs within agribusiness industries are described as the "engines" of emerging economies as demonstrated by their innovations, job and wealth creation, usage of local or domestic resources, facilitation of industrialization and promotion of rural and urban development (Christy et al., 2015). Opportunities in agribusiness in Africa's emerging economies have expanded tremendously due to globalization, regional trade agreements and urbanization. With increased globalization, consumer trends such as changes in food preferences and increased food purchases at supermarkets are occurring across many African countries. While globalization is changing consumers' preferences, urbanization is changing the way food is produced and marketed in Africa. As agribusiness entrepreneurs offer products and services to urban areas, new technologies are emerging to develop value-added food products that meet the demands of African consumers. At the same time, forces such as competition from global business players and climate change pose new challenges to owners and managers of African agribusiness companies.

Despite those market opportunities and challenges, capacity strengthening programmes for agricultural development in Africa traditionally have focused on production directed towards achieving food security objectives. For agribusiness companies to remain competitive, new capacity strengthening strategies and training programmes are needed to address the challenges faced by managers. With the steady growth of the food value-addition and service sectors, more investments are needed to strengthen the capacity of local agribusiness managers so that they may effectively contribute to the sustainable growth and development of African economies. Addressing this capacity gap is a primary goal of the Making Markets Matter (MMM) training programme that was started in 2001 at Stellenbosch University.

This chapter will provide a conceptual understanding of entrepreneurship as a driver for agricultural transformation and economic development in southern Africa as well as an overview of alternative capacity strengthening programmes for business development for small firms in emerging economies.

The role of SMEs in economic development: theory and conceptual frameworks

Despite general support for SMEs programmes, until recently, scholars had given little attention to understanding entrepreneurship as the basis for a comprehensive economic development strategy. Over the years, scholars and development practitioners have accumulated a body of knowledge about the economic development process, but this literature fails to provide a coherent conceptual understanding of how to generate entrepreneurship and sustainable economic progress. Clarity regarding the very role of entrepreneurship and its impact on economic development is also lacking.

Schumpeter's *The Theory of Economic Development*, first published in 1911, links innovation and the entrepreneur, claiming that successful innovation is the source of private profits, which in turn lead to economic growth. The entrepreneur creates new economic combinations by: 1) introducing new products; 2) introducing new production functions that decrease inputs needed to produce a given output; 3) opening new markets; 4) exploiting new sources of materials; and 5) reorganizing an industry (Schumpeter, 1961; Nafziger, 1997). Given Schumpeter's conceptual claims, many empirical gaps are apparent in entrepreneurship research. First, a standard definition for an entrepreneur does not exist. Second, data is unavailable to allow researchers to compare entrepreneurial activity between nations and the data that was available lacked information about a population's entrepreneurial qualities. Third, for international comparisons, data was unavailable about the business start-up process. Until the past decade or so, scholars were unable to make international comparisons of entrepreneurial activity rates or offer a framework for evaluating their conditions.

More recently, researchers have started giving more attention to the connection between entrepreneur or small business development and economic growth. The Global Entrepreneurship Monitor (GEM), a joint research initiative of Babson College in Wellesley and the London Business School, was one group that investigated this new discipline of "entrepreneurial academics". The GEM collected and analyzed data about entrepreneurship and business development globally. The initiative aimed to determine the variance of entrepreneurial activity between countries, the reasons some countries have higher entrepreneurship rates than others, the types of national policies that can increase entrepreneurial activity and the connection between entrepreneurship and economic growth. At the country level, the GEM created a model that identifies nine determinants of entrepreneurial opportunities including: financial support, government policies, government programmes, education and training, research and development transfer, commercial/professional infrastructure, barriers to market entry, access to physical infrastructure and cultural and social norms.

Many development scholars and professionals believe supporting small businesses in emerging economies can be an effective tool to alleviate poverty and sustain a healthy economy (Lichtenstein and Lyons, 2001; Acs and Malecki,

2003). A small but critical portion of new businesses bring knowledge, products or ways of producing goods or services to a community. A greater number of new businesses, in turn, widen the distribution of new goods and services developed by other entrepreneurs. Recent studies have found also that the impact of new business development differs depending on the stage of a country's economic development. One study, which examined data from 36 countries found a U-shaped relationship between new entrepreneurship and the level of economic development (Wennekers et al., 2005). A country's entrepreneurship rate initially declines as its economy develops but then levels off or rises again after it reaches a certain level of development.

Arenius and DeClercq (2005) found that people with higher levels of education are more likely to identify entrepreneurial opportunities than those with lower education levels. Therefore, one policy option could be to foster an environment that facilitates the exchange of information among people who are interested in or knowledgeable about entrepreneurship. This strategy would have to consider a community's entrepreneurial history and specific context to be effective.

Entrepreneurs sometimes have advantages facilitated by wealth or status in society, which are influenced by factors such as: 1) access to more information than competitors; 2) access to training and education; 3) local elites or family connections; 4) access to government; and 5) agreements to restrict entry or output. Entrepreneurial activity is a means for moving up the economic ladder allowing the socioeconomic status of the entrepreneurs to be higher than their parents (Nafziger, 1997). Most studies have shown a higher level of education among entrepreneurs relative to the population at large. Achievements in verbal, mathematical, written and problem-solving skills have contributed to entrepreneurial success, though the time it takes to obtain these, and lengthy apprenticeships, can be costly. In areas where educational opportunities are lacking, individuals tend to accept "secure positions" with high earnings and low risk. Alternatively, areas with a surplus of university graduates tend to encourage entrepreneurship in avoidance of unemployment or low-paying work. Further, cultural norms in transition economies dictate expectations for men and women in the business setting. Those traditional norms often inhibit women's access and success. The characteristics of a successful entrepreneur are inconsistent with what some societies expect of women. Refusal of credits by banks and input suppliers may further hinder women entrepreneurs.

Very often it is the case that business development in emerging economies, especially in low-income communities, is challenged by the very circumstances it aims to eradicate. Poverty coincides with, and most likely contributes to, a deficient state of enterprise development. Increasingly, however, development experts argue that small enterprise development in those settings has a greater likelihood of sustainability than traditional, more costly approaches that seek to attract external businesses to the area. The major contribution of business development to increased community welfare is the generation of new jobs and additional income. A substantial body of evidence suggests that the small business sector has yielded the bulk of new jobs (Acs, 1999). Given this evidence, entrepreneurship as an economic development strategy has continued to gain credibility as governments and donor agencies have expanded funding for entrepreneurship development programmes. This development strategy is perhaps the first major economic development paradigm to be simultaneously applied to low-income areas in both developed and developing economies.

Due to globalization and the competition from lower transaction costs in other countries, efforts to attract businesses from other regions has not always worked. Increasingly, community activists and policymakers are calling for development strategies that focus on homegrown small firms, regional trade associations and local entrepreneurs. They argue that local entrepreneurs are often neglected as agents of development and support a shift in focus to supporting entrepreneurship from "within their communities" rather than trying to attract outside businesses. This notion is backed by a growing body of theory and research that reexamines the "bigger is better" model and emphasizes the organizational embeddedness of small-scale, locally controlled economic enterprises (Robinson et al., 2002). Moreover, it suggests that the establishment of more entrepreneurship-centred economic development may enable economically disadvantaged communities to reverse stagnant economic conditions by creating wealth and jobs through locally owned businesses.

Heated debates have long existed in development economics about the appropriate roles of government, business, individual innovators and civil society in fostering long-term economic growth and poverty reduction. A clear understanding of development policy objectives and strategies are needed as successful economic development is predicated on the design of "institutions" that will establish both effective public policies and successful private strategies.

An alternative capacity strengthening model for SME development

To foster development in economically depressed areas, it is important to have public policies that support healthy entrepreneurial ecosystems.[2] Programmes should focus on retaining financial capital, encouraging business development and providing incentives to curb out-migration and encourage in-migration of skilled people. Examples of instruments or "policy levers" to achieve those objectives include capital subsidies, infrastructure improvements, tax breaks and relaxed regulations for businesses, human capital development and management training. Beyond those traditional strategies, enlightened public policy must also build upon the existing "social capital" in economically depressed regions. Analysts contend that governments can create a favourable growth climate or "enabling environment' by using public funds to provide infrastructure to support genuinely profitable businesses. The Porter (1995) model of economic development offers principles that underline a sound small business-centred strategy:

- Economic rather than social focus that emphasizes the creation of wealth instead of the redistribution of wealth in economically disadvantaged communities

- Emphasis on the private sector with recognition of the supporting and complementary roles played by the government and social service sectors
- Emphasis on engaging skilled and experienced individuals in entrepreneurial activities.

The goal of Porter's model is to identify and exploit the competitive advantages of regions that will translate into truly profitable businesses. For this model to be effective, governments must support the private sector in new economic initiatives, shifting focus from direct intervention to the facilitation of a favourable business environment (or ecosystem).

Since 2001, Stellenbosch and Cornell University have partnered to organize the MMM workshop, an alternative approach to entrepreneurial capacity strengthening based on a contemporary conceptualization of programme design and programme evaluation to monitor transformative change. As we engage African agribusinesses, recognizing that learning will vary across cultures and contexts, we consider all aspects of an empowering educational experience – context, method, philosophy, content, process and the role of the facilitators. Because lasting and sustainable change often comes through a society's educational system, our approach is centred on the participants' needs, questions and curiosity. To ensure effectiveness and impact, the MMM workshop focuses on learning based on African case studies (Mabaya et al., 2011) that provide the participants the opportunity to discover solutions to the problems and challenges they often face in their own businesses. We therefore developed a framework with an emphasis on new and innovative approaches to entrepreneurial capacity strengthening as distinct from the traditional model of thinking about executive programmes. The key features the MMM workshop are outlined in Table 31.1.

What makes the MMM approach unique? Simply put, it is the combination of the key aspects that will affect a company's bottom line, viewed through the lens of both industry and academia. We look beyond the individual and focus on the company as a unit to ensure that change and benefits are systemic for agribusinesses to function effectively. Ultimately, it is the individuals, working

Table 31.1 MMM approach to SME capacity strengthening framework

Description	Traditional approach	Integrated training programme
Unit of analysis	Individual	Company/organizational capacity
Methods	Lectures	Case studies/team building/presentations
Strategy	Clusters	Network development
Problem focus	Market access	Technology, market and capital access
Partners	Ngos	Business Schools, ngos, Corporations and Governments
Facilitators	International	Hybrid – International, Regional and Local
Monitoring and evaluation	External indicators	Participatory assessment

Source: Authors

as a team committed to the same mission and goals, who will expand and grow the company. Additionally, to be successful, African agribusiness must keep up with industry changes and technological innovation, stay ahead of competitors, build and manage knowledge, empower and retain employees, among many things. Thus, our focus on the organizational capacity, as a whole, adds more value and impact. Sometimes, the focus on individual capacity and mentoring, in high turnover environments, can be also a poor use of resources.

Our methods are learner-centred. The use of case studies (Mabaya et al., 2011; Christy et al., 2018) helps participants learn from practical, real life examples and experiences. This approach moves them from focusing on the problems exclusively to engaging with the questions, solutions and applications. The use of other participatory methods such as role-playing, small group discussions and team building activities is very effective in this process of learning.

The emphasis on business networks as opposed to clusters grows out of MMM's years of engagement with African entrepreneurs. We have found that "unlike clusters – which are defined within and often limited by physical location – networks are not bound by geography" (Da Silva and Mhlanga, 2011). Our MMM target audience comes from southern Africa (and beyond) and has a strong desire to connect to and access global markets. Thus, the business networks are more relevant and beneficial for the participants in connecting to local, regional and global networks to access diverse markets.

This focus on business networks also builds on MMM's ability to build collaborative approaches between industry and academia. Traditionally, executive education programmes have tilted towards nongovernmental organizations (NGOs). However, bringing together the best thinking in top business schools, industry, governments and NGOs ensures that the gaps in skillsets and knowledge within African agribusinesses are addressed.

Finally, the MMM workshop has a rigorous monitoring and evaluation to ensure impact and effectiveness. MMM's integrated training programme ensures that the process of learning (inputs, activities and outputs) can augment company performance and better the quality of life for agribusinesses owners, employees and their families as well as communities across Southern Africa. MMM's four primary objectives are to:

- **Enhance** the business management capacity of African entrepreneurs through the combination of short-term, intensive, capacity strengthening and longer-term leadership coaching
- **Facilitate** business and strategic linkages between companies and the broader agribusiness sector in Africa
- **Provide** an opportunity to network by showcasing and marketing products and services or sharing information materials
- **Integrate** industry analysis and case studies of selected companies.

The programme combines a structured learning environment featuring conceptual frameworks and analytical techniques for decision making, contemporary

African agribusiness-specific case studies and the opportunity to share ideas with and learn from peers representing the African continent. As part of establishing business networks and market access, participants are also encouraged to participate in the MMM Product Expo. At the end of the official program, each participant has the opportunity to visit an agribusiness firm as part of a company site visit which includes engagement with management. With all those integrated components, the program's ultimate goal is to improve socioeconomic conditions in communities by strengthening emerging entrepreneurs in Africa through an innovative and multifaceted initiative that seeks to improve the performances of their businesses.

Conclusion

Entrepreneurship refers to both owning and managing a business at one's own risk to take advantage of an economic opportunity. Global competition and corporate restructuring have prompted development scholars to increasingly focus on entrepreneurship as an area of policy and practice. Development literature has historically focused on the roles of the market and the state in the economic development process. More recent literature highlights the role of civil society in development. The potential for civic organizations to facilitate revitalization opportunities beyond what the market and political institutions can offer is gaining increasing recognition. Development practitioners have begun to realize the importance of incorporating and building upon local civic organizations for economic, social and political activities. Sustainable economic growth strategies can no longer separate enterprise innovation from advances in government and civic institutions. Innovations must be reinforced in and complemented by all three sectors to advance a supporting ecosystem for entrepreneurs.

Business development services, important for creating private sector capacity in emerging markets, require experiential knowledge that should be available to employees as well. Many entrepreneurs in Africa face a scarcity of skilled individuals within their local areas. Due to the limited amount of educational opportunities and economic growth potential, depressed areas often experience a "brain drain", in which skilled and educated individuals leave in search of a more lucrative environment. Cultural norms oftentimes prohibit women from achieving the same level of skill and professionalism that is available to men in their societies. To overcome internal and external challenges facing African entrepreneurs, capacity strengthening programmes must create training workshops that are innovative, practical and highly relevant.

Notes

1 The terms entrepreneur, small business, small and medium enterprise (SME), micro-business, and agri-preneur have been used interchangeably in many contexts depending on which country is defining the term. Some measures that are used to define this term include number of people employed, capital employed, and sales turnover.

2 An entrepreneurial ecosystem is "a set of interdependent actors and factors coordinated in such a way that they enable productive entrepreneurship within a particular territory" (Stam and Spigel, 2016). This approach to understanding the environment around entrepreneurs as well as entrepreneurship in an economy builds upon ideas from the regional development literature as well as the strategy literature. For a brief overview, see: Stam and Spigel, 2016; Acs, Stam, Audretsch, and O'Connor, 2017.

References

Acs, Zoltan J. (Ed.) (1999), *Are Small Firms Important? Their Role and Impact*. Boston, MA: Kluwer Academic Publishers and U.S. Small Business Administration.

Acs, Zoltan J. and Edward J. Malecki (2003), *Entrepreneurship in Rural America: The Big Picture*. Kansas City, MO: Federal Reserve Bank of Kansas City.

Acs, Zoltan J., Erik Stam, David B. Audretsch, and Allan O'Connor (2017), "The Lineages of the Entrepreneurial Ecosystem Approach," *Small Business Economics*, 49(1), 1–10.

Arenius, Pia, and Dirk De Clercq (2005), "A Network-Based Approach on Opportunity Recognition," *Small Business Economics*, 24(3), 249–65.

Christy, Ralph, Joselito Bernardo, Aimée Hampel-Milagrosa, and Lin Fu (Eds.) (2018), *Asian Agribusiness Management: Case Studies in Growth, Marketing, and Upgrading Strategies*. Singapore: World Scientific.

Christy, Ralph, Mohammad Karaan, Edward Mabaya and Krisztina Tihanyi (Eds.) (2015), *From Principles to Best Practices: A "Making Markets Matter" Guide to Managing African Agribusinesses*. Ithaca, NY: Market Matters Inc.

Da Silva, Carlos A. and Nomathemba Mhlanga (2011), *Innovative Policies and Institutions to Support Agro-Industries Development*. Rome: Food and Agriculture Organizations (FAO).

Lichtenstein, Gregg A. and Thomas S. Lyons (2001), "The Entrepreneurial Development System: Transforming Business Talent and Community Economies," *Economic Development Quarterly*, 15(1), 3–20.

Mabaya, Edward, Krisztina Tihanyi, Mohammad Karaan and Johan van Rooyen (Eds.) (2011), *Case Studies of Emerging Farmers and Agribusinesses in South Africa*. Stellenbosch, South Africa: Sun Press.

Nafziger, Wayne (1997), *The Economics of Developing Countries*. Upper Saddle River, NJ: Prentice Hall.

Porter, Michael E. (1995), "The Competitive Advantage of the Inner City," *Harvard Business Review*, 73(3), 55–71.

Robinson, Kenneth L., Thomas A. Lyson, and Ralph D. Christy (2002), "Civic Community Approaches to Rural Development in the South: Economic Growth with Prosperity," *Journal of Agriculture and Applied Economics*, 34(2), 327–338.

Schumpeter, Joseph. (1961), *The Theory of Economic Development: An Inquiry into Profits, Capital, Credit, Interest, and the Business Cycle* (Redvers Opie, Trans.). New York: Oxford University Press.

Stam, Erik, and Ben Spigel (2016), "Entrepreneurial Ecosystems," *Utrecht University School of Economics (USE) Discussion Paper Series Nr: 16–13*.

Wennekers, Sander, André Van Wennekers, Roy Thurik, and Paul Reynolds (2005), "Nascent Entrepreneurship and the Level of Economic Development," *Small Business Economics*, 24(3), 293–309.

32 Changing farm structure in Africa

Implications on agricultural transformation in Southern Africa

Milu Muyanga and Thomas S. Jayne

Introduction

National development policy strategies within the region (including the Comprehensive Africa Agriculture Development Programme – CAADP) officially regard the smallholder farming sector as an important vehicle for achieving agricultural growth, food security and poverty reduction objectives. In sub-Saharan Africa, smallholder farmers constitute the bulk of agricultural producers and the majority remains mired in low productivity and poverty. A major lesson for Southern African, and sub-Saharan Africa in general, from the experience of smallholder-led Asia, is that if we want agricultural growth to reduce poverty, it must be inclusive such that a large percentage of the rural smallholder population is able to participate in the process. The key elements of agricultural transformation process include increased productivity and sustainability followed by commercialization. Commercialization is the transition from subsistence to market-oriented patterns of production and input use.

The productivity of farming is a key driver of real incomes and productivity in the rest of the economy. While expansion of area under cultivation, *agricultural extensification*, has been the major source of growth in agricultural production for many decades in this region, the scope for continued agricultural extensification to drive agricultural production growth is increasingly limited in light of growing land scarcity as a result of mounting population growth. *Agricultural intensification*, or raising productivity on existing farmland, has been touted as crucial strategy to improve the continent's agricultural growth. However in a number of recent applied studies, agricultural *intensification* is found to rise with population density up to a point; beyond this threshold, rising population density is associated with sharp declines in output per unit of land (Muyanga and Jayne, 2014; Ricker-Gilbert et al., 2014; Josephson, et al., 2014). These unsustainable agricultural productivity trends are being attributed to factors such as shortened fallows, deterioration in soil quality and land fragmentation. Willy et al. (2019) find presence of a "silent" form of soil degradation as a result of dwindling soil organic carbon and critical soil micronutrients as well as increased soil acidity due to continued use of inorganic fertilizers on tiny pieces of land.

Rapid growth of medium-scale farms in sub-Saharan Africa

Meanwhile, evidence is emerging showing a changing structure of land ownership in Africa, a major trend that is likely to affect agri-food systems in Southern Africa region and sub-Saharan Africa in general. Africa has witnessed a rise in the number of commercialized medium-scale farmers. This refers to farmers operating between 5 and 100 hectares of land. This group has little in common with large-scale commercial farmers in terms of farm size, access to finance, input application rates and farm management strategies. In Southern Africa, Zambia, for example, while the overall population of smallholders has increased by 33.5%, the number of medium-scale farmers has grown by 103% (Sitko and Jayne, 2014).

Considering Zambia, Tanzania, Kenya and Ghana, only in Kenya is a substantial majority of national farmland, about two-thirds, under small-scale farms. In Tanzania, Zambia and Ghana, the percentages of farmland held by medium-scale farms are, 39, 53 and 32%, respectively (Table 32.1). In every case, including Kenya, land controlled by medium-scale farms exceeds that under the control of large-scale farms above 100 ha (Jayne et al., 2016).

Within the past decade, the amount of agricultural produce that these farms contribute to countries' national output has also risen rapidly. In some Southern African countries, like Zambia and Tanzania, medium-sized farms now account for roughly 40% of the country's marketed agricultural produce (Figure 32.1). This investment appears to be most common in countries with relatively abundant land: Zambia, Tanzania and Ghana clearly fall into this category. Meanwhile, the sector is much smaller in more densely settled countries such as Nigeria, where it accounted for less than 20% of marketed output in 2016, and Rwanda, where its share was well under 5% in 2014. Note, however, that even in these densely populated countries, this sector's share of output has risen substantially since the mid-2000s. This is exactly what we would expect over the course of agri-food system transformation, as long as policy does not create major impediments.

Patterns of medium-scale participation by crop suggest that grains and oilseeds are major focuses for these farmers (Figure 32.1). The sector's highest share across crops is in grains in Nigeria and Rwanda, and grains are a close second to oilseeds in Southern African Tanzania. Oilseeds lead in Tanzania. Involvement in horticulture varies across countries: these crops show the sector's second-highest share in Ghana but the lowest share in Tanzania.[1] Given this sector's apparent growing importance, much more needs to be known about Southern African cropping patterns, sales behaviour and the amount of on-farm employment they generate.

Table 32.1 Changes in farm structure in Zambia, Tanzania and Ghana

Farm size "category"	Number of farms		% of farms in Period II	% growth in number of farms between two periods	% of total cultivated area		% growth in area operated between two periods
	Period I (year)	Period II (year)			Period I (year)	Period II (year)	
Zambia	2001	2012			2001	2012	
0–5 ha	797,157	1,167,315	83	46.4	79.1	47.9	−39.4
5–10 ha	20,832	165,129	12	692.7	14.3	25	74.8
10–20 ha	2,352	53,454	4	2,172.7	6.6	15	127.3
20–100 ha	–	13,839	1	–	–	12.1	–
Total	820,341	1,399,737	100	70.6	100	100	–
Tanzania	2008	2012			2008	2012	
0–5 ha	5,454,961	6,151,035	91	12.8	62.4	56.3	−9.8
5–10 ha	300,511	406,947	6	35.4	15.9	18	13.2
10–20 ha	77,668	109,960	2	41.6	7.9	9.7	22.8
20–100 ha	45,700	64,588	1	41.3	13.8	16	–
Total	5,878,840	6,732,530	100	14.5	100	100	–
Ghana	1992	2013			1992	2013	
0–5 ha	2,037,430	2,580,685	84	26.7	60.7	45.5	−25.0
5–10 ha	116,800	320,411	10	174.3	17.2	22.8	32.6
10–20 ha	38,690	117,722	4	204.3	11	16.1	46.4
20–100 ha	18,980	37,421	1	97.2	11.1	12.2	–
> 100 ha	–	1,740	0	–	–	3.5	–
Total	2,211,900	3,057,978	100	38.3	100	100	–

Sources: Zambia– Zambia MAL Crop Forecast Surveys, 2001 and 2012; Tanzania – LSMS/National Panel Surveys, 2008 and 2012; Ghana GLSS Surveys, 1992 and 2013

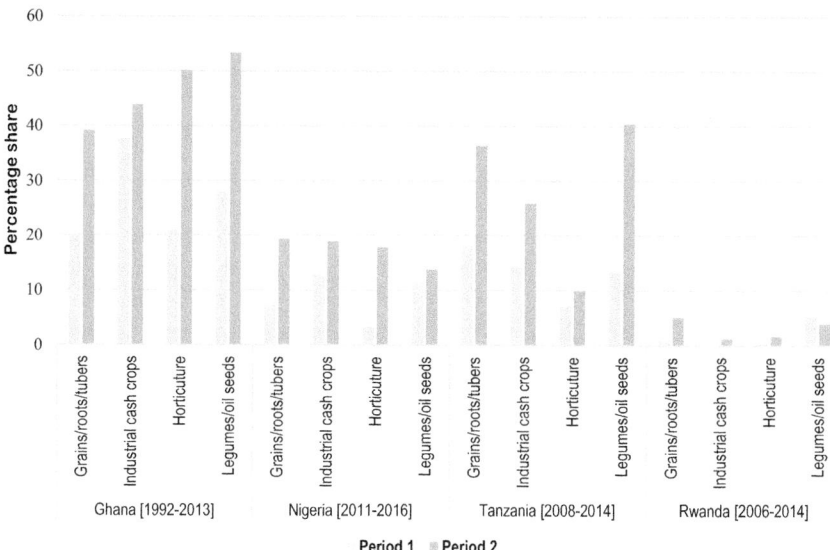

Figure 32.1 Medium-scale (5–100 ha) farms share of national value marketed crop output

Sources: Own presentation; data from Zambia- Zambia MAL Crop Forecast Surveys; Tanzania, Nigeria and Rwanda – LSMS/National Panel Surveys; and Ghana GLSS Surveys

Pathways into medium-scale farming

Much about the processes behind this group's growth has remained unclear. Is this growth driven by land accumulation by relatively productive small-scale farmers who are transitioning to a larger scale production through capital and assets accumulation – a precursor to the smallholder-led agricultural transformation that Johnston and Kilby (1975) and Mellor (1995) talked about? If this is the case, is this evidence of farm consolidation and successful smallholder-led agricultural transformation? Is the growth driven by market-oriented land institutions and policies that encourage investment in land based on willingness to pay and, hence, largely by individuals from outside the small-scale farming sector? Understanding the factors driving the growth of this class of medium-scale farmers has never been more important as many African countries seek to modernize their agricultural sector by transitioning from a subsistence-based to market-driven rural economies.

Ongoing work under APRA-Nigeria Project[2] shows that slightly more than half of the medium-scale farmers in Ogun and Kaduna states used incomes generated outside of farming to enter this size (Muyanga et al., 2019). From a sample of 1,010 medium-scale farms, 47% of them started as small-scale farms and graduated into medium-scale farming status. In some countries, like in Kenya and Zambia, we also find that many current medium-scale farmers started out with capital generated from nonfarm sources. For example results from medium-scale farms' data collected by the Tegemeo Institute and Michigan State University in Kenya show that about 41% of the surveyed 200 farms in Western and Rift Valley provinces of Kenya followed farm-led pathways into medium-scale farming status (Muyanga, 2013).

It is important to note that the smallholder farms that grew organically from small-scale into medium-scale status were not typical smallholders. While most rural farm households operated about two hectares of land when they started farming in Nigeria, those who transitioned into medium-scale status owned about four hectares of land when they started. Less than 10% percent of them owned less than one hectare of land. Most of them (52%) owned between two and five hectares while about 28% had over five hectares, even though they farmed less. In Kenya, only about 6% of the group that followed farm-led pathway into medium-scale farming started out owning less that one hectare of land. The results further show that the group that followed the nonagricultural led pathway in to medium-scale farming in Kenya and Nigeria was relatively younger and had post-secondary education attainments. Similar trends will most likely be encountered in many of the countries in Southern Africa.

Factors contributing to the growing medium-scale farming

Jayne et al. (2016) identify four reasons behind the striking growth of medium-scale farms:

First, high world food prices. For about ten years, there has been a prolonged surge in global food prices. This ushered in major, and much publicized, investment in African farmland by foreign investors. What happened largely under the radar was very large, in aggregate, farmland investments by African professionals, entrepreneurs and civil servants. The amount of land acquired by these medium-scale African farmers since 2000 far exceeds the amount acquired by foreign investors. Many appear to have started out as small-scale farmers who successfully accumulated land and expanded their agricultural operations. Others are relatively wealthy and influential, often professionals, entrepreneurs or retired civil servants. Many accumulated wealth from nonfarm jobs, invested in land and became either part-time or full-time farmers. Many are based in rural areas and have political or social influence with local traditional authorities. Others are urban "telephone farmers" who retain jobs in the cities, hire managers to attend to their farms and occasionally visit on weekends.

Second, key policy reforms. Reforms in the 1990s, including the removal of restrictions on the private movement of agricultural food commodities across district borders and the related demise of government grain marketing parastatals, improved the conditions for private investment in African agri-food systems. The effects of these reforms exploded after world food prices suddenly skyrocketed. They enabled thousands of small-, medium- and large-scale private firms to respond rapidly to profitable incentives.

Third, key complementary policy reform related to land markets. Rise of land markets and capture of agricultural policies by urban/rural wealthy persons and farm lobbies/unions. Unlike some 20 years ago, now land sales and rental transactions are mostly legal, even in areas of customary tenure. Agricultural policies have also become more favourable to medium- and large-scale farms interests. Most national farmer unions and lobbies support policies that raise food prices, promote the conversion of land from customary tenure to statutory land to promote access to land through market transactions and input and credit subsidy programmes that allow bigger farms to participate in the programmes. Increased government spending on agriculture, where benefits are related to size, is disproportionately captured by large farms. Common rhetorical themes used to justify this position are that public support should go to "progressive" farmers who view "farming as a business" and have adequate access to capital (Jayne et al., 2016).

Fourth, rapid urbanization Urban growth coupled with rising incomes, the diet transformation and the continuing effects of the commodity price surge of 2007–08 have made farming a more attractive commercial opportunity in Africa.

Medium-scale farms are a source of dynamism, technical change and commercialization

We identify three channels through which medium-scale farms are likely to bring new sources of capital and know-how to African agriculture. First, the rise of this sector is associated with increased large-scale investment in grain wholesaling, often but not only by multinational grain trading companies such as Cargill. The rise of urban areas increases demand for food, but this demand can be satisfied from imports or from domestic production. Investment response by private traders in domestic value chains is at least due to the growing surpluses of medium-scale farms, which dramatically reduce the unit cost of rural assembly. Very high costs of such assembly in systems dominated by many dispersed smallholder farmers are a major impediment to change in scale and technology further down the chain.

Second, by attracting private investment around them, these farmers may improve input- and output market access for surrounding smallholder farmers. Quantitative evidence on this topic is not yet strong, but personal observation of major changes abounds. In one of the only quantitative studies, Chamberlin and Jayne (2017) find some evidence that a higher concentration of farms of 5–10 ha (at the bottom end of the medium-scale farm class) leads to higher rural incomes in Tanzania. Large trading firms are setting up buying depots in areas with many medium-scale farms, which improves output market access for surrounding smallholder farmers. Medium-scale farms attract tractor rental providers who now provide mechanization services to smallholders that they otherwise would not, since making a trip for a few smallholder farmers would not be profitable. This allows smallholders to farm their land with much less labour input, freeing up opportunities to work off the farm. Van der Westhuizen (2018) shows large increases in smallholder farmer use of tractor rental services in Tanzania between 2008–09 and 2014–15, from an average of roughly 5% to about 15%. Similar synergies might be expected in markets for variable inputs such as fertilizer and plant protection chemicals, though data are lacking on this effect. Evidence from Nigeria show that medium-scale farms are offering extension guide/ services to smallholders, sell farm inputs to smallholders, purchase farms inputs together with smallholders and rent tractors and farm machinery to smallholders (Muyanga et al., 2019). Finally, there is much casual information about medium-scale farmers engaging smallholder farmers in outgrowing schemes, whether on their own or in the context of donor-funded development programmes. If such arrangements prove profitable for these medium-scale farmers, this could become another channel through which they enhance opportunities for smallholder farmers.

Third, to the extent that these farmers spend in the local economy – hiring labour, purchasing food and services – they will stimulate off-farm employment opportunities for rural people formerly dependent on subsistence farming. Again, quantitative evidence is limited, but Poulton (2018), for example, notes local shop owners in the Kilombero Valley of Tanzania indicating that most of their household goods are purchased by "medium scale farmers who have moved into the area" in response to rural electrification and improved roads. Chamberlin and Jayne's (2017) finding that it is the smaller end of the medium-scale segment that has the biggest effect on incomes may be consistent with the idea that these farmers are more likely to make most of their expenditures locally, compared to farmers at the upper end of that distribution or large-scale farmers.

There are some negatives associated with growing medium-scale farming

While there are many positives associated with growing medium-scale farming, there are some negative changes too. The rapidly growing share of land in the medium-scale holdings is leading to concentration of landownership and use, as represented by the Gini coefficient, is rising in many sub-Saharan countries. Jayne et al. (2014) report that Gini coefficients on land cultivation from the early 1990s to the mid-2000s rose from 0.51 to 0.54 in Kenya, and from 0.54 to 0.65 in Ghana. In Zambia in Southern Africa, the Gini on landholdings rose from 0.42 to 0.49 between 2001 and 2012. They report that these levels of concentration "far exceed those of most Asian countries in the 1980s". The rising concentration of land is driving increased land scarcity that may be constraining the growth of small-scale farm holdings and raising entry barriers into farming to new entrants, like the youth.

Medium-scale farms get their land from traditional chiefs or by purchasing land from others, including small-scale farm households. Though data are scarce, a strong trend over the past ten to 15 years in Africa has been engaging in programmes to move land from customary to statutory tenure, sometimes leading to the loss of land by smallholder farmers.

Policy implications

Medium/large-scale farm investment in Africa is injecting important sources of capital and expertise into underperforming current farming systems. There are significant potential positive spillover benefits from medium- to large-scale cropping activities to adjacent smallholder communities (assuming appropriate institutional arrangements exist or are designed). One such benefit is increasing access by smallholder communities to agricultural technologies, credit, extension and marketing services, thus improve the food security and welfare of smallholders in those communities. Medium/large holders may also provide a valuable source of off-farm agricultural wage employment (and thus additional income) for an adjacent smallholder community.

In terms of differences in agricultural productivity, evidence is emerging indicating that medium-scale farms are in fact more (land and labour) productive than smallholdings (Muyanga and Jayne, 2019). Medium-scale farmers are more capitalized and educated compared to their smallholder counterparts. This productivity advantage is largely driven by mechanization and input use intensity. Consequently, improving medium-scale farms' access to land can help the countries increase their domestic production of key staple crops.

The much-awaited smallholder-led agricultural transformation in sub-Saharan Africa seems uncertain with ever-shrinking farm sizes. Smallholder farms have become "too small" to generate meaningful production surpluses and participate in broad-based inclusive agricultural growth processes given existing on-shelf production technologies. Thus, the smallholder-led agricultural transformation process is unlikely to occur, and structural transformation will require sources of vigour that can generate money flows in rural areas, enabling smallholder households to find viable employment in off-farm activities. This chapter emphasizes the difficulty for farmers, starting with small landholding sizes, to expand their scale of production and eventually graduate into medium-scale and more commercialized systems of farming. Landholding size is a critical policy issue given that a majority of smallholders in sub-Saharan Africa own or control less than one hectare of land. Land availability and access to capital are among the most important factors that enabled transition from small to medium-scale farming.

Even though much remains unknown and the story is still unfolding in Southern Africa, medium-scale farms may be one of the important sources of rural dynamism. We believe that medium-scale farms are an important driver of rural transformation in Southern Africa just like in other regions of sub-Saharan Africa. Medium-scale farms have contributed to Africa's 4.6% annual rate of agricultural production growth between 2000 and 2015 (Jayne et al., 2018). Similar trends will most likely be observed in many of the relatively land abundant countries in Southern Africa.

It is true that the rise of land markets is creating a new class of landless workers who are dependent on the local nonfarm economy for their livelihoods. Land administration policies in Southern Africa seem to be lagging behind to support more sustainable and inclusive land dynamics in particular, agriculture and rural transformation in general. Policymakers will need guidance on how to minimize these hardships –protecting those who are most vulnerable as the processes of economic transformation gradually raise living standards for the majority of the population.

Notes

1   The analysis uses LSMS/NPS panel data and the agricultural census data. Even though these datasets are nationally representative, datasets based on population-based sampling miss information on urban households that are likely to be engaged in less land intensive and high value enterprises such as horticulture. The surveys are also conducted during the short and long rain seasons, meaning they are likely to miss the bulk of horticultural production that tends to take place in the dry season.
2   www.future-agricultures.org/apra/

References

Chamberlin, J. and T.S. Jayne. (2017) *Does Farm Structure Matter? The Effects of Farmland Distribution Patterns on Rural Household Incomes in Tanzania. Food Security Policy Innovation Lab Research Paper 94.* East Lansing, Michigan.

Jayne, T.S., Jordan Chamberlin, Lulama Traub, Nicholas Sitko, Milu Muyanga, F.K. Yeboah, Ward Anseeuw, Antony Chapoto, Ayala Wineman, Chewe Nkonde, and Richard Kachule (2016) Africa's changing farm size distribution patterns: The rise of medium-scale farms. *Agricultural Economics* 47: 197–214.

Jayne, T.S., Jordan Chamberlin, and Rui Benfica (2018) Africa's unfolding economic transformation. *The Journal of Development Studies* 54: 777–787.

Jayne, T.S., Antony Chapoto, Nicholas Sitko, Chewe Nkonde, Milu Muyanga, and Jordan Chamberlin (2014) Is the scramble for land in Africa foreclosing a smallholder agricultural expansion strategy? *Journal of International Affairs* 67(2).

Johnston, B.F. and P. Kilby (1975) *Agriculture and Structural Transformation: Economic Strategies in Late Developing Countries.* New York: Oxford University Press.

Josephson, A.L., J. Ricker-Gilbert, and R.J.G.M. Florax (2014) How does population density influence agricultural intensification and productivity? Evidence from Ethiopia. *Food Policy* 48: 142–152.

Mellor, J.W. (1995) *Agriculture on the Road to Industrialization.* Baltimore, MD: International Food Policy Institute, Johns Hopkins University Press.

Muyanga, M.C., A. Aromolaran, S. Liverpool-Tasie, T.S. Jayne, and T. Awokuse. (2019) *Changing Farm Structure and Agricultural Commercialization in Nigeria: Livelihoods and Policy Implications.* APRA Working Paper 6/2019. Available here https://opendocs.ids.ac.uk/opendocs/bitstream/handle/123456789/14576/APRA_WP26_Changing_Farm_Structure_and_Agricultural_Commercialisation_in_Nigeria.pdf?sequence=1&isAllowed=y.

Muyanga, M.C. (2013) *Smallholder Agriculture in the Context of Increasing Population Densities in Rural Kenya*. A PhD Dissertation submitted to Michigan State University.

Muyanga, M.C. and T.S. Jayne (2014) Effects of rising rural population density on smallholder agriculture in Kenya. *Food Policy* 48: 98–113.

Muyanga, M.C. and T.S. Jayne (2019) Revisiting the farm size-productivity relationship based on a relatively wide range of farm sizes: Evidence from Kenya. *American Journal of Agricultural Economics* 101(4) (July): 1140–1163.

Poulton, C. (2018) *Kilombero Stories, Futures Agriculture*. www.future-agricultures.org/blog/kilombero-stories/.

Ricker-Gilbert, J., C. Jumbe, and J. Chamberlin. (2014) How does population density influence agricultural intensification and productivity? Evidence from Malawi. *Food Policy* 48: 114–128.

Sitko, N.J. and T.S. Jayne (2014). Structural transformation or elite land capture? The growth of "emergent" farmers in Zambia. *Food Policy* 48: 194–202.

van der Westhuizen, D., T.S. Jayne, and F.H. Meyer (2018) *Rising Tractor Use in sub-Saharan Africa: Evidence from Tanzania*. Paper presented at the Agricultural & Applied Economics Association (AAEA) meetings, Washington, DC, August 5, 2018.

Willy, K.D., M. Muyanga, and T.S. Jayne (2019) Can economic and environmental benefits associated with agricultural intensification be sustained at high population densities? A farm level empirical analysis. *Land Use Policy* (81): 100–110.

33 Food security in Africa

A complex issue requiring new approaches to scientific evidence and quantitative analysis

Mario Giampietro

Introduction

The problem with the Cartesian dream of prediction and control

The disappointing results achieved so far in relation to sustainability goals (e.g., reduction of greenhouse gas emissions, protection of biodiversity or circularization of the economy) flag the necessity to improve the effectiveness of sustainability policies and the models used to inform the process of decision making. When dealing with complex sustainability issues ("wicked problems"), the conventional approach to purposeful modelling (the Cartesian dream of prediction and control) might not be the most suited approach. Existing quantitative approaches (conventional approaches based on economic narratives) do not allow an integrated analysis of the different factors determining: 1) material standard of living; 2) food, energy, water security; and 3) environmental security because these factors can only be observed across different dimensions and scales of analysis. For this reason, today, relevant sustainability issues can only be analyzed in quantitative terms "one at the time" using nonequivalent models. This is what generates the "silo governance syndrome" – i.e., solving a given problem by setting targets that ignore negative side effects related to other problems. This fact can explain, for example, why many policies dealing with sustainability of agriculture (both in developed and in developing countries) have been so far ineffective and even contradictory with each other.

Quantitative storytelling: a complexity revolution in sustainability analysis

The process of producing numbers is inseparable from the process of defining the meaning and the relevance of the numbers. For this reason, we always need to know: 1) why and how the numbers came into being in the first place – i.e., they depend on the problem definition (decided by whom?) determining the choice of preanalytical narratives driving the choice of models; and 2) why and how they get back in the form of knowledge claims used to guide policy. The meaningful use of numbers requires transdisciplinary research. Unfortunately, at the moment much of the sustainability discussions are based on numbers that are produced by simplistic analytical tools (mono-scale and mono-dimensional indicators), inept to address complex issues, that are thrown at each other (divorced from the original context) by the discussants. The complexity revolution proposed here entails the combination of two distinct changes:

1   *Quantitative Storytelling* – a novel philosophy in the use of scientific evidence for the governance of sustainability
2   *Multi-Scale Integrated Assessment of Societal and Ecosystem Metabolism* (MuSIASEM) – a novel accounting scheme used for the structuring of the perception (qualitative choices) and representation (quantitative choices) of the sustainability predicament. This novel accounting scheme based on the concept of the metabolic pattern of social-ecological systems, can be used in a process of coproduction of knowledge claims in decision making to radically change the nature of the information space used as "evidence".

A different "WHY" for the generation of quantitative information

Quantitative Storytelling is based on the consideration of the following questions: 1) "what if it is not possible to identify the optimal solution to the problem?" (whose problem?); 2) "what if it is not possible to identify an uncontested intervention that will provide an optimal improvement?" (for whom? for how long? at which costs?); 3) "what if it is not possible to eliminate large doses of uncertainty both from our analysis of the existing situation and from our predictions of future events?" If we admit that it is impossible to make reliable predictions and to provide full control about desired futures, what is then the possible role of quantitative analysis? In quantitative storytelling, scientific analysis is applied to answer questions such as: "what cannot happen if" or "what could go wrong if we try to implement this policy?" (in order to eliminate implausible or risky policies!) or "can we define the consequences of proposed policies in terms of winners and losers?" (in order to avoid unfair policies). To implement this new philosophy, we need an analytical framework based on a complex representation of the metabolic pattern of social-ecological systems integrating different types of quantitative information into a salient characterization of sustainability issues.

A different "HOW" for the generation of quantitative information

MuSIASEM accounting scheme allows people to generate a *sustainability sudoku* by preserving coherence across quantitative analyses (integrating indicators) based on different metrics carried out across different levels and dimensions of analysis. This is an extremely important achievement because of the unavoidable existence of side effects (negative/positive consequences of an improvement according to an indicator over other indicators) to be expected when considering the different dimension of the nexus. The "silo-governance syndrome" is exactly generated by an excessive importance given to the achievement of a specific target at the time, when side effects in relation to other targets are not considered: solving a problem in relation to food, we may generate a different problem in relation to water and so on.

Food security in Africa is definitely a complex issue

In Africa, the continuous increase in the population pressure associated with a generalized raise in expectation for better economic conditions – new parents want a better future for them and their children – is generating the need for a dramatic expansion of economic activities. The aspirations for a quick increase in the level of personal monetary income in the time span of a generation translates into a "mission impossible" due to the huge requirement of infrastructure, capital and resources that this expansion would require. Therefore, the dramatic process of transformation taking place in Africa is at risk of jeopardizing human well-being, the environment, the richness of cultures and the fragile sociopolitical institutions all over Africa. The problem is especially clear in relation to food security because traditional agricultural production faces important external constraints such as limited land available, poor soils and scarcity of water. In this situation an attempt to boost the traditional yields above their specific benchmarks can only be obtained through high impact on the environment and a very low economic gain for the farmers. As a result of this situation, we witness a massive migration away from rural areas. However, differently from what happened in the industrial revolution in Europe, this migration is not driven by the availability of better jobs in the cities. Rather, rural migrants are running away from unacceptable living conditions in the countryside. This mass of migrant population is generating another critical situation in shanty towns where the poor settle around mushrooming urban agglomerates. In this urban context the access to food is strictly determined by market laws: the food security of the urban poor depends on cheap food supply. This situation translates into an internal tension in the society because cheap food supply for the cities means low economic revenues for the farmers feeding the city (or vice versa). To make things worse, in many areas of Africa, the problems associated with a scarcity of resources are worsened by: 1) a generalized collapse of social fabric (due to an excessive speed of cultural changes or the occurrence of wars), making it impossible to guarantee the satisfaction of basic needs; 2) the effect of climate change that are disrupting the traditional patterns of agricultural activity; and 3) land grabbing from foreign investors. In this situation the use of complex analytical tools capable of generating a more robust and articulated diagnostic and anticipation of problems is a must because:

1   The systems of agricultural production in Africa are extremely diverse reflecting the diversity of cultural contexts and environmental conditions in which farmers operate. This diversity represents an additional problem because it requires an "ad hoc" tailoring of the management of plantations,

pastures, agricultural or horticultural crops to specific conditions and a continuous adjusting of the original choices in response to the effects of climate change. This challenge calls for an analytical approach capable of identifying the relevant factors (referring to different scales and dimensions) determining the sustainability of food production across the mosaic of situations experienced in Africa.

2   The sustainability of food security of Africa depends on the ability to handle when deciding policies complex dilemmas associated with direct tensions/ trade-offs: 1) "intensification of agriculture" vs. "impact on the environment"; 2) "development of urban areas" vs. "development of rural areas"; 3) the implications of the nexus between water-energy-food (the various flows are entangled and it is impossible to optimize just one of them without affecting the others); 4) "modernization" vs. "protection of the existing culture" – i.e., the war on poverty ends up eliminating the poor.

The next section shortly illustrates a holistic toolkit capable of identifying and characterizing a set of factors relevant for an informed policy deliberation.

Analyzing food security from multiple perspectives, dimensions and scales

The Drivers–Pressure–State–Impact–Response (DPSIR) framework

This section presents an application of a toolkit developed in the EU project MAGIC (Moving Towards Adaptive Governance in Complexity: Informing Nexus Security – https://magic-nexus.eu/) that applies insights of complexity to the analysis of food security, addressing the implications of the nexus. It is based on the causal Drivers–Pressure–State–Impact–Response (DPSIR) framework proposed by the European Environment Agency in 1999, to study the interactions of socioeconomic systems with their environment (EEA, 2018). The DPSIR framework allows the integration of different "types of indicators": 1) indicators referring to the "state" are those referring to what is going on inside the society (the technosphere – referring to the set of processes under human control); 2) indicators referring to the "pressure" are those referring to the pressure exerted on the environment (the biosphere – referring to the set of processes outside human control). It should be noted that the indicators of pressure do not translate directly into indicators of impact. Because of the openness of modern society based on trade, imported food can be associated with an environmental pressure that is generating impact elsewhere (in the ecosystems of the agricultural systems producing the imported food). In the same way exported food does generate an environmental pressure inside the system without contributing to the local food security. For this reason, when applying the state-pressure analysis it is necessary to address another key aspect of the analysis of food security: how much open the system is, in relation to imports and exports of food commodities.

The MuSIASEM toolkit

An accounting system called Multi-Scale Integrated Analysis of Societal and Ecosystem Metabolism (MuSIASEM) – Giampietro et al., 2012, 2014 – makes it possible to establish a bridge across different indicators defined across different levels and dimensions.

After having characterized the level of openness of the system under analysis it becomes possible to characterize four key aspects of the system:

1   the local *end use matrix* – the quantity of inputs required in the process of food production taking place in the technosphere (labour, energy, fertilizer, power capacity, etc.) inside the considered system

2   the local *environmental pressure matrix* – the quantity of primary sources (land, soil, green water, abstraction from ground water) and the sink capacity (natural processes absorbing pollution) that would be required to maintain the metabolic pattern without affecting the health of ecosystems

3   *the externalize end use matrix* – the quantity of inputs required in the process of food production taking place in the technosphere (labour, energy, fertilizer, power capacity, etc.) but outside the border of the investigate system; they are embodied in the commodities crossing the border of the system

4   the *externalized environmental pressure matrix* – the quantity of primary sources (land, soil, green water, abstraction from ground water) and the capacity of sinks (absorbing pollution) required for the production of the traded food commodities; they are embodied in what is crossing the border.

A visualization based on the rationale proposed by the European Environment Agency is given in Figure 33.1.

In order to carry out an analysis of environmental impact, one has to contextualize the various typologies of pressures – e.g., flow of nitrogen per hectare in the soil, abstraction of blue water per hectare for irrigation, requirement of land in production determining destruction of local habitat – against the characteristics (qualitative and quantitative) of the local ecological funds. Therefore, the analysis of impact requires the use of spatial analysis (information organized in GIS layers) because each type of pressure has to be assessed in relation to a specific relevant information layer – e.g., the impact of the abstraction of water for irrigation on the water table depends on the relation between the volume of the abstracted flow and the pace of recharge of a specific aquifer. The two have to be compared on a GIS layer focusing on the location of aquifers. As mentioned earlier in the case of Africa it is important to apply the toolkit to study the dynamic relation of food supply and food demand between urban and rural areas. That is, we can define for both an urban system (a city) and a rural system (identified by an administrative area):

1   The desirability of existing situations, identifying the factors associated with a given level of bio-economic pressure in the society: 1) the minimum

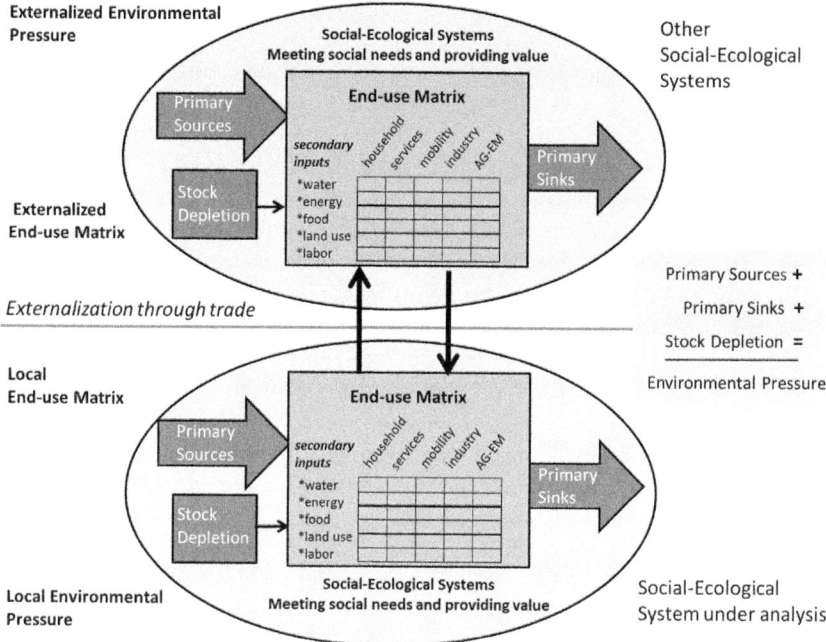

Figure 33.1 The state-pressure relations in an open Social-Ecological System

Source: Own representation

return of labour expected in a given community (operating below this threshold would imply the crumbling of the social fabric) – the productivity of labour can be assessed both in terms of monetary value and in biophysical terms, depending on what is produced in the various tasks; 2) the minimum level of services expected in a given community (operating below this threshold would imply the crumbling of the social fabric) – this factor can be measured in "hours of paid work per capita per year" in the community for services (health care, education, police and other administrative tasks).

2   The viability of existing situations, identifying the factors associated with the given level of productivity of production factors: 1) what are the factors determining the economic viability of farmers, the combination of land available, productivity of land (yields) with the structure of costs and revenues; 2) what is the level of intensification of agricultural production (the profile of inputs and outputs per hectares) at the moment? – are we dealing with "high external input agriculture" (market agriculture), "low external input agriculture" (subsistence agriculture) or a mix of the two?

3   The feasibility of existing situations determined by the severity of external biophysical constraints. Is the existing demographic pressure generating

problem of limited access to land or rather the limited access to land is due to institutional settings? Is the requirement of land compatible with the preservation of terrestrial habitats? Is the pressure associated with existing production activity representing a threat to the health of ecological funds when considering existing drivers? Can we characterize the limits imposed by biophysical processes as distinct from those imposed by institutional settings?

4   What is the level of openness of the systems of agricultural production: are they producing food to feed rural people (and themselves)? Are they producing food to feed the cities? Are they producing food for international exports? Answering these questions is essential if we want to understand the nature of the trade-offs over intensification of agriculture, food security, human well-being and environmental impact. In fact, as discussed earlier there is a direct relation and tension between urban and rural areas. Basically, cities import all the food they consume from their surroundings: the supply of food to the city has to come from local rural areas or has to be imported from other countries. The chosen solution makes a huge difference in the dynamic of rural development. In addition to that, today in Africa, land grabbing is implying that some of the natural resources available in rural areas are used to produce food that will leave the country. In general, the decisions about the origins and the destinations of food flows in Africa are not taken by the local farmers, even though they are affected by these decisions that have an overwhelming effect on the resulting trade-offs. Foreign investors may be willing to use the best land for commercial activities, reducing in this way the options of local farmers for local production. We can expect that a strategy having the goal of increasing as much as possible the supply of food to the local people while reducing the impact on the environment and boost the well-being of rural communities against a very heterogeneous set of external biophysical constraints can only result in a mosaic of production systems: subsistence farming (integrated by a multifunctionality of the landscape in marginal areas), mixed farming and high external input farming (on the best land) depending on the specificity of geographic and economic situations. If we accept this hypothesis it is really important to control the overall effect of the demographic pressure on the availability of land and the distortion that foreign investors can generate, in order to evaluate the pros and cons of increasing or reducing the trade of agricultural commodities.

5   Last, but certainly not least, it is essential to address the issue of the nexus over water, energy, food and landscape both in diagnostic mode – what is the level of nexus impact associated with the existing state (the giving mix and quantity of production systems)? – and in anticipation mode – what types of impact can be anticipated because of changes in population, changes in the mix of activities or changes in climatic conditions? Can we track the flows of consumed inputs and produced outputs from and to the technosphere (having an economic relevance in terms of costs and revenues)? Can we track the flows of required inputs and resulting wastes and

emissions from and to the biosphere (having an environmental relevance in terms of constraints and impact)? Can we fractionate this analysis across levels and scale of analysis using GIS layers?

Practical applications

Sound policies for agricultural transformation will require a holistic contextualization of the variety of different inputs of scientific information used in the decision process. The worrisome trends experienced across the planet of political instability, the progressive shortage of resources and environmental deterioration leading to geopolitical turmoil entail an increased responsibility for decision makers.

It is time to guarantee a better quality of policies in terms of robustness, fairness and transparency, and this requires a revolution in the production and use of quantitative analysis in agriculture and food production.

A new generation of quantitative information has to be used to address simultaneously the four challenges to the governance of sustainability as it relates to agriculture:

1   *How to characterize, in biophysical and quantitative terms, the factors that can be used to deliberate about a desirable standard of living* – how to define a GOOD QUALITY OF LIFE? What should be considered as a desirable standard of living when considering the various attributes of sustainability?

     This is an issue very often raised in sustainability discussions – e.g., prosperity without growth, post-growth economy, sufficiency economy – and it has been proposed by several governments as a working concept – e.g., *"buen vivir"* (Ecuador), *happiness index* (Nepal) and *Det Goda Livet* (Sweden).

2   *How to assess the level of food and energy security of a country.* The analysis of the metabolic pattern of countries should be used to identify dangerous dependences on imports in relation to: 1) energy security and food security – when the internal demand cannot be met by local supply because it exceeds external biophysical constraints (available primary sources and sinks in the country); and 2) the effects of the exports of a country on the rest of the economy (not only in terms of natural resources but in terms of technology and capital).

3   *How to track the impact of the various pressures across scales using GIS.* In order to be able to geo-localize the effect of the environmental pressure against the characteristics of local ecological funds – assess the level of environmental loading – one has to use the lens of the *microscope*. Different type of impacts can be calculated in relation to:

     1)   the need of protecting habitats and biodiversity
     2)   GHG emissions (direct and indirect associated to changes in land uses)
     3)   water resources, both on the supply side (over drafting of aquifers) and the sink side (leakage of NP in the aquifers)

4)   pesticides (effect on pollinators) and other dangerous pollutants
5)   waste management (plastic and urban wastes), etc.

The possibility of mapping the pressures generated by the technosphere onto the characteristics of natural ecosystems across scales to calculate the resulting impact is a natural feature of the MuSIASEM accounting scheme.

4   *How to establish an interactive interface capable of visualizing and communicating the results of the analysis in the form of an effective decision support.* The simultaneous analysis across different scales and dimensions translates into an excessive density of the quantity of information to be processed (too many results).

What needs to be done by policymakers?

It is essential for policymakers to:

1   Support the development of better decision support tools based on soft-ware to build an interface with those creating polices for transformation of agriculture (using dashboards, visualization tools and interactive features). These tools have to be used when deliberating over policies regarding the future of agriculture and food production.

2   Give financial support for capacity building to produce local scientists that can develop such toolkits using, for example, the MuSIASEM accounting scheme.

3   Strengthen in-country scientific capacity for working with these tool kits in *diagnostic mode*: based on the current situation in agriculture and food production in the country they should identify: 1) the quantitative and qualitative characteristics of the constituent components of the agricultural food system in terms of benchmarks; 2) the important functional and structural relations to be considered in policies; and 3) bottlenecks, critical situations and potential factors that can hamper the implementation of policy in relation to changes in interrelated conditions.

4   Strengthen in-country scientific capacity for working with these tool kits in *anticipatory mode*: they should explore, using contingent analysis over "what if" scenarios, the internal and/or external limits – i.e., the option space – in order to identify policies for agricultural food production systems avoiding not feasible, not viable or not desirable outcomes in relation to the goal of agricultural improvement of food production.

In conclusion, the MuSIASEM toolkit gives decision makers the ability to identify and study the factors determining the:

1   Feasibility = compatibility with external constraints: processes outside human control

2   Viability = compatibility with internal constraints: processes under human control

3 Desirability = compatibility with aspirations and social norms required to preserve the social fabric of proposed policies to improve agricultural production.

References

EEA. (2018) www.eea.europa.eu/help/glossary/eea-glossary/dpsir.

Giampietro, M., Mayumi, K. and Sorman, A.H. (2012) *The Metabolic Pattern of Societies: Where Economists Fall Short*. London: Routledge.

Giampietro, M., Aspinall, R.J., Ramos-Martin, J. and Bukkens, S.G.F. (Eds.) (2014) *Resource Accounting for Sustainability Assessment: The Nexus Between Energy, Food, Water and Land Use*. London: Routledge.

Part VI

Conclusions

34 The way forward

The editors

Richard A. Sikora, Eugene R. Terry,
Paul L. G. Vlek and Joyce Chitja

This book contains short and concise chapters written by experts coming from a wide spectrum of agricultural disciplines on a broad array of important constraints, technologies and policy options needed to transform agriculture. The book is divided into five parts:

1 the status of agriculture in Southern Africa
2 drivers and constraints
3 current technologies
4 emerging technologies
5 policies and processes.

These parts describe the main problems facing the farming community and suggest means to improve access to these technologies and thereby improve agriculture and food production in Southern Africa.

Farming is often seen as a simple straight forward process that includes, among other factors, soil cultivation, sowing of seed, fertilization, weeding, pest management, harvesting and grain storage. Agriculture is not that simple but is a highly complex system of interactions between man and nature. There are many forms of agricultural production and each has specific approaches and requirements to be effective. If certain tools are not available when needed, crop production suffers. The tools for improved levels of production, to name a few, are quality seed, high yielding varieties, fertilizer, mechanization, plant and animal pest management and access to markets. The complexity of agricultural transformation is shown in the word cloud in Figure 34.1.

To produce the food we eat, farmers must rely on the availability of natural resources i.e., rainfall, temperature, soil fertility and pest management. In addition, farmers need access to a large number of building blocks or production tools, that when used properly, can lead to a good harvest without negative environmental impact – or a sustainable form of agricultural production.

For the vast majority of small- and medium-size family farmers in Southern Africa these tools are either: unknown, not available or too expensive. Conversely, large family and commercial farms have access to these tools, and this is the reason for the discrepancy in yield between the two groups of farmers. The

Figure 34.1 The complexity of agricultural transformation

Source: Richard A. Sikora, made with permission of wordclouds.com, Zygomatic 2015

lack of land ownership, access to credit, proper agricultural extension services, training, access to markets and government support strongly limits their ability to farm effectively and help improve the food security issues impacting the region.

The authors of the chapters in this book outline specific problems that farmers experience during and after production across the entire food value chain. They then present current and/or emerging technologies that can be used to offset the impact of the biological and/or physical constraints that they face on a daily, seasonal and yearly basis.

Most important, the authors also outlined what policies need to be modified or newly developed by decision makers to make these technologies available to the farming community.

Therefore, a very important part of the book deals with improving policies and processes that are needed in order to stimulate the transformation of agriculture and thereby improve food production and food availability within a country.

The recommendations given by the 67 authors who compiled the chapters in this book are of course the opinions and visions of the authors themselves. They wrote these chapters with the objective of improving agriculture in Southern Africa. We, the editors, hope the recommendations in this book are examined by decision makers and that they have an impact in transforming agriculture in the future.

Any attempt by the editors to try and synthesize the information presented in these chapters into recommendations for decision makers was considered extremely difficult – and not what the editors had in mind from the start.

Instead, the book is seen as a handbook outlining the most important building blocks needed for improving agriculture at the small- and medium-size farm level over what is presently the state of the art.

If these tools are selected and combined in a logical framework that fits the agriculture uniqueness of each farming system of the countries of Southern Africa, and policies and processes are devised to support local farmers, then the concepts and tools presented in the book will lead to the improved food production levels that are needed in the future.

Index

Note: Numbers in *italics* indicate figures and numbers in **bold** indicate tables on the corresponding page.

applications 305–306; quantitative storytelling: 298–299; renewable energy and 202; role of policymakers 306–307; small farms and 232; small rural markets 253–254; state-pressure relations in an open Social-Ecological System 302, *303*; sustainability of 301
foot and mouth disease (FMD) 62, 132, 133
fruit: citrus 58, 60–61; climatic conditions 170–171; food safety 172–173; improving market access 175–176; increasing consumer demand 174–175; local low-input and regional high-input production 172; markets 56; postharvest losses 171–172; quality standards 174–175; supermarket impacts 171, 173; urbanization impacts 171; value of 170
fuelwood 209–210

Gender and Development (GAD) 224
gender equity: factors in gender policy change **228**; non-transformative gendered agricultural contexts 225–227; policies 227–229; strengthening 227–229; transformation 227–229
gender inequality 11
genetically modified (GM) crops: banana 165; benefits 167; *Bt* cowpea 164; countries with commercialized 161–162; crops using CRISPR-type technologies 165–166; Event 709A 164; government policies 166–167; impacts 167; national variety performance trials 167; TELA Maize Project 163
genomics 127–128
genomic tools 99
German Corporation for International Cooperation GmbH (GIZ) 122
Ghana: changes in farm structure **290**; GM crop field trials 162, 164; medium-scale farms 289; QualiTrace 182
gig economy 181–182
giveaway 48, 53
Global Entrepreneurship Monitor (GEM) 281
governmental extension systems 175
grain markets 84
Green Climate Fund, 203
Greenpeace 162
green revolution 80–81, 84–85, 145
groundwater 41

harmonization 109–110
Hello Tractor 182
hermetic storage structures 118–121, **122**
highly pathogenic avian influenza (HPAI) 133

high value commodities 56
HIV/AIDS 11–12, 133
horticulture: biomass transfer for production 208; climatic conditions 171; cold chain technologies/infrastructure 171–172; food safety 172–173; implementing postharvest-related technologies at scale. 173; improving market access 175–176; increasing consumer demand 174–175; local low-input and regional high-input production 172; postharvest losses 171–172; quality standards 174–175; seeds 174; shift in agricultural research towards 176; supermarket impacts 171, 173; traditional African vegetables 176–177; urbanization impacts 171
Human Development Index (HDI) 17

ichneumonid wasp (*Diadegma semiclausum*) 147
income inequality 12
indigenous farm animal genetic resources (AnGR) 126, 128
indigenous knowledge systems (IKS) 111, 122
infectious diseases 133
information and communication technologies (ICT) 73, 134, 137, 179, 182–185
infrastructure 56, 236–237, 263–264
innovation 281
insecticides 147
integrated pest management (IPM): agrochemicals 105–106; application of 106–109, *108*; biocontrol 106, *107*, 109–110; bioprospecting policies 111; cultural methods 105; efficacy trials 110; host-plant resistance 104–105; lack of regional harmonization 109; model of 106; policy issues 109–111; registration costs 110; toxicology testing 110
Integrated Soil Fertility Management 71–74, 145
intercropping 91
internet *180*
"Internet of Things" 181
irrigation 40–42, 61, 201–202

Joburg Market 244, 248

kaolin 117
Kenya: emergence of ICT sector 185; financial technology 186; food safety 175; GM crop field trials 162; growth in non-hydropower renewable energies 200; medium-scale farms 289, 291–292; SHEP project 235